GLOSSARY
for the
WORLDWIDE
TRANSPORTATION
of
DANGEROUS GOODS
and HAZARDOUS
MATERIALS

Malcolm A. Fox

 Springer

Acquiring Editor:	Ken McCombs
Project Editor:	William Heyward
Cover design:	Dawn Boyd
Manufacturing:	Carol Slatter

Library of Congress Cataloging-in-Publication Data

Fox, Malcolm (Malcom A.)
  A glossary of terms for the worldwide transportation of dangerous goods
and hazardous materials / Malcom Fox. — 1st ed.
        p.   cm.
    Includes bibliographical references and index.
    ISBN 978-3-662-11892-4   ISBN 978-3-662-11890-0 (eBook)
    DOI 10.1007/978-3-662-11890-0

    1. Hazardous substances — Transportation Dictionaries. I. Title.
  T55.3.H3 F69 1999
  604.7—dc21   99-36315
                                                                                CIP

(Orders from outside the U.S.A. and Canada to Springer-Verlag)
ISBN 978-3-662-11892-4

© 1999 by Springer-Verlag Berlin Heidelberg
Originally published by CRC Press LLC in 1999

No claim to original U.S. Government works
International Standard Book Number 3-540-64822-4
Library of Congress Card Number 99-36315

Printed on acid-free paper

To Odile

The tension before an explosion (as you already eknowsion)
Can cause a lot of commotion
And when the explosion occurs
Bang, crash, flash, and smash
There will be a lot of erosion

Hannah Fox, age 8

Table of Contents

Acknowledgements..ix
Introduction..x
Regulatory Background xii
How to Use the *Glossary for the Worldwide Transportation of Dangerous
    Goods and Hazardous Materials* xiv

Adhesives...1
Aerosols 3
Alcohols...5
Alkaloids..7
Ammunition..8
Ammunition, Toxic..19
Antifreeze 22
Antiknock Compounds 23
Asbestos...25
Batteries 27
Bituminous Products..32
Bleach 34
Carbon..36
Castor Beans 38
Catalysts...39
Chemical Kits and Samples 40
Cleaning Liquids..42
Coal..44
Corrosives and Class 8...47
Cosmetics and Perfumes..52
Dangerous Goods and Hazardous Materials..54
Dangerous When Wet Materials and Division 4.3.....................................58
Drugs and Medicines 61
Dyes and Pigments..63
Elevated Temperature Materials 65
Environmentally Hazardous Substances...67
Explosive Articles..69
Explosives and Class 1...74
Extracts 86
Fertilizers 88
Fibres and Fibrous Products..90
Fire Extinguishers 94
Flammable Liquids and Class 3 96
Flammable Solids and Division 4.1 99
Gases and Class 2..104
Halogenated Polyphenyls 111

Hazardous Waste 113
Infectious Substances and Division 6.2 115
Initiating Explosives 119
Inorganic Compounds..128
Leather 132
Lifesaving Equipment..134
Lighters 137
Magnetized Materials...139
Marine Pollutants...141
Matches..143
Mercury..145
Metallurgical By-Products 147
Metals, Alloys..151
Metals, Elemental Products 153
Metals, Inorganic Compounds...156
Miscellaneous Dangerous Goods (including Class 9) 158
Nitrocellulose Products..161
Organic Compounds 164
Organometallics and Related Compounds....................................167
Oxidizers and Class 5...170
Oxygen Generators 173
Paints and Coatings..175
Pesticides 179
Petroleum...183
Pine Products 188
Polymers and Resins...189
Pressurized Articles 193
Pyrotechnics and Signals 194
Radioactive Materials and Class 7 200
Refrigerants and Halocarbons..215
Refrigerating Machines..218
Self-Propelled Vehicles 219
Solid Bulk Materials 221
Solvents..224
Spontaneously Combustible Materials and Division 4.2 226
Terminology 229
Tires 254
Toxic Substances and Division 6.1 255
Units of Measure...258

References...263
Index of UN Numbers...275
Index of Entries...285

Acknowledgements

The following international transportation authorities kindly granted permission to reprint portions of their standards related to dangerous goods classification and identification:

□ The United Nations, extracts from the *Recommendations on the Transport of Dangerous Goods*.

□ The International Maritime Organization, extracts from the *International Maritime Dangerous Goods Code*.

□ The International Atomic Energy Agency, extracts from the *Regulations for the Safe Transport of Radioactive Material*.

□ The International Civil Aviation Organization, extracts from the *Technical Instructions for the Safe Transport of Dangerous Goods by Air*.

□ The International Air Transport Association, extracts from the *Dangerous Goods Regulations*.

A number of individuals generously spared time to review and comment on the text including Dick Elbourne, International Air Transport Association; Marion Fox; Ralph Kummler, PhD, Associate Dean for Research and Director, Hazardous Waste Management Programs, College of Engineering, Wayne State University; Jonathan Swindell, McLaren/Hart Inc.; and Dick Wentworth, Hazardous Materials Coordinator, NAO Logistics, General Motors Corporation. Thanks also to my students at Wayne State University and elsewhere. From their questions I drew the inspiration for this book.

Labelmaster in Chicago, Illinois, gave me access to the latest revisions of the above texts, and the following companies provided additional research information:

□ Bristol-Myers Squibb Company

□ Morton International, Inc.

□ The Ensign-Bickford Company

In order that readers of subsequent editions may benefit from your suggestions, comments related to omissions and shortcomings will be most welcome.

Introduction

Worldwide, 500,000 shipments of materials which pose chemical, physical, or biological risks to human health, property, or the environment are made each day by air, rail, road, sea, and inland waterways totalling over 3.6 billion metric tons each year.[1] To ensure safety during transportation, the means by which these dangerous goods and hazardous materials[2] are packaged and handled is prescribed by international authority including the United Nations, the International Maritime Organization, the International Atomic Energy Agency, the International Civil Aviation Organization, and the International Air Transport Association, as well as national authorities such as the Department of Transportation in the United States. In fact, the United Nations establishes model regulations that function as recommendations addressed to international organizations and national governments.

At the core of regulation lies *hazard identification*: once accurately identified, the hazards of dangerous goods may be communicated and the material safely packaged, segregated, transported, and handled by qualified personnel. Incorrectly identified materials increase greatly the risk of explosion, fire, poisoning, or some other mishap. To aid identification, each authority maintains a list of the articles, substances, and materials it regulates comprising thousands of entries including chemical names, industry-specific terms, tradenames, generic descriptions, and other specialized terms common to the language of transportation. While much of this language is recognizable, some is less well understood even to transportation, environmental, and health professionals.

The *Glossary for the Worldwide Transportation of Dangerous Goods and Hazardous Materials* (the *Glossary*) explains these specialized terms using simple language, understandable to shippers around the world. Thereby, it functions as a guide to all those with the fundamental responsibility to identify the hazards and proper shipping names of regulated materials for domestic or international shipment. Specifically, the *Glossary*

▫ Describes around 1400 entries[3] in the lists of regulated dangerous goods and hazardous materials maintained by major international and national transportation authorities.

---

[1] KOE.

[2] The term *dangerous goods* is used internationally; *hazardous materials* describes the same materials and is used in the United States and other countries. The terms are used interchangeably in the *Glossary*.

[3] The *Glossary* does not explain standard chemical names for which many excellent chemical dictionaries are available (although many chemicals are used as examples to explain certain generic entries).

□   Sorts these entries into 75 related subjects to discuss them in context and bring out subtle but critical differences between similar entries.

□   Explains why certain articles, materials, and substances are considered dangerous.

□   Explains the language used in the regulatory lists to describe chemical groups, chemical and physical phenomena, chemical processes, and chemical nomenclature.

□   Presents side-by-side and for comparison regulatory definitions provided by the major transportation authorities.

Finally, the language of international transportation is English. The *Glossary* is particularly intended to support the compliance and safety efforts of those for whom English is not their first language.

As those involved in transportation understand, it is among the shipper's responsibilities to ensure that dangerous goods and hazardous materials are properly identified; the *Glossary* is not a substitute for the regulations in this endeavour. Rather, the *Glossary* may serve to confirm a classification or as a pointer for additional research. Ultimately, it is left to the discretion of the reader to determine whether the descriptions and properties described in this text match those of any particular material.

Regulatory Background

HAZARD CLASSES

Based on the United Nations' scheme, regulatory transportation authorities[4] have identified nine classes of materials regulated in transportation, of which some are further categorized into divisions:

- Class 1: Explosives (six divisions)
- Class 2: Gases (three divisions)
- Class 3: Flammable liquids (three divisions[5])
- Class 4: Flammable solids; substances liable to spontaneous combustion; substances which, on contact with water, emit flammable gases (three divisions)
- Class 5: Oxidizing substances and organic peroxides (two divisions)
- Class 6: Toxic and infectious substances (two divisions)
- Class 7: Radioactive material
- Class 8: Corrosives
- Class 9: Miscellaneous dangerous goods

REGULATORY LISTS

Each regulatory authority maintains a list of around 3000 regulated articles, materials, and substances that meet one or more hazard classes or divisions.[6] While individual entries on these lists may be unique to the regulatory authority, many, with minor modifications, are common to all being based on the United Nations' *Recommendations on the Transport of Dangerous Goods, Model Regulations* (ROT). The *Glossary* covers entries from the following:

- United Nations (UN): *Dangerous Goods List*, Part 3; Annex: Model Regulations on the Transport of Dangerous Goods, Recommendations on the Transport of Dangerous Goods, 10th revised edition; UN: Geneva, 1997.

- International Maritime Organizaton (IMO): *General Index (alphabetical) of Dangerous Goods*; International Maritime Dangerous Goods Code, including Amendment 29-98, 11 to 20 May, 1998; IMO: London, 1997.

---

[4] Certain national authorities have hazard class schemes that are not harmonized with that of the United Nations.

[5] Unlike other authorities, the International Maritime Organization maintains three divisions for Class 3.

[6] The International Atomic Energy Agency limits its list of entries to 25 descriptions of radioactive materials derived from the United Nations' recommendations.

- International Atomic Energy Agency (IAEA): *Table VIII. Excerpts from List of United Nations Numbers, Proper Shipping Names and Descriptions, Subsidiary Risks and Their Relationship to the Schedules*, Section V; Regulations for the Safe Transport of Radioactive Material, 1996 Edition, IAEA Safety Standards Series; IAEA: Vienna, 1996.

- International Civil Aviation Organization (ICAO): Table 2-14 *Dangerous Goods List*, Part 2, 11.5; Technical Instructions for the Safe Transport of Dangerous Goods by Air, 1999-2000; ICAO: Montreal, 1998.

- United States Department of Transportation (DOT): *Hazardous Materials Table*, Sec. 172.101; Title 49 of the Code of Federal Regulations, Revised as of Oct 1, 1998 (including revisions published in Federal Register through Feb 1, 1999); Office of the Federal Register, United States National Archives and Records Administration: Washington, 1998.

- International Air Transport Association (IATA): *List of Dangerous Goods*, Subsection 4.2; Dangerous Goods Regulations, 40th Edition; IATA: Montreal, 1998.

REGULATORY LIST ENTRIES

No list can include by name the infinite number of dangerous goods warranting regulation; consequently four types of entries, each of decreasing specificity are used to represent all regulated materials:

1. Single entries that identify well-defined materials such as *Methanol*.
2. Generic entries based on the use or application of the material such as *Adhesives* or *Paint*.
3. Generic entries based on the chemical family of the material such as *Alcohols, n.o.s.* (The suffix 'n.o.s.' means *not otherwise specified*.)
4. Generic entries based on the hazard of the material such as *Flammable liquid, n.o.s.* These are to be used if the dangerous goods cannot be described by a more precise entry.

The correct choice of entry is the basis of all subsequent hazard communication, packaging, and handling regulation; in fact, the entry must be chosen that describes the goods with the highest degree of specificity.

How to Use the *Glossary for the Worldwide Transportation of Dangerous Goods and Hazardous Materials*

The following key explains the seven elements that make up each chapter in the *Glossary* (see figure on opposite page).

1. CHAPTER TITLE

The *Glossary* is divided into 75 chapters, around which approximately 1400 entries from the regulatory lists are centred. There are exceptions to this rule:

- The *Terminology* Chapter explains chemical and physiochemical language used in multiple entries rather than the entries themselves. Such terms include *anhydrous, flash point, glacial, liquid, metal, tertiary, unphlegmatized*, etc.

- The *Units of Measure* Chapter explains the meaning of the units used in the transportation regulations to measure temperature, pressure, volume, magnetic strength, radioactivity, etc., and the interrelationship between similar measures.

2. ENTRIES

Each chapter deals with one or more related entries extracted from the lists of dangerous goods maintained by the regulatory authorities. All generic regulatory entries (i.e., those related to use, application, chemical family, or hazard) are included; standard chemical names are for the most part not included.

The related entries along with their hazards, where applicable, are listed alphabetically and as they appear in the regulations. They form the basis of the main index at the back of the *Glossary*. The following points should be noted:

- Slight wording, punctuation, or grammatical differences between the different regulatory lists account for many entries that appear similar; e.g., both "P.c.b.s.," and "PCBs," acronyms for polychlorinated biphenyls from ICAO and IATA, respectively, are included in the Halogenated Polyphenyls chapter. Differences in singular or plural entries are generally ignored.

- Entries appear regardless of whether they are legal shipping names; e.g., "Activated carbon" appears in many regulatory lists and references "Carbon, activated," the proper shipping description. Both are included in the *Glossary*.

- Entries appear even if they represent materials that are forbidden in or not restricted from transportation; e.g., "Aluminium dross, wet" (forbidden) and "Cinnabar" (not restricted).

1. CHAPTER TITLE   2. ENTRIES

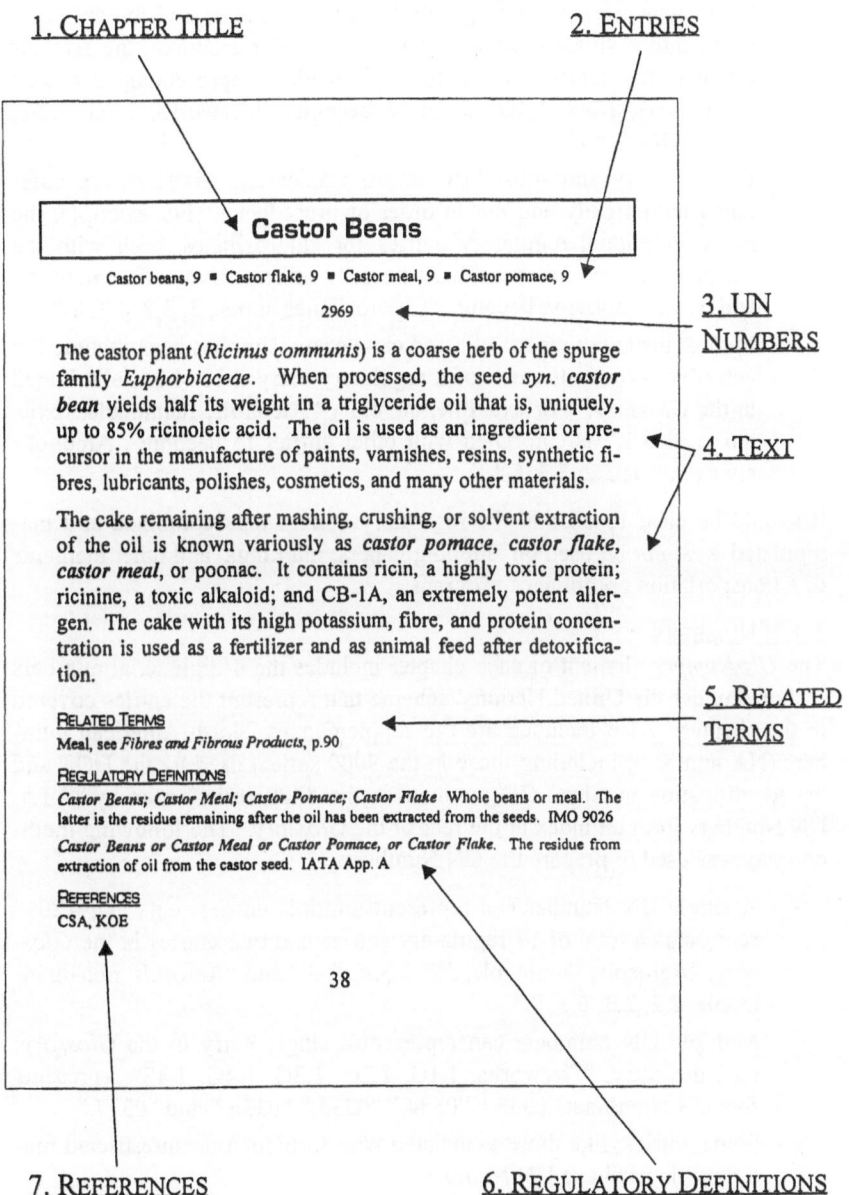

Castor Beans

Castor beans, 9 ▪ Castor flake, 9 ▪ Castor meal, 9 ▪ Castor pomace, 9

2969

3. UN NUMBERS

The castor plant (*Ricinus communis*) is a coarse herb of the spurge family *Euphorbiaceae*. When processed, the seed *syn. castor bean* yields half its weight in a triglyceride oil that is, uniquely, up to 85% ricinoleic acid. The oil is used as an ingredient or precursor in the manufacture of paints, varnishes, resins, synthetic fibres, lubricants, polishes, cosmetics, and many other materials.

4. TEXT

The cake remaining after mashing, crushing, or solvent extraction of the oil is known variously as *castor pomace*, *castor flake*, *castor meal*, or poonac. It contains ricin, a highly toxic protein; ricinine, a toxic alkaloid; and CB-1A, an extremely potent allergen. The cake with its high potassium, fibre, and protein concentration is used as a fertilizer and as animal feed after detoxification.

RELATED TERMS
Meal, see *Fibres and Fibrous Products*, p.90

5. RELATED TERMS

REGULATORY DEFINITIONS
Castor Beans; Castor Meal; Castor Pomace; Castor Flake Whole beans or meal. The latter is the residue remaining after the oil has been extracted from the seeds. IMO 9026

Castor Beans or Castor Meal or Castor Pomace, or Castor Flake. The residue from extraction of oil from the castor seed. IATA App. A

REFERENCES
CSA, KOE

38

7. REFERENCES   6. REGULATORY DEFINITIONS

- Regulatory entries differing only by hazard class or division are combined into a single entry in the *Glossary*. For example, the five individual regulatory entries for "Fireworks" representing divisions 1.1G, 1.2G, 1.3G, 1.4G, and 1.4S become, "Fireworks, 1.1G, 1.2G, 1.3G, 1.4G, 1.4S."

- Both primary and subsidiary hazard classes and divisions are combined numerically and *not* in order of precedence. For example, the many individual regulatory entries for chlorosilanes, each with the various primary and subsidiary hazards of corrosivity, flammability, and water-reactivity, become, "Chlorosilanes, n.o.s., 3, 3.2, 4.3, 8."

- In most instances where a hazard class or division has been included in narrative form in the text of a regulatory entry it has been eliminated in the *Glossary*. For example, the entry "Alcohols, flammable, toxic, n.o.s., 3, 6.1," is combined with other entries to become, "Alcohols, n.o.s., 3, 3.1, 3.2, 3.3, 6.1."

It should be clear to all that the regulatory entries thus combined and manipulated *must not* be used on shipping papers, markings, and other elements of a transportation compliance program.

3. UN NUMBERS

The *UN Number* element of each chapter includes the 4-digit serial numbers assigned under the United Nations' scheme that represent the entries covered in the chapter. Also included are the supplementary North American numbers (NA numbers, including those in the 9000 series) used by the DOT and the identification numbers (ID numbers in the 8000 series) used by IATA. UN Numbers form an index at the rear of the *Glossary*. The following methodology was used to prepare the UN Numbers:

- A single UN Number can represent multiple entries; e.g., "UN1201" represents a total of 17 regulatory entries and two entries in the *Glossary*: "Aerosols, flammable, 2.1, 2.3, 6.1, 8," and "Aerosols, non-flammable, 2.2, 2.3, 6.1, 8."

- Multiple UN Numbers can represent a single entry in the *Glossary*; e.g., the entry, "Fireworks, 1.1G, 1.2G, 1.3G, 1.4G, 1.4S" represents five UN Numbers: "0333," "0334," "0335," "0336," and "0337."

- Some entries, like those associated with forbidden or unrestricted materials, do not have UN Numbers.

- The prefixes "UN," "ID," or "NA" are omitted but may be required on use.

4. TEXT

The *Text* element of each chapter describes each of the entries in context presenting relationships and differences between similar terms.[7] As each regulatory entry is described for the first time in its chapter, it is italicized and bolded; e.g., the regulatory entry "Petroleum products, n.o.s., 3" becomes **petroleum products** in the Petroleum chapter. Similarly, the regulatory entry "Extracts, flavouring, liquid, 3" is represented by the key words **extracts** and **flavouring** in the Extracts chapter and **liquid** in the Terminology chapter. Elsewhere in the *Glossary*, the same key word or phrase may occur; in these instances, the page on which it appears is included in the index but the key word or phrase is not highlighted. The following points should be noted:

- The term *syn.* is used often to indicate that one key word or phrase is synonymous with another; e.g., **gunpowder syn. black powder.**

- Adjectives used to describe physical, chemical, or toxic hazards are not necessarily used in a regulatory context. For example, a sodium hydroxide solution described in the text as *corrosive* does not mean that it is sufficiently corrosive to meet the regulatory definition of a corrosive material (Class 8).

- CAS Registry Numbers,[8] which take the form: "00000-00-0" are used to identify the chemical compounds that are intimately related to the entries covered by the chapter.

5. RELATED TERMS

The *Related Terms* element of each chapter provides references in the *Glossary* to key words or phrases that

- Are used in an entry in the current chapter but are explained elsewhere; e.g., the key word "corrosive" from the entry "Coal tar dye, corrosive, liquid, n.o.s." in the Dyes and Pigments chapter is explained in the Corrosives and Class 8 chapter.

- Are associated with the entries as they appear in the text of the regulations; e.g., the entry "Insecticide gas, n.o.s. (aerosols in boxes), 2.2" appearing in IATA's *List of Dangerous Goods* becomes "Insecticide gas, n.o.s., 2.2" in the Pesticides chapter supplemented with the Related Term "aerosol" referenced to the Aerosols chapter.

6. REGULATORY DEFINITIONS

The *Regulatory Definitions* element in each chapter includes relevant definitions from the regulatory authorities. In particular, definitions that cover

---

[7] Even after extensive research the author was unable to find references to a small number of entries. Footnotes indicate the occurrence of those terms.

[8] Chemical Abstract Service (CAS) Registry Number, American Chemical Society. The *Dictionary of Chemical Names and Synonyms* (DOCN) was the major source of CASRNs.

identification and classification from glossaries and those associated with hazard classification are reproduced. Readers are encouraged to read through these definitions to highlight the frequent differences between regulatory authorities and to round out their understanding of a particular term.

All regulatory definitions are referenced to their sources. Rather than the usual method of citation (e.g., TI 2.2.1.2 for section 2.2.1.2 of ICAO's *Technical Instructions*), the following system is used:

- "UN" for the United Nations' *Recommendations on the Transport of Dangerous Goods* (ROT); e.g., "UN 2.1.1.1."
- "IMO" for the IMO's *International Maritime Dangerous Goods Code* (IMDG); e.g., "IMO Class 7, 1.1.5."
- "IAEA" for the IAEA *Regulations for the Safe Transport of Radioactive Material* (RFT); e.g., "IAEA Para. 222."
- "ICAO" for ICAO's *Technical Instructions for the Safe Transport of Dangerous Goods by Air* (TI); e.g., "ICAO 2-7.1."
- "US" for the United States' *Title 49 of the Code of Federal Regulations* (49CFR); e.g., "US 171.8."
- "IATA" for IATA's *Dangerous Goods Regulations* (DGR); e.g., "IATA 3.3.2."

Where two or more authorities use the same definition with only differences in format or punctuation, the definition is not duplicated; it is reproduced from its source according to the following hierarchy: UN, IMDG, ICAO, US, and IATA. IAEA takes precedence for all definitions related to radioactive materials. Definitions differing in wording, spelling, or pluralization, however small, are considered unique. Corrections to regulatory definitions appear within brackets: [ ].

It should be understood that these definitions may change slightly from year-to-year. It is the responsibility of the reader to confirm the current status of any particular definition.

7. REFERENCES

The *Reference* element of each chapter includes codes to the texts and other resources used to compile this text. The full titles of each document are listed in the *References* section in the back of the *Glossary*.

Adhesives

Adhesives, containing a flammable liquid, 3, 3.1, 3.2, 3.3 ■ Adhesives containing flammable liquid, 3 ■ Cement, flammable ■ Cement, liquid ■ Pyroxylin cement

1133

Adhesives (a term generally synonymous with *glues, mastics, mucilages,* and *pastes*) produce relatively permanent surface bonds between two substances (paper, glass, wood, metal, etc.) that develop by chemical reaction or as the adhesive cools or dries. The enormous variety of adhesives limits any system of nomenclature, but classification by binder, the major active ingredient, yields the following:

□ Animal, e.g., albumen, gelatines, casein, shellac, and beeswax.
□ Vegetable, e.g., natural resins such as gum arabic, oils, waxes, and mucilages.
□ Mineral, e.g., inorganic materials such as silicates, mineral waxes and resins, and bitumen.
□ Elastomeric, i.e., natural and synthetic rubbers.
□ Synthetic thermoplastics which soften on heating, e.g., cellulose derivatives, vinyl polymers, saturated polyesters, polyacrylates, polyethers, and polysulphones.
□ Synthetic thermosets which cure on heating to solids, e.g., amino plastics, epoxides, phenolic resins, unsaturated polyesters, polyaromatics, and furanes.

The constituency of traditional adhesives is limited to the single binders listed above; contemporary adhesives are likely to combine one or more binders with some combination of the following additives:

□ Diluents to carry the components and provide viscosity control.
□ Hardeners to cure the binder.
□ Catalysts to increase the rate and improve the efficiency of chemical reactions (cure time and cross-linking).
□ Accelerators, inhibitors, and retarders to control the curing rate.
□ Modifiers such as fillers, extenders, thinners, plasticizers, stabilizers, and wetting agents.

In the context of adhesives, *cement* is a natural rubber- or silicone-based elastomeric. Rubber cements contain a suitable solvent such as naphtha or aromatic hydrocarbons. *Pyroxylin cements* are adhesives based on solutions of nitrocellulose in alcohol, ether, or another solvent. Hydraulic cements used in construction, such as portland or pozzolana cement, are nonhazardous mixtures composed of some combination of lime, alumina, and silica which sets into a hard product (concrete) when water is added (the term *portland*

1

derives from the resemblance borne by the first synthetic hydraulic cement to a building stone quarried from the Isle of Portland off the coast of England).

HAZARDS

Certain binders are flammable and other additives may be hazardous, but it is the common use of flammable solvents as thinners or diluents that presents the primary hazard in transportation.

RELATED TERMS

Flammable; Flammable liquid, see *Flammable Liquids and Class 3*, p.96

Liquid, see *Terminology, Liquid*, p.241

Pyroxylin, see *Nitrocellulose Products*, p.161

REGULATORY DEFINITIONS

Cement. The fine grey powder composed of lime, alumina and silica which sets to a hard product when water is added. Also known as hydraulic cement or Portland cement. It is used to make concrete. This product is not restricted for transport by air. IATA App. A

Cement, flammable. This product, properly called an adhesive, usually contains rubber or a rubber-like substance and a solvent. It is used to bond other substances together such as paper or leather. The solvent may be flammable. IATA App. A

REFERENCES

AH, HCC, KOC, MEOS, MH14, THS

Aerosols

Aerosols, 2 ■ Aerosols, flammable, 2.1, 2.3, 6.1, 8 ■ Aerosols, non-flammable, 2.2, 2.3, 6.1, 8 ■ Pressurized products

1950

An *aerosol* is a suspension of liquid or solid particles in a gas. Natural aerosols include smoke, mist, and fog, all particles suspended in air. Aerosols may be mechanically generated by making a solution, emulsion, or suspension of a liquid or solid in a compressed or liquefied gas in a sealed container. As the gas (the propellant) is released through a valve an aerosol or aerogel (foam) is generated. The pressurized containers holding these mixtures are known as *aerosols* or *pressurized products*.

There are a large number of propellant-product combinations. Common propellants include halocarbons (although the use of CFCs is widely abandoned), compressed air, carbon dioxide, nitrous oxide, butane, and propane. Aerosols are used in industrial applications and consumer articles to dispense or apply cosmetics and perfumes, paints and coatings, medicines and drugs, starches, pesticides, disinfectants, deodorants, and many other materials.

HAZARDS
Both the compressed gas propellant and the product propelled (an often inert but possibly toxic, flammable, or corrosive material) present a potential hazard in transportation.

RELATED TERMS

Biological products, see *Infectious Substances and Division 6.2*, p.115

Class 8, see *Corrosives and Class 8*, p.47

Division 6.1, see *Toxic Substances and Division 6.1*, p.255

Engine starting fluid, see *Self-Propelled Vehicles*, p.219

Flammable, see *Gases and Class 2*, p.104

Medicinal preparations, see *Drugs and Medicines*, p.61

Non-flammable, see *Gases and Class 2*, p.104

Pepper spray, see *Ammunition, Toxic*, p.19

Poison, see *Toxic Substances and Division 6.1*, p.255

Self-defense spray, aerosol, see *Ammunition, Toxic*, p.19

Tear gas devices, see *Ammunition, Toxic*, p.19

Toxic gas, see *Gases and Class 2*, p.104

REGULATORY DEFINITIONS
For the purpose of these instructions an *aerosol* means any non-refillable receptacle made of metal, glass or plastic and containing a gas compressed, liquefied or dissolved under pressure, with or without a liquid, paste or powder, and fitted with a self-closing release device allowing the contents to be ejected as solid or liquid particles in suspension in a gas, as a foam, paste or powder, or in a liquid or gaseous state. ICAO 2.5.1

Aerosol means any non-refillable metal receptacle containing a gas compressed, liquefied or dissolved under pressure, the sole purpose of which is to expel a nonpoisonous (other than a Division 6.1 Packing Group III material) liquid, paste, or powder and fitted with a self-closing release device allowing the contents to be ejected by the gas. US 171.8

Aerosol. Means any non-refillable receptacle made of metal, glass or plastic and containing a gas compressed, liquefied or dissolved under pressure, with or without a liquid, paste or powder, and fitted with a self-closing release device allowing the contents to be ejected as solid or liquefied particles in suspension in a gas, as a foam, paste or powder, or in a liquid or gaseous state. IATA App. A

REFERENCES
DOSATT, HCC, KOC, MEOS, MH14, POG

Alcohols

Alcohol, 3.2 ▪ Alcohol, denatured ▪ Alcohol, denatured, 3.2 ▪ Alcohol, denatured solutions, 3.3 ▪ Alcoholic beverages, 3, 3.2, 3.3 ▪ Alcohol, industrial ▪ Alcohol, industrial, 3.2 ▪ Alcohol, industrial, solutions, 3.3 ▪ Alcohols, n.o.s., 3, 3.1, 3.2, 3.3, 6.1 ▪ Alcohol solutions, 3.3 ▪ Denatured alcohol, 3 ▪ Fermentation amyl alcohol ▪ Fusel oil, 3, 3.2, 3.3 ▪ Liquor ▪ Methylated spirit ▪ Tertiary alcohol

1170 ▪ 1201 ▪ 1986 ▪ 1987 ▪ 3065

Alcohols are a class of organic compounds containing one or more hydroxyl groups (-OH) attached to a carbon atom that exists in one of four configurations:

1. Attached to no other carbon atom, i.e., methanol.
2. Bonded to one other carbon atom, a primary alcohol.
3. Bonded to two other carbon atoms, a secondary alcohol.
4. Bonded to three carbon atoms, *tertiary alcohols*, such as *tert*-amyl alcohol.

Alcohols include phenol, the sterols, glycols, diols, and glycerols.

ALCOHOLIC BEVERAGES

Alcoholic beverages syn. liquor describe beverages for human consumption containing ethanol (64-17-5) which is miscible in water, the other principal ingredient. Together they hold in solution or suspension a wide and sometimes complex variety of flavours, sugars, and other compounds characteristic of the beverage. Ethanol may be added to a beverage, or more often produced by the fermentation (the enzymatic decomposition of sugars and starches) of carbohydrate vegetable matter:

❑ Beer: based on the fermentation of starches from barley malt, rice, corn, wheat, potatoes, and other sources. While low-alcohol beers are made, most beers contain around 3 to 5% ethanol by volume.

❑ Wine: based on the fermentation of sugars from grapes as well as other fruits, rice, and leaves. Table wines range up to around 14% ethanol by volume; dessert or aperitif wines, usually sweeter, range from around 14 to 24% ethanol.

❑ Distilled spirits: beverages distilled from the fermentation products of characteristic ingredients that impart distinctive and prized flavours: whiskeys from grains, e.g., rye, barley, and corn; gin from juniper berries; brandy from fruit; rum from sugar cane; tequila from the cactus *Agave tequilana weber*; and vodka from purified grain distillates. Spirits are usually bottled at ethanol concentrations around 40% or higher by volume.

OTHER ALCOHOLS

Denatured alcohol is ethanol rendered unfit for drinking or internal medicinal use by the addition of a small fraction of some malodorous or obnoxious substance. Once denatured, ethanol is no longer subject to beverage tax although it is still widely acceptable commercially as a solvent, raw material, fuel, fluid, etc. The most effective denaturants have a boiling point sufficiently close to that of ethanol to make separation by distillation difficult. They include methanol, brucine, brucine sulphate, quassin, n-butanol, petroleum, and wood distillates. *Methylated spirits* and *industrial alcohol* are used synonymously with *denatured alcohol*, although the first specifically applies to those alcohols denatured by methanol (67-56-1) while the latter may have been denatured only by dilution with water.

Fusel oil (8013-75-0) is a mixture of the eight possible isomers of amyl alcohol and smaller amounts of other alcohols resulting from the fermentation of certain carbohydrates including grains and potatoes, hence *fermentation amyl alcohol* (123-51-3) which is principally 3-methyl-1-butanol. Fusel oil is a widely used solvent.

HAZARDS

Many alcohols including ethanol, methanol, and the amyl alcohols are flammable and exhibit varying degrees of toxicity. The often desired inebriating effect of alcoholic beverages is one manifestation of the potential toxicity of ethanol. Almost all denaturants are toxic by definition. Other alcohols may have other hazards.

RELATED TERMS

By Volume, see *Terminology, By Volume,* p.233

Flammable, see *Flammable Liquids and Class 3,* p.96

Solutions, see *Terminology, Solutions,* p.247

Toxic, see *Toxic Substances and Division 6.1,* p.255

REGULATORY DEFINITIONS

None

REFERENCES

27CFR21, HCC, KOE, MH12, SSD

Alkaloids

Alkaloid salts, liquid, n.o.s., 6.1 ■ Alkaloid salts, solid, n.o.s., 6.1 ■ Alkaloids, liquid, n.o.s., 6.1 ■ Alkaloids, solid, n.o.s., 6.1 ■ Nicotine compound, liquid, n.o.s., 6.1 ■ Nicotine compound, solid, n.o.s., 6.1 ■ Nicotine preparation, liquid, n.o.s., 6.1 ■ Nicotine preparation, solid, n.o.s., 6.1

1544 ■ 1655 ■ 3140 ■ 3144

Alkaloids encompass a range of crystalline solid or liquid organic substances primarily of vegetable origin and usually derived from cyclic nitrogen compounds. Many exhibit physiological activity including some well-known and sometimes extremely toxic medicinal and recreational drugs:

- Atrophine, used in medicines, pesticides, and nerve gas.
- Caffeine, extracted from coffee beans, tea leaves, and the kola nut and used in medicine and beverages.
- Cocaine, extracted from *Erythroxylon coca* and used as a nerve stimulant and anaesthetic.
- Codeine and morphine, narcotics derived from opium.
- Mescaline, a hallucinogen derived from peyote, a small cactus.
- *Nicotine* (54-11-5), a thick oil derived from tobacco and used as an insecticide and fumigant.
- Quinine, an antimalarial drug and flavouring extracted from cinchona bark.
- Strychnine, a poison extracted from the seeds of *Nux vomica*.

Alkaloids are usually basic and combine with acids to form *alkaloid salts*, a property often exploited to extract them from their source. Other alkaloids occur naturally as salts of organic acids. Common salts include hydrochlorides, salicylates, sulphates, nitrates, acetates, and tartrates such as morphine acetate, cocaine hydrochloride, and strychnine nitrate. Water, alcohol, and ether solutions of alkaloids and their salts are often used to administer or carry the alkaloid, particularly for medicinal purposes. *Nicotine preparations* can include a variety of liquid and solid mixtures of nicotine (soluble in alcohol, chloroform, ether, and water), nicotine salts, and many other *nicotine compounds* (e.g., nicotine sulphate and nicotine tartrate).

RELATED TERMS

Alkaloid and alkaloid salts (pesticides), see *Pesticides*, p.179

Compound, see *Terminology, Compound*, p.234

Liquid, see *Terminology, Liquid*, p.241

Poisonous, see *Toxic Substances and Division 6.1*, p.255

Salts, see *Terminology, Salts*, p.247

Solid, see *Terminology, Solid*, p.247

REGULATORY DEFINITIONS

None

REFERENCES

HCC, KOC, KOE

Ammunition

Ammunition, blank ∎ Ammunition, fixed, semi-fixed or separate loading ∎ Ammunition, incendiary liquid or gel, with burster, expelling charge or propelling charge, 1.3J ∎ Ammunition, incendiary (water-activated contrivances) ∎ Ammunition, incendiary (water-activated contrivances) with burster, expelling charge or propelling charge ∎ Ammunition, incendiary, white phosphorus with burster, expelling charge or propelling charge, 1.2H, 1.3H ∎ Ammunition, incendiary with or without burster, expelling charge or propelling charge, 1.2G, 1.3G, 1.4G ∎ Ammunition, practice, 1.3G, 1.4G ∎ Ammunition, proof, 1.4G ∎ Ammunition, rocket ∎ Ammunition, SA (small arms) ∎ Ammunition, sporting ∎ Bag charges ∎ Bangalore torpedoes ∎ Bombs with bursting charge, 1.1D, 1.1F, 1.2D, 1.2F ∎ Bombs with flammable liquid with bursting charge, 1.1J, 1.2J ∎ Bursters, explosive, 1.1D ∎ Cartridges for weapons, blank, 1.1C, 1.2C, 1.3C, 1.4C, 1.4S ∎ Cartridges for weapons, inert projectile, 1.2C, 1.3C, 1.4C, 1.4S ∎ Cartridges for weapons, with bursting charge, 1.1E, 1.1F, 1.2E, 1.2F, 1.4E, 1.4F ∎ Cartridges, safety ∎ Cartridges, safety, blank ∎ Cartridges, small arms, 1.2C, 1.3C, 1.4C, 1.4S ∎ Cartridges, small arms, blank, 1.3C, 1.4C, 1.4S ∎ Cartridges, sporting ∎ Charges, bursting, plastics bonded, 1.1D, 1.2D, 1.4D, 1.4S ∎ Charges, depth, 1.1D ∎ Charges, propelling, 1.1C, 1.2C, 1.3C, 1.4C ∎ Charges, propelling, for cannon, 1.1C, 1.2C, 1.3C ∎ Depth charges ∎ Engines, rocket ∎ Grenades, empty primed, 1.4S ∎ Grenades, hand or rifle, with bursting charge, 1.1D, 1.1F, 1.2D, 1.2F ∎ Grenades, practice, hand or rifle, 1.2G, 1.3G, 1.4G, 1.4S ∎ Jet thrust unit (Jato) ∎ Mines with bursting charge, 1.1D, 1.1F, 1.2D, 1.2F ∎ Missiles, guided ∎ Model rocket motor, 1.4C, 1.4S ∎ Projectiles with burster or expelling charge, 1.2D, 1.2F, 1.2G, 1.4D, 1.4F, 1.4G ∎ Projectiles, inert with tracer, 1.3G, 1.4G, 1.4S ∎ Projectiles with bursting charge, 1.1D, 1.1F, 1.2D, 1.2F, 1.4D ∎ Propellant, liquid, 1.1C, 1.3C ∎ Propellant, solid, 1.1C, 1.3C ∎ Rifle grenade ∎ Rocket motors, 1.1C, 1.2C, 1.3C ∎ Rocket motors, liquid fuelled, 1.2J, 1.3J ∎ Rocket motors with hypergolic liquids with or without expelling charge, 1.2L, 1.3L ∎ Rockets, liquid fuelled with bursting charge, 1.1J, 1.2J ∎ Rockets with bursting charge, 1.1E, 1.1F, 1.2E, 1.2F ∎ Rockets with expelling charge, 1.2C, 1.3C, 1.4C ∎ Rockets with inert head, 1.3C ∎ Torpedoes, liquid fuelled with inert head, 1.3J ∎ Torpedoes, liquid fuelled with or without bursting charge, 1.1J ∎ Torpedoes with bursting charge, 1.1D, 1.1E, 1.1F ∎ Warheads for guided missiles ∎ Warheads, rocket with burster or expelling charge, 1.4D, 1.4F ∎ Warheads, rocket with bursting charge, 1.1D, 1.2D, 1.1F ∎ Warheads, torpedo, with bursting charge, 1.1D

0005 ∎ 0006 ∎ 0007 ∎ 0009 ∎ 0010 ∎ 0012 ∎ 0014 ∎ 0033 ∎ 0034 ∎ 0035 ∎ 0043 ∎ 0056 ∎ 0110 ∎ 0136 ∎ 0137 ∎ 0138 ∎ 0167 ∎ 0168 ∎ 0169 ∎ 0180 ∎ 0181 ∎ 0182 ∎ 0183 ∎ 0186 ∎ 0221 ∎ 0238 ∎ 0242 ∎ 0243 ∎ 0244 ∎ 0247 ∎ 0250 ∎ 0271 ∎ 0272 ∎ 0276 ∎ 0279 ∎ 0280 ∎ 0281 ∎ 0284 ∎ 0285 ∎ 0286 ∎ 0287 ∎ 0291 ∎ 0292 ∎ 0293 ∎ 0294 ∎ 0295 ∎ 0300 ∎ 0318 ∎ 0321 ∎ 0322 ∎ 0323 ∎ 0324 ∎ 0326 ∎ 0327 ∎ 0328 ∎ 0329 ∎ 0330 ∎ 0338 ∎ 0339 ∎ 0344 ∎ 0345 ∎ 0346 ∎ 0347 ∎ 0348 ∎ 0349 ∎ 0362 ∎ 0363 ∎ 0369 ∎ 0370 ∎ 0371 ∎ 0372 ∎ 0395 ∎ 0396 ∎ 0397 ∎ 0398 ∎ 0399 ∎ 0400 ∎ 0412 ∎ 0413 ∎ 0414 ∎ 0415 ∎ 0417 ∎ 0424 ∎ 0425 ∎ 0426 ∎ 0427 ∎ 0434 ∎ 0435 ∎ 0436 ∎ 0437 ∎ 0438 ∎ 0446 ∎ 0447 ∎ 0449 ∎ 0450 ∎ 0451 ∎ 0452 ∎ 0457 ∎ 0458 ∎ 0459 ∎ 0460 ∎ 0488 ∎ 0491 ∎ 0495 ∎ 0497 ∎ 0498 ∎ 0499

Ammunition describes devices containing some combination of explosive, biological, nuclear, or chemical material for use in military, law-enforcement, or sporting applications. An individual unit of ammunition is called a *round*. Biological weapons (those which project, disperse, or disseminate biological agents) were banned by the *Geneva Protocol* of 1925 (COTB). The *Chemi-*

cal Weapons Convention (COTC) outlaws the use of chemical weapons. Ammunition may be categorized by its means of delivery:

□ *Projectile ammunition* is ejected from the chamber or barrel of a weapon toward the target by the gases generated on ignition of an explosive propelling or expelling charge. *Propelling charges* are integral to the round and may continue to provide propulsion during flight, while *expelling charges* are independent of the round. The traditional use of gunpowder for these charges is now almost completely superseded by the use of smokeless powders: one of the single-, double-, or triple-base propellants. (Propellants also burn at the base of some projectiles during flight to counteract the drag generated by air currents at that point.)

□ Nonprojectile ammunition is delivered by mechanical or manual projection, or by dropping from an aircraft, by awaiting the arrival of the target, or other means.

Ammunition may simply deliver high velocity inert projectiles that cause injury and damage, or it may deliver warheads that contain *bursters* (charges that break open the projectile and scatter its contents: incendiaries, tear gases, and pyrotechnics), high explosive *bursting charges* that explode and scatter shrapnel, or powerful explosive charges.

PROJECTILE AMMUNITION

Small Arms Ammunition

Small arms ammunition syn. SA ammunition or *small arms cartridges* used in pistols, rifles, revolvers, machine guns, and shotguns consist of a metal case (cartridge case or shell) often of brass or steel containing a bullet, a propelling charge, and a percussion initiating device (primer or cap). The diameter of the bullet (a spire- or ogival-pointed slug of metal wholly or partially covered by a metal jacket) defines the calibre of the ammunition, ranging up to 15 or 20mm for small arms. Armour-piercing bullets contain lead, whose density provides high momentum, surrounding an extremely hard steel core. Other bullets may be hollowed to carry detonating explosives, incendiaries, or tracers.

As it strikes the primer, the shock of a weapon's firing pin or hammer ignites the initiating explosive it contains which in turn ignites the propelling charge. The breech (loading and fire unit) of the weapon is momentarily sealed by the expanded cartridge case as the hot gases accelerate the bullet through the chamber. The spent cartridge is automatically or manually expelled before the weapon is reloaded.

Sporting ammunition describes rifle and handgun ammunition for hunting and target practice. Commercial hunting bullets, ranging in size from around 6 to 12mm, may differ from the conical shape of military bullets in that they may be sharply pointed or blunt. Shotguns, used in the military and law-enforcement as well as hunting, use cylindrical shells holding the propelling charge, primer, and multiple metal balls varying from 1 to 5mm in diameter.

Artillery Ammunition

Artillery ammunition for guns, howitzers, cannons, and mortars ranges upwards in calibre from the 20mm limit that generally divides it from small-arms ammunition. Artillery ammunition is divided into three types:

- *Fixed ammunition*, that containing the projectile and propelling charge in a single cartridge.
- *Semi-fixed ammunition*, that in which a round of ammunition is supplied in two units: the shell containing the payload and fuze, and an independent cartridge containing the propelling charge and primer. The propelling charge may be adjusted to change velocity and range before the two units are fitted together prior to loading.
- *Separate loading ammunition*, that in which the projectile and propelling charge are loaded separately. They are used in large-calibre weapons where the combined weight of the projectile and charge would make handling difficult, to permit adjustment of the propelling charge, or when *bag charges* are in use.

Bag charges are waterproofed cloth bags containing a propelling charge and sometimes a primer that are designed to be consumed during firing so that the weapon can be reloaded without the need to eject a spent cartridge. When used in separate loading ammunition, bag charges are inserted into the chamber of the weapon with the projectile placed on top. In other designs, combustible cartridge cases are used that add to the ballistic performance of the weapon. Bag charges are also inserted into the cartridges of semi-fixed ammunition to adjust the propelling charge.

Large-calibre artillery ammunition permits a more complex payload including high explosives, bursters, shrapnel, bursting charges, fuzes, incendiaries, etc.

Self-Propelled Ammunition

In contrast to small-arms or artillery ammunition, self-propelled ammunition incorporates a *rocket* in which the propelling charge called a *propellant* (a mixture of fuel and an oxidizer) is burned with sufficient thrust to project the rocket to its target. Liquid-fuelled rockets are those in which fuel and oxidizer are injected from separate storage tanks into the thrust chamber of a *rocket engine* allowing for great control over combustion. Solid-fuelled rockets are those in which the thrust chamber of a *rocket motor* is already charged with an intimate solid mixture of fuel and oxidizer which burns until exhausted. Advantageously, solid rockets require less pumping, maintenance, and handling and, because the propellant cannot slosh around in flight, are more stable than liquid rockets. Hybrid liquid and solid rockets exist.

In addition to catalysts, stabilizers, opacifiers (to reduce heat radiation ahead of the flame), flash depressants, plasticizers, and binders, the main fuel and oxidizers for rocket propellants include

- Liquid rockets: liquefied oxygen, fluorine, hydrogen, boron hydride, hydrazine, hydrogen peroxide, nitric acid, and JP4.
- Solid rockets: carboxy- or hydroxy-terminated polybutadienes, ammonium perchlorate, ammonium nitrate, HMX, RDX, polysulphides, nitrocellulose, nitrate esters, nitramines, waxes, asphalts, oils, and resins.

While the term *missile* can apply to any projectile ammunition, *guided missiles* describes those rockets whose course or trajectory is controlled electronically by computers and communications equipment.

Rockets may also be used for aircraft or spacecraft boosters, steering, and braking systems. *Jet thrust units syn. JATO* (*jet-assisted take-off*) are auxiliary devices containing solid fuel rocket motors which provide thrust for the takeoff and initial acceleration of aircraft or missiles. They were introduced for use in heavily laden aeroplanes and seaplanes. The more accurate term may be *RATO*, for *rocket-assisted take-off*, because jet engines draw oxygen for fuel combustion from the atmosphere (rocket engines carry their own oxidizers). *Model rocket motors* are toys containing black powder in tubes which is ignited to provide thrust as the combustion gases are discharged through a clay nozzle.

Torpedoes are tubular rockets or missiles designed to operate underwater. They are the principal weapons of submarines but may be dropped from aircraft or surface vessels. Torpedoes may be expelled by compressed air or propelled by a propelling charge, steam, or electricity.

Warheads

Warheads contain the agent intended to inflict damage in any projectile ammunition, be it inert, explosive, nuclear, biological, or chemical, although the term is most often applied to self-propelled missiles, rockets, and torpedoes. Warheads include casings, destructive agents, and a power supply. They may be loaded into the ammunition just before use. With the addition of fuzes, safety, and arming mechanisms, warheads become *armament systems*.

NONPROJECTILE AMMUNITION

Many types of ammunition are not projected toward their target and, therefore, do not contain a propelling or expelling charge. They do, however, include any combination of bursters, bursting charges, and other destructive payloads:

- *Bombs*, ammunition that is dropped by aircraft, although bombs may be equipped with rockets to give them added velocity and penetrating power. They range in size up to many thousands of kilograms. In some countries, bombs describe projectiles fired from mortars.
- *Grenades*, small, short-range munitions that are projected by hand and are filled with bursters or bursting charges. They disperse chemicals (tear gases and incendiaries) or shrapnel as they explode on impact or by a time fuze. Grenades are also projected from specially designed grenade launchers or from rifles using grenade cartridges (grenades with an integral cartridge) or bullet catchers that allow them to be projected with a normal round of ammunition.
- *Depth charges*, ammunition dropped or catapulted into water and which explode on contact with or in the vicinity of their targets through the action of water pressure or by a time fuze.
- *Mines* are distributed in the subsurface of minefields or landing strips, suspended in the air with balloons, or floated in water. They work by fuzes or remote control, or by being triggered on contact with or by the influence (pressure or magnetic fields) of the target itself (personnel, boats, vehicles, aircraft).
- *Bangalore torpedoes* are simple metal pipes or tubes a few centimetres in diameter and one or two metres in length. They are packed with high explosives and slipped into barbed wire fencing or wire obstacles to cut a path a few metres wide, or laid on the ground to detonate buried mines.

Incendiary ammunition damages property, equipment, and personnel by localized fire and burning. It includes mines and flamethrowers (in which a compressed gas may project the fuel) as well as grenades, bombs, artillery shells, and other projectiles that have been used on areas as large as $20km^2$ to destroy cities by firestorm and conflagration. Many chemical compositions are used including petroleum fuels, white phosphorus, finely divided metals (zirconium is often used as is depleted uranium), certain organometallics (e.g., triethylaluminium or triethylmagnesium), and organic chemicals. A typical mixture for an incendiary bomb may include 86% benzene, 10% rubber as a thickener, and 4% white phosphorus. Thermite, a common incendiary which burns at 2000°C is a 73:27% iron oxide-magnesium mixture.

Depending on the composition, incendiaries include an ignition device or ignite spontaneously when exposed to the moisture or oxygen in the atmosphere. Bursters and bursting charges often scatter and ignite the fuel or expose the pyrophoric material. Pyrogels (gels) are pyrophoric incendiaries that consist of a petroleum product (e.g., gasoline, other distillates, or asphalt) thickened with some agent such as isobutyl metacrylate or natural rubber. Metals, including magnesium, and oxidizers, such as sodium nitrate, are added to increase the temperature of combustion.

BLANK, PRACTICE, AND PROOF AMMUNITION

Many types of ammunition are used for testing, training, and ceremonial purposes:

- *Proof ammunition* is designed to test the ammunition or weapon. It may be fully or partially functional.
- *Blank ammunition syn. safety cartridges* refers to small-arms cartridges containing a charge but no bullet which are used for ceremonial or training purposes. Wads holding the charge in place are ejected powerfully enough to cause harm at short distances. Blank cartridges may also be made of plastic. These split under the explosion but eject no solid material.
- *Practice ammunition* can refer to blank ammunition, but is usually applied to fully operational large-calibre ammunition in which an otherwise extremely expensive payload has been substituted with less damaging or less complex payloads; e.g., practice involving armour-piercing ammunition may include replicas. Small-arms practice ammunition may involve plastic bullets suitable for short-range target practice.
- Drill ammunition contains no explosive and is used for loading practice.

RELATED TERMS

Ammunition, illuminating, see *Pyrotechnics and Signals*, p.194

Ammunition, industrial, see *Explosive Articles*, p.69

Ammunition, lachrymatory, see *Ammunition, Toxic*, p.19

Ammunition, smoke, see *Pyrotechnics and Signals*, p.194

Ammunition, tear-producing, see *Ammunition, Toxic*, p.19

Ammunition, toxic, see *Ammunition, Toxic*, p.19

Bombs, illuminating, see *Pyrotechnics and Signals*, p.194

Bombs, photo-flash, see *Pyrotechnics and Signals*, p.194

Bombs, smoke, non-explosive, see *Pyrotechnics and Signals*, p.194

Bombs, target identification, see *Pyrotechnics and Signals*, p.194

Cartridge case, see *Explosive Articles*, p.69

Cartridges, flash, see *Pyrotechnics and Signals*, p.194

Cartridges, illuminating, see *Pyrotechnics and Signals*, p.194

Cartridges, signal, see *Pyrotechnics and Signals*, p.194

Cases, cartridge, see *Explosive Articles*, p.69

Cases, combustible, see *Explosive Articles*, p.69

Charges, shaped without detonator, see *Explosive Articles*, p.69

Charges, shaped, flexible, linear, see *Explosive Articles*, p.69

Charges, supplementary, explosive, see *Initiating Explosives*, p.119

Combustible, see *Terminology, Combustion*, p.233

Contrivances, water-activated, see *Explosive Articles*, p.69

Detonators for ammunition, see *Initiating Explosives*, p.119

Explosive, see *Explosives and Class 1*, p.74

Flammable liquid, see *Flammable Liquids and Class 3*, p.96

Fuses, tracer, see *Pyrotechnics and Signals*, p.194

Grenades, illuminating, see *Pyrotechnics and Signals*, p.194

Grenades, smoke, see *Pyrotechnics and Signals*, p.194

Hypergolic, see *Terminology, Hypergolic*, p.240

Igniters see *Initiating Explosives*, p.119

Jet thrust igniters, for rocket motors or Jato, see *Initiating Explosives*, p.119

Liquid, see *Terminology, Liquid*, p.241

Plastics bonded charge, see *Explosives and Class 1*, p.74

Primer, see *Initiating Explosives*, p.119

Primers, cap type, see *Initiating Explosives*, p.119

Primers, small arms, see *Initiating Explosives*, p.119

Projectile, illuminating, see *Pyrotechnics and Signals*, p.194

Propellant, single, double or triple base, see *Explosives and Class 1*, p.74

Rockets, line-throwing, see *Lifesaving Equipment*, p.134

Shaped charges, see *Explosive Articles*, p.69

Sounding devices, see *Explosive Articles*, p.69

Tear gas cartridges, see *Ammunition, Toxic*, p.19

Tear gas grenades, see *Ammunition, Toxic*, p.19

Tracer, see *Pyrotechnics and Signals*, p.194

Tracers for ammunition, see *Pyrotechnics and Signals*, p.194

Water-activated contrivances, see *Explosive Articles*, p.69

REGULATORY DEFINITIONS

Ammunition Generic term related mainly to articles of military application consisting of all kind[s] of bombs, grenades, rockets, or mines, projectiles and other similar devices or contrivances. UN App. B, ICAO A2, IATA App. A

Ammunition. Generic term related mainly to articles of military application consisting of all types of bombs, grenades, rockets, mines, projectiles and other similar devices or contrivances. US 173.59

Ammunition, Incendiary Ammunition containing incendiary substance which may be a solid, liquid or gel including white phosphorus. Except when the composition is an explosive per se, it also contains one or more of the following: a propelling charge with primer and igniter charge; a fuze with burster or expelling charge. The term includes: Ammunition, Incendiary liquid or gel, with burster, expelling charge or propelling charge; Ammunition, Incendiary with or without burster, expelling charge or propelling charge; Ammunition, Incendiary, White Phosphorus with burster, expelling charge or propelling charge. UN App. B, ICAO A2, IATA App. A

Ammunition, incendiary. Ammunition containing an incendiary substance which may be a solid, liquid or gel including white phosphorus. Except when the composition is an explosive per se, it also contains one or more of the following: a propelling charge with primer and igniter charge, or a fuze with burster or expelling charge. The term includes: Ammunition, incendiary, liquid or gel, with burster, expelling charge or propelling charge; Ammunition, incendiary with or without burster, expelling charge or propelling charge; and Ammunition, incendiary, white phosphorus, with burster, expelling charge or propelling charge. US 173.59

Ammunition, Practice Ammunition without a main bursting charge, containing a burster or expelling charge. Normally it also contains a fuze and a propelling charge. The term excludes the following articles which are listed separately: Grenades, Practice. UN App. B, ICAO A2, US 173.59

Ammunition, Proof Ammunition containing pyrotechnic substances, used to test the performance or strength of new ammunition, weapon component or assemblies. UN App. B, ICAO A2, US 173.59

Ammunition, Proof. Ammunition containing pyrotechnic substance(s) used to test the performance or strength of new ammunition or weapon components or assemblies. IATA App. A

Bombs Explosive articles which are dropped from aircraft. They may contain a flammable liquid with bursting charge, a photo-flash composition or a bursting charge. The term excludes torpedoes (aerial) and includes: Bombs, Photo-Flash; Bombs with bursting charge; Bombs With Flammable Liquid with bursting charge. UN App. B, ICAO A2, IATA App. A

Bombs. Explosive articles which are dropped from aircraft. They may contain a flammable liquid with bursting charge, a photo-flash composition or bursting charge. The term excludes torpedoes (aerial) and includes bombs, photo-flash; bombs with bursting charge; bombs with flammable liquids, with bursting charge. US 173.59

Bursters, explosive Articles consisting of a small charge of explosive used to open projectiles, or other ammunition in order to disperse their contents. UN App. B, ICAO A2, IATA App. A

Bursters, explosive. Articles consisting of a small charge of explosive to open projectiles or other ammunition in order to disperse their contents. US 173.59

Cartridges, blank Articles which consist of a cartridge case with a centre or rim fire primer and a confined charge of smokeless or black powder but no projectile. Used for training, saluting or in starter pistols, etc. UN App. B, ICAO A2, IATA App. A

Cartridges, blank. Articles which consist of a cartridge case with a center or rim fire primer and a confined charge of smokeless or black powder, but no projectile. Used in training, saluting, or in starter pistols, etc. US 173.59

Cartridges for Weapons (1) Fixed (assembled) or semi-fixed (partially-assembled) ammunition designed to be fired from weapons. Each cartridge includes all the components necessary to function the weapon once. The name and description shall be used for small arms cartridges that cannot be described as "cartridges, small arms". Separate loading ammunition is included under this name and description when the propelling charge and projectile are packed together (see also "Cartridges, blank"). (2) Incendiary, smoke, toxic and tear-producing cartridges are described in this Glossary under Ammunition, Incendiary etc. UN App. B

Cartridges for Weapons. 1) Fixed (assembled) or semi-fixed (partially-assembled) ammunition designed to be fired from weapons. Each cartridge includes all the components necessary to function the weapon once. The name and description shall be used for small arms cartridges that cannot be described as "cartridges, small arms". Separate loading ammunition is included under this name and description when the propelling charge and projectile are packed together (see also "Cartridges, blank"). 2) Incendiary, smoke, toxic and tear-producing cartridges are described in this Attachment under "ammunition, incendiary" etc. ICAO A2

Cartridges for weapons. (1) Fixed (assembled) or semi-fixed (partially assembled) ammunition designed to be fired from weapons. Each cartridge includes all the components necessary to function the weapon once. The name and description should be used for military small arms cartridges that cannot be described as cartridges, small arms. Separate loading ammunition is included under this name and description when the propelling charge and projectile are packed together (see also Cartridges, blank). (2) Incendiary, smoke, toxic, and tear-producing cartridges are described under ammunition, incendiary, etc. US 173.59

Cartridges for Weapons. (a) fixed (assembled) or semi-fixed (partially-assembled) ammunition designed to be fired from weapons. Each cartridge includes all the components necessary to function the weapon once. The name and description shall be used for small arms cartridges that cannot be described as "Cartridges, small arms". Separate loading ammunition is included under this name and description when the propelling charge and projectile are packed together (see also "Cartridges, blank"); (b) incendiary, smoke, toxic and tear-producing cartridges are described in this appendix under "Ammunition, incendiary" etc. IATA App. A

Cartridges for Weapons, Inert Projectile Ammunition consisting of a projectile without bursting charge but with a propelling charge. The presence of a tracer can be disregarded for classification purposes provided that the predominant hazard is that of the propelling charge. UN App. B, ICAO A2, IATA App. A

Cartridges for weapons, inert projectile. Ammunition consisting of a casing with propelling charge and a solid or empty projectile. US 173.59

Cartridges, Small Arms Ammunition consisting of a cartridge case fitted with a centre or rim fire primer and containing both a propelling charge and a solid projectile. They are designed to be fired in weapons of calibre not larger than 19.1 mm. Shot-gun cartridges of any calibre are included in this description. The term excludes: Cartridges, Small Arms, Blank listed separately in the Dangerous Goods List; and some small arms cartridges which are listed under Cartridges for Weapons, Inert Projectile. UN App. B, ICAO A2

Cartridges, small arms. Ammunition consisting of a cartridge case fitted with a center or rim fire primer and containing both a propelling charge and solid projectile(s). They are designed to be fired in weapons of caliber not larger than 19.1 mm. Shotgun cartridges of any caliber are included in this description. The term excludes: Cartridges, small arms, blank, and some military small arms cartridges listed under Cartridges for weapons, inert projectile. US 173.59

Cartridges, Small Arms Ammunition consisting of a cartridge case fitted with a centre or rim fire primer and containing both a propelling charge and solid projectile(s). They are designed to be fired in weapons of calibre not larger than 19.1 mm. Shot-gun cartridges of any calibre are included in this description. The term excludes: Cartridges, small arms, blank which are listed separately and some small arms cartridges which are listed under Cartridges for weapons, inert projectile. IATA App. A

Charges, bursting Articles consisting of a charge of detonating explosive such as hexolite, octolite or plastics bonded explosive designed to produce effect by blast or fragmentation. UN App. B, ICAO A2, US 173.59, IATA App. A

Charges, Depth Articles consisting of a charge of detonating explosive contained in a drum or projectile. They are designed to detonate under water. UN App. B, ICAO A2, US 173.59, IATA App. A

Charges, expelling A charge of deflagrating explosive designed to eject the payload from the parent articles without damage. UN App. B, ICAO A2, US 173.59, IATA App. A

Charges, Propelling Articles consisting of a propellant charge in any physical form, with or without a casing, for use as a component of rocket motors or for reducing the drag of projectiles. UN App. B, ICAO A2, IATA App. A

Charges, propelling. Articles consisting of propellant charge in any physical form, with or without a casing, for use in cannon or for reducing drag for projectiles or as a component of rocket motors. US 173.59

Charges, Propelling for Cannon Articles consisting of a propellant charge in any physical form, with or without a casing, for use in a cannon. UN App. B, ICAO A2, US 173.59, IATA App. A

Expelling Charge. An explosive charge designed to eject the projectile from the parent article without damage. IATA App. A

Grenades, hand or rifle Articles which are designed to be thrown by hand or to be projected by a rifle. The term includes: Grenades, hand or rifle, with bursting charge; Grenades, Practice, hand or rifle. The term excludes grenades, smoke which are listed under Ammunition, Smoke. UN App. B, ICAO A2, IATA App. A

Grenades, hand or rifle. Articles which are designed to be thrown by hand or to be projected by rifle. The term includes: grenades, hand or rifle, with bursting charge; and grenades, practice, hand or rifle. The term excludes: grenades, smoke. US 173.59

Mines Articles consisting normally of metal or composition receptacles and a bursting charge. They are designed to be operated by the passage of ship, vehicles or personnel. The term includes "Bangalore torpedoes". UN App. B, ICAO A2, IATA App. A

Mines. Articles consisting normally of metal or composition receptacles and bursting charge. They are designed to be operated by the passage of ships, vehicles, or personnel. The term includes Bangalore torpedoes. US 173.59

Projectiles Articles such as a shell or bullet which are projected from a cannon or other artillery gun, rifle or other small arm. They may be inert, with or without tracer, or may contain a burster or expelling charge or a bursting charge. The term includes: Projectiles, inert, with tracer; Projectiles with burster or expelling charge; Projectiles with bursting charge. UN App. B, ICAO A2, IATA App. A

Projectiles. Articles, such as a shell or bullet, which are projected from a cannon or other artillery gun, rifle, or other small arm. They may be inert, with or without tracer, or may contain a burster, expelling charge or bursting charge. The term includes: projectiles, inert, with tracer; projectiles, with burster or expelling charge; and projectiles, with bursting charge. US 173.59

Propellant, liquid. Substances consisting of a deflagrating liquid explosive, used for propulsion. UN App. B, US 173.59

Propellant, Liquid. A substance consisting of a deflagrating liquid explosive, used for propulsion ICAO A2

Propellant, Liquid. A substance consisting of a deflagrating liquid explosive, used for propulsion or for reducing the drag of projectiles. IATA App. A

Propellants Deflagrating explosive used for propulsion or for reducing the drag of projectiles. UN App. B, ICAO A2, US 173.59, IATA App. A

Propellant, solid. Substances consisting of a deflagrating solid explosive, used for propulsion. UN App. B, US 173.59

Propellant, Solid A substance consisting of a deflagrating solid explosive, used for propulsion. ICAO A2

Propellant, Solid. A substance consisting of a deflagrating solid explosive, used for propulsion or for reducing the drag of projectiles. IATA App. A

Rocket Motors Articles consisting of a solid, liquid or hypergolic fuel contained in a cylinder fitted with one or more nozzles. They are designed to propel a rocket or a guided missile. The term includes: Rocket Motors; Rocket Motors With Hypergolic Liquids with or without expelling charge; Rocket Motors, Liquid Fuelled. UN App. B, ICAO A2, IATA App. A

Rocket motors. Articles consisting of a solid, liquid, or hypergolic propellant contained in a cylinder fitted with one or more nozzles. They are designed to propel a rocket or guided missile. The term includes: rocket motors; rocket motors with hypergolic liquids with or without an expelling charge; and rocket motors, liquid fuelled. US 173.59

Rockets Articles consisting of a rocket motor and a payload which may be an explosive warhead or other device. The term includes guided missiles and: Rockets, Line-Throwing; Rockets, Liquid Fuelled with bursting charge; Rockets with bursting charge; Rockets with expelling charge; Rockets with inert head. UN App. B, ICAO A2, IATA App. A

Rockets. Articles containing a rocket motor and a payload which may be an explosive warhead or other device. The term includes: guided missiles; rockets, line-throwing; rockets, liquid fuelled, with bursting charge; rockets, with bursting charge; rockets, with expelling charge; and rockets, with inert head. US 173.59

Torpedoes Articles containing an explosive or non-explosive propulsion system and designed to be propelled through water. They may contain an inert head or a warhead. The term includes: Torpedoes, Liquid Fuelled with inert head; Torpedoes, Liquid Fuelled with or without bursting charge; Torpedoes with bursting charge. UN App. B, ICAO A2, US 173.59, IATA App. A

Warheads Articles consisting of detonating explosives. They are designed to be fitted to a rocket, guided missile or torpedo. They may contain a burster or expelling charge or bursting charge. The term includes: Warheads, Rocket with burster or expelling charge; Warheads,

Rocket with bursting charge; Warheads, Torpedo with bursting charge. UN App. B, ICAO A2, IATA App. A

Warheads. Articles containing detonating explosives, designed to be fitted to a rocket, guided missile, or torpedo. They may contain a burster or expelling charge or bursting charge. The term includes: warhead rocket with bursting charge; and warheads, torpedo, with bursting charge. US 173.59

REFERENCES
29CFR1910, 32CFR552, CCO, COTB, COTC, D, DODT, DOST, DOW, E3, HG, IW, JDOMS, JDOMT, KOE, PFT, POG, SG, TEA, TIE, WAT

Ammunition, Toxic

Ammunition, lachrymatory ■ Ammunition, tear-producing, non-explosive with neither burster nor expelling charge, 6.1, 8 ■ Ammunition, tear-producing, non-explosive with neither burster nor expelling charge, non-fuzed, 6.1 ■ Ammunition, tear-producing, non-explosive, without burster, expelling charge or propelling charge, non-fuzed, 6.1, 8 ■ Ammunition, tear-producing, with burster, expelling charge or propelling charge, 1.2G, 1.3G, 1.4G, 6.1, 8 ■ Ammunition, toxic, non-explosive, without burster or expelling charge, non-fuzed, 6.1 ■ Ammunition, toxic (water-activated contrivances), with burster, expelling charge or propelling charge ■ Ammunition, toxic with burster, expelling charge or propelling charge, 1.2K, 1.3K, 6.1 ■ Irritating agents ■ Irritating material ■ Pepper spray ■ Self-defense spray, aerosol ■ Self-defense spray, non-pressurized, 9 ■ Tear gas candles, 4.1, 6.1 ■ Tear gas candles, non-explosive, 4.1, 6.1 ■ Tear gas cartridges ■ Tear gas devices ■ Tear gas devices containing tear gas substances ■ Tear gas grenades ■ Tear gas grenades, non-explosive, 4.1, 6.1 ■ Tear gas substance, liquid, n.o.s., 6.1 ■ Tear gas substance, solid, n.o.s., 6.1

0018 ■ 0019 ■ 0020 ■ 0021 ■ 0301 ■ 2016 ■ 2017 ■ 1693 ■ 1700 ■ 3334

Toxic ammunition contains chemical agents which, through their adverse physiological effects, are in use by military or law enforcement organizations to kill, injure, or incapacitate humans. Less powerful agents are used by the public for self-defense. While the use of lethal chemical agents like mustard gas and nerve agents to kill or maim has been banned by the *Chemical Weapons Convention* (COTC), the use of toxic chemicals as suppressive methods to control or subdue crowds and riots is widespread.

Underlying any categorization of chemical agents is the fact that all are toxic to a greater or lesser degree. It is the concentration of the agent and the period of exposure that determines whether it produces reversible irritation or a range of incapacitating effects that can lead, sometimes promptly, to death of the victim. For law enforcement purposes, chemical agents must be non-lethal under normal use with rapidly reversible effects, but be used in sufficient quantities to have immediate and decisive results. Only a few chemicals meet these criteria, and even fewer when storage, decomposition, temperature stability, and economy are considered.

LACHRYMATING AGENTS

The term *tear gas* is used widely and without precision. Strictly speaking it applies to *lachrymatory ammunition* syn. *tear-producing ammunition* which contains chemical agents that are absorbed by epithelial surfaces of the body (outer layers of mucous membranes) and irritate the eyes by causing smarting, swelling, and closure. Reduced vision and disorientation is caused as profuse tears generated by the lachrymal gland change the refractive power of the eye. Nerves in other areas of the skin may be similarly affected.

Tear gas agents include ethylbromoacetate, the first chemical police ordnance, used in Paris; bromobenzylcyanide, the strongest lachrymator; and chloroacetophenone (CN gas), almost as strong, but widely used. Chloro-acetophenone is also called by its trademark *Mace*,[1] or *chemical mace* to distinguish it from the oleoresin derived from nutmeg and used as a spice.

PEPPER SPRAY

Self-defense spray is sold in small aerosols and other devices for self-defense. The chemical agent employed is often *capsaicin* (404-86-4), the oleoresin extracted from cayenne peppers (*Capisicum annum*), hence **pepper spray**. Aerosol products contain 10 to 15% capsaicin. It is an inflammatory which causes temporary blindness and restricted breathing for up to 45 minutes. Self-defense sprays may come as key rings, or be disguised as flashlights or fountain pens, and be used against wildlife in addition to human offenders. Mace is also used in some self-defense sprays.

NAUSEATING AGENTS

Nauseating agents are violent irritants used in those uncommon instances where the mental attitude of a crowd has rendered the use of tear gas ineffectual. Often based on arsine, these agents enter and lodge in the lungs, nose, throat, and other mucous membranes where they attack nerve endings to produce acute pain, involuntary muscular reflexes, and secretion of bodily fluids. Severe exposure fully incapacitates a victim by uncontrollable coughing, sneezing, lachrymating, nose discharge, vomiting, urination, and defecation, along with headaches, chest pains, cramps, and difficulty breathing. Nauseating agents include diphenylaminechloroarsine, the most commonly used agent, and diphenylchloroarsine.

IRRITATING AGENTS

True *irritating agents syn. irritating materials* lie in effect between lachrymatory and nauseating agents, although the distinction is a matter of exposure and can encompass both, as it does in United States Army terminology. On use, irritants act as lachrymatory agents with some of the debilitating effects of nauseating agents. They include chloropicrin and o-chlorobenzalmalononitrile, known as CS gas.

AMMUNITION

Many types of fuzed or nonfuzed projectile ammunition (tear gas grenades, shells, small arms cartridges, bombs, and other devices) are used to deliver chemical agents. Ammunition may be ejected from a weapon using a propelling or expelling charge, thrown into place by hand, or sprayed from an aerosol. If projected, ammunition may employ an explosive burster or burning device to vaporize or gasify the solid or liquid toxic agent. Alternatively, the agent may be sufficiently volatile to evaporate once exposed to the atmosphere or be volatilized by the heat of the propelling charge.

---

[1] Trademark, Mace Security International.

Tear gas candles incorporate a fuel in which the toxic agent is mixed. They include fuses or other initiating devices that ignite the mixture which, on burning, generates a gas that carries the agent as it escapes from the container. They are designed to burn at temperatures that do not decompose the chemical agent and deliver a long, continuous cloud covering a large area, but not too coolly to permit the grenade to be picked up and thrown back at the police or troops.

RELATED TERMS

Aerosols, see *Aerosols*, p.3

Ammunition, see *Ammunition*, p.8

Burster, see *Ammunition*, p.8

Cartridges, see *Ammunition*, p.8

Chemical sample, toxic, liquid or solid, see *Chemical Kits and Samples*, p.40

Expelling charge, see *Ammunition*, p.8

Flammable, see *Flammable Solids and Division 4.1*, p.99

Fuzed, see *Initiating Explosives*, p.119

Gas, see *Terminology, Gas*, p.239

Grenades, see *Ammunition*, p.8

Irritating material, see *Toxic Substances and Division 6.1*, p.255

Liquid, see *Terminology, Liquid*, p.241

Propelling charge, see *Ammunition*, p.8

Solid, see *Terminology, Solid*, p.247

Toxic, see *Toxic Substances and Division 6.1*, p.255

REGULATORY DEFINITIONS

Ammunition, Tear-Producing with burster, expelling charge or propelling charge Ammunition containing tear-producing substance. It also contains one or more of the following: a pyrotechnic substance; a propelling charge with primer and igniter charge; a fuze with burster or expelling charge. UN App. B, ICAO A2, IATA App. A

Ammunition, tear-producing with burster, expelling charge or propelling charge. Ammunition containing tear-producing substance. It may also contain one or more of the following: a pyrotechnic substance, a propelling charge with primer and igniter charge, or a fuze with burster or expelling charge. US 173.59

Ammunition, toxic. Ammunition containing toxic agent. It may also contain one or more of the following: a pyrotechnic substance, a propelling charge with primer and igniter charge, or a fuze with burster or expelling charge. US 173.59

Ammunition, Toxic with burster, expelling charge or propelling charge Ammunition containing toxic agent. It also contains one or more of the following: a pyrotechnic substance; a propelling charge with primer and igniter charge; a fuze with burster or expelling charge. UN App. B, ICAO A2, IATA App. A

Self-defense spray means an aerosol or non-pressurized device that: (1) Is intended to have an irritating or incapacitating effect on a person or animal; and (2) Meets no hazard criteria other than for Class 9 (for example, a pepper spray; see 173.140(a) of this subchapter) and, for an aerosol, Division 2.1 or 2.2 (see 173.115 of this subchapter), except that it may contain not more than two percent by mass of a tear gas substance (e.g., chloroacetophenone (CN) or 0-chlorobenzylmalo[no]nitrile (CS); see 173.132(a)(2) of this subchapter.) US 171.8

Tear Gas Candles, Non-Explosive; Tear Gas Grenades, Non-Explosive Devices containing tear-producing substances which, in minute quantities dispersed in air, cause extreme eye irritation and profuse tears. IMO 6261

REFERENCES

CTD, MH14, TGM, WAT

Antifreeze

Anti-freeze, liquid ■ Compound, anti-freeze liquid ■ De-icing fluid

Water has excellent heat-transfer properties. This and its almost universal availability supports its general use to keep machinery cool by removing excess heat. Disadvantageously, it promotes corrosion and its operating range is curbed by its freezing point of $0°C$ and boiling point of $100°C$, temperatures easily reached in winter conditions or in hot machinery, like internal combustion engines. To abate this problem, chemical *antifreezes* are added which depress the freezing point of water. For example, a 50% solution by weight of ethylene glycol in water drops the freezing point to $-22°C$. If the antifreeze chemical alone boils at a higher temperature than water, then it will also raise the boiling point of the mixture and increase the operating range still further.

Antifreeze chemicals, such as ethylene and propylene glycol, brines (e.g., calcium chloride solution), methanol, and ethanol are combined with water, corrosion inhibitors (e.g., phosphates, borates, and thiazoles), antifoam agents, and dyes to make *antifreeze compounds*.

Antifreezes function similarly when used as *de-icing fluids* for aircraft, airport runways, locks, and other applications. Here, the antifreeze melts the ice crystals by forming a mixture that freezes lower than the ambient temperatures. For aircraft de-icing and other critical applications, antifreezes are often concentrated glycols, water being considered too corrosive and flammable solvents too dangerous.

HAZARDS
Antifreezes and de-icers contain a variety of chemicals with possible toxic effects, but the flammability of alcohol-based mixtures is of primary concern in transportation.

RELATED TERMS
Compound, see *Terminology, Compound*,
p.234
Flammable liquid, n.o.s., see *Flammable*

Liquids and Class 3, p.96
Fluid, see *Terminology, Fluid*, p.238
Liquid, see *Terminology, Liquid*, p.241

REGULATORY DEFINITIONS
De-Icing Fluids. Frequently contain large proportions of alcohol or other flammable liquids. IATA App. A

REFERENCES
HCC, KOC, KOE, MEOS, MH14

Antiknock Compounds

Antiknock compound, mixture ■ Motor fuel anti-knock compounds ■ Motor fuel anti-knock mixture, 3, 6.1

1649

Pistons inside spark-ignition internal combustion engines compress air-fuel mixtures which are ignited by an electrical spark. The combustion gases expand and push back against the piston generating useful mechanical work. Higher compression results in greater reaction efficiency but also higher temperatures. If the temperature exceeds the autoignition temperature of the mixture, premature ignition generates a force that works against the piston cycle, reduces overall power, causes overheating, and produces an audible knocking sound. Any given fuel-engine combination has an upper limit to the compression ratio that can be used without knock. Compression-ignition engines (those without spark plugs) manifest the same tendency.

Premature ignition is reduced or eliminated by the addition to the fuel of *antiknock compounds*, the most common of which is tetraethyl lead (ethyl fluid), an organometallic compound consisting of lead surrounded by four ethyl groups ($-C_2H_5$). Other antiknock compounds include mixtures of tetraethyl lead and its homologues in which the ethyl groups are replaced by some number of methyl groups ($-CH_3$): tetramethyl lead (i.e., methyl lead), ethyl trimethyl lead, triethyl methyl lead, and diethyl dimethyl lead. Other antiknock compounds include methyl-tert-butyl ether (MTBE), methanol, and iron pentacarbonyl.

To compare the knocking tendency of fuels, they are compared to that of iso-octane in a standard iso-octane/n-heptane mixture which is arbitrarily assigned an octane rating of 100. It is against this number that other spark-ignition engine fuels are rated. Compression-ignition engines use a similar system based on cetane.

HAZARDS
Common antiknock compounds exhibit flammability and toxicity.

RELATED TERMS
Compound, see *Terminology, Compound*, p.234

Mixture, see *Terminology, Mixture*, p.243

REGULATORY DEFINITIONS
Motor Fuel Anti-Knock Mixture. A mixture of one or more organic lead components such as tetraethyl lead, triethylmethyl lead, diethyldimethyl lead, ethyltrimethyl lead, and tetramethyl lead, with one or more halogen compounds such as ethylene dibromide and ethylene dichloride. ICAO A2, IATA App. A

Motor fuel antiknock mixtures are: a. Mixtures of one or more organic lead mixtures (such as tetraethyl lead, triethylmethyl lead, diethyldimethyl lead, ethyltrimethyl lead, and tetramethyl

23

lead) with one or more halogen compounds (such as ethylene dibromide and ethylene dichloride), hydrocarbon solvents or other equally efficient stabilizers; or b. tetraethyl lead.   US 172.102(c)(1)14

REFERENCES
FAF, HCC, KOE, MH12

Asbestos

Actinolite ■ Amosite ■ Anthophyllite ■ Asbestos ■ Asbestos, blue, 9 ■ Asbestos, brown ■ Asbestos, white, 9 ■ Blue asbestos ■ Blue asbestos (crocidolite), 9 ■ Brown asbestos (amosite, mysorite), 9 ■ Chrysotile ■ Crocidolite ■ Mysorite ■ Talcum with tremolite and/or actinolite ■ Tremolite ■ White asbestos ■ White asbestos (chrysotile, actinolite, anthophyllite, tremolite), 9

2212 ■ 2590

Asbestos (1332-21-4) is the generic term used to describe over 30 different types of naturally occurring hydrated silicates representing two mineral groups (serpentine and amphiboles) that separate into fibres on mechanical processing. While the exact taxonomy continues to evolve, the sole serpentine variety is called *chrysotile* (12001-29-5) while the amphiboles include five forms:

□ Riebeckite or glaucophane, including *crocidolite* (12001-28-4) *syn. blue asbestos.*

□ Grunerite or cummingtonite-grunerite, including grunerite asbestos known as *amosite* (12172-73-5) *syn. mysorite.*

□ *Anthophyllite* or gedrite, including anthophyllite asbestos (77536-67-5).

□ *Tremolite* or ferroactinolite, including tremolite asbestos (77536-68-6).

□ *Actinolite* asbestos (77536-66-4).

White asbestos (12001-29-5) refers to the degree to which the mineral surface achieves complete reflectance over the visible spectrum rather than to the actual colour or hue of the mineral; consequently, *white* describes multiple asbestos types, although it is most commonly applied to actinolite, chrysotile, and sometimes anthophyllite and tremolite. Similarly, while *brown asbestos* principally refers to amosite, it actually describes the degree to which the mineral deviates from colourlessness toward yellow, tan, or brown.

Asbestos is resistant to heat, moisture, chemicals, microorganisms, wear, deformation, and decay and insulates against noise, heat, and electricity. It has been widely used commercially in fabrics, paper, filters, fillers, insulating boards, cements, fireproof garments, curtains, shields, brake linings, shingles, pipe coverings, and moulded products.

Unfortunately, inhalation of asbestos fibres can lead to asbestosis, bronchogenic cancer, and mesothelioma. Other diseases and cancers have been reported, including those of the gastrointestinal tract. Less risk of exposure is presented by those products in which asbestos is embedded in a cement, plastic, asphalt, resin, mineral, or other binder in such a way that the fibres are only released and become airborne if the product is cut, abraded, damaged, or otherwise worked.

Talcum (talc, 14807-96-6) is a natural hydrous magnesium silicate that can be associated with deposits of crude asbestos. Industrial talc, used extensively as a filler in plastics, cosmetics, coatings, lubricating compositions, and polishes may contain or be strengthened with up to 2 particles per 100 of asbestos fibre.

RELATED TERMS

None

REGULATORY DEFINITIONS

Asbestos. Asbestos is a generic name for naturally occurring mineral silicate fibres of the Serpentine and Amphibole series. In the Serpentine series is Chrysotile, commonly known as white asbestos. In the amphibole series are Actinolite, Amosite or Mysorite (commonly known as brown asbestos), Anthophyllite, Crocidolite (commonly known as blue asbestos) and Tremolite. All types of asbestos can be hazardous to health, blue and brown asbestos being the more dangerous types. ICAO A2

Asbestos, blue, brown or white, includes each of the following hydrated mineral silicates: chrysotile, crocidolite, amosite, anthophyllite asbestos, tremolite asbestos, actinolite asbestos, and every product containing any of these materials. US 173.216

Asbestos. Is a generic name for naturally occurring mineral silicate fibres of the Serpentine and Amphibole series. In the Serpentine series is Chrysotile, commonly known as white asbestos. In the Amphibole series are Actinolite, Amosite or Mysorite (commonly known as brown asbestos), Anthophylite, Crocidolite (commonly known as blue asbestos) and Tremolite. All types of asbestos can be hazardous to health, blue and brown asbestos being the more dangerous types. IATA App. A

REFERENCES

DFA, HCC, HOHM, KOC, KOE, MEOS, MH14, SSD, STRA, STRPV, WNN

Batteries

Accumulators, electric ■ Aircraft hydraulic power unit fuel tank ■ Alkaline corrosive battery fluid ■ Batteries, containing lithium ■ Batteries, containing sodium, 4.3 ■ Batteries, dry ■ Batteries, dry, containing potassium hydroxide, solid, electric storage, 8 ■ Batteries, electric storage ■ Batteries, lithium type ■ Batteries, wet, filled with acid, electric storage, 8 ■ Batteries, wet, filled with alkali, electric storage, 8 ■ Batteries, wet, non-spillable, electric storage, 8 ■ Batteries, wet, without electrolyte and fully discharged ■ Battery acid ■ Battery fluid, acid, 8 ■ Battery fluid, alkali, 8 ■ Battery-powered equipment, 9 ■ Battery-powered vehicle, 9 ■ Battery, wet, filled with acid or alkali with automobile (or named self-propelled vehicle or mechanical equipment containing internal combustion engine) ■ Battery, wet, with wheelchair ■ Cells containing sodium, 4.3 ■ Corrosive battery fluid ■ Electric storage batteries ■ Electrolyte (acid) for batteries ■ Electrolyte (acid or alkali) for batteries ■ Electrolyte (alkali) for batteries ■ Heat producing article, battery operated equipment, 9 ■ Lithium batteries, 9 ■ Lithium batteries contained in equipment, 9 ■ Lithium batteries packed with equipment, 9 ■ M86 fuel, 3.2 ■ Mobility aids ■ Self-propelled vehicle ■ Storage batteries, wet ■ Vehicles, self-propelled ■ Wheelchair, electric (spillable or non-spillable type batteries), 9 ■ Wheelchair, electric with batteries

2794 ■ 2795 ■ 2796 ■ 2797 ■ 2800 ■ 3028 ■ 3090 ■ 3091 ■ 3165 ■ 3171 ■ 3292 ■ 8038

Batteries are comprised of one or more *cells* connected in series or parallel and assembled with terminals or contacts. When these contacts are connected to a device electrical current flows through the completed circuit providing power. Cells are the basic electrochemical unit used to generate or store electrical energy; they consist of

□ Two electrically conductive electrodes, one positive and one negative. Battery systems are often described by their electrode combinations, of which there are a large number that vary in energy and power density, discharge and temperature stability, size and weight, environmental effect, cost, ease of manufacture, physical resistance, and safety.

□ An *electrolyte*, a semi-solid or liquid (*syn. battery fluid*) that allows conductivity (transport of ions) between the electrodes of the cell.

Between the electrodes lies an electrical potential difference. Once a circuit is made by connecting the electrodes, and until the potential difference is eliminated, electrons flow through the connection (an electric current) as the electrolyte ions (charged atoms or molecules) transport the current in the electrolyte. During the chemical reaction that takes place, the negative electrode is oxidized as the positive electrode is reduced.

Electrochemical batteries can be divided into two types:

□ Primary batteries, in which the conversion of chemical to electrical energy is not reversible and the battery is nonrechargeable.

□   Secondary batteries, in which the conversion is reversible by supplying electrical energy to the cell and reversing the chemical changes.
Secondary batteries are also called *electric accumulators syn. electric
storage batteries* or *storage batteries*.

WET AND DRY BATTERIES

Batteries that require a liquid electrolyte are called *wet batteries*. *Corrosive
battery fluid* refers to either acid electrolytes *syn. battery acid*, like the common lead-acid automobile battery which uses a solution of sulphuric acid, or
alkali electrolytes *syn. alkaline corrosive battery fluid*, like potassium hydroxide (1310-58-3) solutions in nickel-cadmium and other alkaline battery
systems. *Dry batteries* or dry cells, like all primary batteries, use electrolytes
immobilized in pastes, gels, or absorbed into separator materials. Some batteries are loaded with a dry, solid chemical (e.g., potassium hydroxide) which
is diluted with water to become a liquid electrolyte. The hazards associated
with handling and transportation prior to use are thereby reduced.

Some reactions at the electrolyte-electrode interface generate gas. In most
car batteries, for example, sulphuric acid (the electrolyte) reacts with lead
(the electrode) to generate hydrogen gas which is vented to avoid the buildup of explosive concentrations. Over time, the electrolyte is depleted and
may require replenishment. In this instance, the battery is not sealed and
presents the possibility of spillage during handling. Other nonspillable designs avoid this problem in a number of ways:

□   The battery contains an excess of electrolyte which cannot be depleted
during normal service life.
□   Safety valves allow the release of gases but not liquids.
□   Hermetic seals close the battery by fusion, gasketing, and crimping to
prevent the escape of accumulated gases and vapours.
□   Using electrode-electrolyte combinations that do not generate gases.

LITHIUM AND SODIUM BATTERIES

Lithium batteries and cells exploit the extremely electropositive nature of
lithium, allowing it to generate the highest of potential differences with another electrode. Lithium is highly reactive, particularly with water, dictating
the use of electrolytes made from molten salts and other non-aqueous organic
liquids and gels. Lithium battery equipment includes military radios, data
communication devices, position locators, surveillance devices, night vision
glasses, sonobuoys, drilling instrumentation, standby power sources, as well
as many nonmilitary items.

Sodium batteries and cells are designed such that the electrolyte is solid and
the electrodes are liquid. In hermetically sealed sodium-sulphur batteries, the
sodium and sulphur electrodes are heated to a molten state (300 to 350°C)

and ions pass between them through a solid electrolyte. The result is that when cold, these batteries are electrically inert, although the elemental sodium represents an extreme reaction hazard.

BATTERY-OPERATED EQUIPMENT

The list of *battery-operated* or *battery-powered equipment* is extensive:

- Communication equipment, e.g., cellular phones, pagers, cordless phones.
- Computers and calculators.
- Radios, tape players, and televisions.
- Razors, massagers, toothbrushes, hand tools, hand vacuums, mowers, blenders, mixers, and knives.
- Toys, games, and novelties.
- Lighting products.
- Hearing aids, heart products, and medical testers.
- Scientific, e.g., nautical, laboratory, geodetic, electronic.
- *Heat producing articles*, e.g., underwater torches which can generate tremendous heat when operated in air, soldering equipment, and some photoflashes.

Mobility aids, such as *electric wheelchairs,* use a variety of battery systems, including lead-acid batteries. *Battery-powered vehicles* compete primarily on environmental grounds with the internal combustion engines of other *self-propelled vehicles* (i.e., cars, trucks, lawnmowers, golf carts, etc.). Commercializing these vehicles beyond specialist functions has meant improvements in battery technology; the development of compact, high-efficiency motors; vehicle body weight and rolling resistance improvements; and the development of a battery charging infrastructure. Thus far, technology has focused on nickel-cadmium, sodium-sulphur, and nickel-iron electrode systems, in addition to the traditional lead-acid battery.

FUEL CELLS

Fuel cells are electrochemical devices in which fuels (e.g., hydrogen, carbon monoxide, hydrocarbons, and alkali metals), oxidants, and reaction products move into and out of a system of electrodes separated by an electrolyte. The reduction-oxidation reactions that take place generate a direct current while the materials are supplied to the cell. A number of transportation and other applications for this technology are being explored, partly because of the environmental benefits the reaction products have over those of fossil fuels. *M86 fuel*, a mixture of anhydrous and methyl hydrazines, is used in fuel cells including those used to generate electricity for some *aircraft hydraulics* systems. These fuel tanks are leak-tight, double-walled aluminium pressure vessels that contain up to 42 litres of M86.

HAZARDS

While generally of a mature technology and abuse-resistant, batteries are subject to many types of failure which can lead to hazards in transportation. The presence of an electrical charge, normally discharged in a safe manner, can lead to sparks that result in fires. Leaks and ruptures can result in the loss of corrosive electrolyte or molten sodium. The electrochemical reactions that occur within the cell are as varied as the systems and conditions, but toxic and flammable gases and chemicals can be involved or generated, possibly in ignitable concentrations. Prolonged, low-resistance short-circuits involving high-rate batteries may generate considerable heat with the potential for rupture and fire. M86 fuel components are flammable, toxic, and corrosive to many metals.

RELATED TERMS

Acid, see *Corrosives and Class 8*, p.47
Alkali, see *Corrosives and Class 8*, p.47
Automobile, see *Self-Propelled Vehicles*, p.219
Corrosive, see *Corrosives and Class 8*, p.47
Fluid, see *Terminology, Fluid*, p.238
Internal combustion engine, see *Self-*

Propelled Vehicles, p.219
Solid, see *Terminology, Solid*, p.247
Vehicle (flammable, gas powered), see *Self-Propelled Vehicles*, p.219
Vehicle (flammable, liquid powered), see *Self-Propelled Vehicles*, p.219

REGULATORY DEFINITIONS

Batteries, containing Sodium; Cells, Containing Sodium Series of hermetically sealed metal cells containing sodium, electrically connected and secured within a metal casing. "Cold" batteries (batteries containing elemental sodium only in the solid state) are electrically inert. Batteries are activated by heating to between 300°C and 350°C before operating to produce electricity. Activated batteries (i.e., "hot" batteries containing liquid elemental sodium) may cause fire through short-circuit of the terminals. IMO 4332-1

Batteries, Containing Sodium. Articles consisting of a series of Cells, Containing Sodium that are secured within, and fully enclosed by a metal casing so constructed and closed as to prevent the release of dangerous goods under normal conditions of transport. Although designed and intended to provide a source of electrical energy, these batteries are electrically inert at any temperature at which the sodium contained in the battery is in a solid state. ICAO A2, IATA App. A

Batteries, Dry. Are sealed, non-vented batteries of the type used in flashlights or for the operation of small apparatus. They contain zinc salts and other solids, or may be of the nickel cadmium type or other combinations of metals. Such batteries must be packed in inner packagings in such a manner as to effectively prevent short circuits and to prevent movement which could lead to short circuits. IATA App. A

Batteries, Dry, Containing Potassium Hydroxide, Solid Series of metal plates immersed in dry potassium hydroxide in a closed receptacle. IMO 8119

Batteries, Dry, Containing Potassium Hydroxide, Solid. Storage batteries filled with potassium hydroxide, solid which are shipped from the factory in their original dry state and filled with the dry alkali. Water would be added to the battery before first being used. ICAO A2, IATA App. A

Batteries, Wet, Electric Storage. Consist of a series of metal plates immersed in an electrolyte. The electrolyte is a dilute sulphuric acid, but for a certain type of battery the electrolyte is a solution of potassium hydroxide. Both of these electrolytes are corrosive liquids. The casing for the acid containing batteries is commonly plastic. Storage batteries of either of these types, when containing electrolyte, are classed as corrosive liquids. Storage batteries in transit may cause damage by leakage of the electrolyte or may produce fire by accidental short circuiting of the terminals. Non-spillable batteries are designed and constructed so as to positively prevent leakage of the electrolyte, irrespective of the position of the battery. This is achieved by the use of jelly type electrolyte or porous absorbent separators or by specially designed filling and venting devices. IATA App. A

Batteries, Wet, Filled with Acid; Batteries, Wet, Filled with Alkali Metal plates immersed in acid or alkaline electrolyte in a glass, hard rubber or plastics receptacle. IMO 8120

Batteries, Wet, Filled with Acid or Alkali. A series of metal plates immersed in an electrolyte, which is usually dilute sulphuric acid, but for a certain type of battery the electrolyte is a solution of potassium hydroxide. Both of these electrolytes are corrosive liquids. The casing for the acid containing batteries is commonly plastic. Storage batteries of either of these types, when containing electrolyte, are classed as corrosive liquids. Storage batteries in transit may cause damage by leakage of the electrolyte or may produce fire by accidental short-circuiting of the terminals. ICAO A2

Batteries, Wet, Non-Spillable Metal plates immersed in gelled alkaline or acid electrolyte in a glass, hard rubber or plastics receptacle of a non-spillable type. IMO 8121

Batteries Wet, Without Electrolyte, and Fully Discharged. Are usually wet type batteries which have been shipped from the factory in their original dry state with the intent that electrolyte would be added just before placing the batteries in service. They may also be wet batteries from which the electrolyte has been removed. In this latter instance the cells should be thoroughly flushed with water and allowed to drain before shipping. IATA App. A

Cells, Containing Sodium. Articles consisting of hermetically sealed, metal casings which fully enclose the dangerous goods and which are so constructed and closed as to prevent the release of the dangerous goods under normal conditions of transport. In addition to sodium, cells covered by this entry may also contain sulphur, but no other dangerous goods. Although designed and intended to provide a source of electrical energy, these cells are electrically inert at any temperature at which the sodium contained in the cell is in a solid state. ICAO A2, IATA App. A

Electrolyte. The term commonly applied to the dilute sulphuric acid used in ordinary lead plate storage batteries. The solution of potassium hydroxide used in some storage batteries is also called electrolyte. ICAO A2

Electrolyte. Is the term commonly applied to dilute sulphuric acid used in the ordinary lead plate storage batteries. The solution of potassium hydroxide used in some storage batteries is also called electrolyte. The term electrolyte is sometimes applied to the strong sulphuric acid which is meant for use in storage batteries after dilution with water. IATA App. A

Lithium Batteries; Lithium Batteries Contained in Equipment; Lithium Batteries Packed with Equipment Electrical batteries containing lithium or lithium alloy encased in a rigid metallic body. Lithium batteries may also be shipped in, or packed with, equipment. IMO 9033

Lithium Batteries or Lithium Cells. A battery is two or more cells which are electrically connected together by a permanent means. A cell is a single encased electromechanical unit which exhibits a voltage differential across its two terminals. ICAO A2, IATA App. A

REFERENCES
ANS, BHA, BRB, BTH, EVT, HCC, HOB, IMDG, LAC, PBP, SBV1, SBV2, TEA, VAA

Bituminous Products

Asphalt, cut back ■ Cut-backs ■ Bitumen ■ Pencil pitch ■ Pitch prill ■ Road asphalt ■ Road asphalt or tar liquids ■ Tars, liquid, 3.2, 3.3 ■ Tars, liquid, including road asphalt and oils, bitumens and cut backs, 3

1999

Bituminous materials are dark brown or black, semi-solid or liquid, thermoplastic mixtures of hydrocarbons derived from natural or synthetic processes in which hydrocarbon mixtures have lost their volatile components leaving a denser residue. Natural bitumens come from exposed and weathered petroleum and rock deposits. Synthetic bitumens come from the residue remaining after the distillation of petroleum, coal tar, and other organic materials like wood and peat. The complexity of the high molecular weight hydrocarbon oils and resins bitumens contain make complete chemical characterization impossible. The terms *bitumen*, *tar* (8007-45-2), *pitch* (61789-60-4), and *asphalt* (8052-42-4) apply to any of these substances, although *pitch* and *tar* also describe the sticky resins that exude from various trees.

Petroleum asphalt is used extensively as a binder in concrete road paving, hence *road asphalt*. Mixed with asphalt, coal tar pitch makes exceptional road surfaces that resist oils, are non-skid, adhere excellently to stone, and provide weathered hard surfaces. *Cutback asphalt* is asphalt dissolved in and made less viscous by a solvent (usually a petroleum distillate) allowing easier application in road repair and waterproofing. *Road oil* is very fluid asphalt used as a dust suppressant.

Prills are small globules of material that are often of high purity either separated from inferior pieces or manufactured into granules, pellets, or cylinders by dropping the melted material from a height or other means. *Pitch prills* are spherical, granular, or cylindrical products (*pencil pitch*) made from bitumens.

HAZARDS
Bitumens contain many chemicals of unknown toxicity and all will burn when heated to a sufficiently high temperature. Flammable solvents are widely added to bitumens to decrease their viscosity and improve handling. Bitumens may also be transported at sufficiently high temperatures to evolve these solvents and other flammable constituents in ignitable concentrations.

RELATED TERMS
Asphalt, see also *Petroleum*, p.183
Dead oil, see *Coal*, p.44
Liquids, see *Terminology, Liquid*, p.241
Oils, see *Terminology, Oil*, p.244

Pencil pitch, see also *Solid Bulk Materials*, p.221
Pitch prill, see also *Solid Bulk Materials*, p.221

REGULATORY DEFINITIONS
None

REFERENCES
HCC, HCD, KOE, MH14, TCAOP, TIR

Bleach

Bleaches are used in the chemical processing of textiles, paper pulp, oils, foodstuffs, coatings, fibres, and other materials to remove or decolourize impurities. They are also used as cleaners and disinfectants. There are two broad classes of bleach, the first of which are the oxidizing bleaches (those that contain chlorine or active oxygen):

□ Chlorine bleaches: chlorine, hypochlorites, nitrogenchlorocompounds, chlorine dioxide.
□ Peroxygen (active oxygen) bleaches: hydrogen peroxide, peroxy, and peroxyhydrate compounds, peracids, and related.

The chlorine or available oxygen destroy chromophoric (colour-bearing) bonds in the molecules of the impurities they oxidize shifting the radiation-absorbing and reflecting properties of the molecule toward the ultraviolet. The new compound absorbs and radiates less light of specific wavelengths and reflects light more evenly, creating a whiter appearance.

The second class of bleaches (i.e., sulphur dioxide, sulphites and bisulphites, dithionites) work by reducing chemical bonds in the impurities (often carbonyl bonds) to achieve similar light shifts. Bleaches may also increase the solubility of dyes, allowing them to be more easily washed out.

While there are many chemicals in use, the term *bleach syn. bleach solution* commonly means a hypochlorite, principally sodium hypochlorite, a common household bleach, but also calcium, magnesium, and lithium hypochlorites. *Bleach liquor* is a solution of calcium hypochlorite and calcium chloride commonly used to bleach paper pulp. Hypochlorite solutions are alkaline, only remaining stable above the pH of 11.

Bleaching powders (7778-54-3) are complex, solid mixtures of calcium hypochlorite, calcium chloride, calcium hydroxide, and their hydrates. They are applied as dry powders to moist or wet environments (e.g., septic tanks, soils, sewers) where they form a solution with the water making chlorine available for oxidation.

In addition to their oxidizing and reducing properties, the use of chlorine bleaches results in chlorinated organic compounds which have many unwanted environmental consequences. Alternatives are being rapidly implemented.

RELATED TERMS

Hypochlorite, see *Inorganic Compounds*, p.128

Leather bleach, see *Leather*, p.132

Mixture, see *Terminology, Mixture*, p.243

Powder, see *Terminology, Powder*, p.246

Solution, see *Terminology, Solutions*, p.247

REGULATORY DEFINITIONS

Hypochlorite Solution. Water solutions containing a soluble hypochlorite varying over a wide range in concentration. The solutions are alkaline and corrosive but are not flammable. If the hypochlorite solution contacts strong acids, a decomposition takes place to produce the noxious chlorine-type gases. ICAO A2

Hypochlorite Solution. Are water solutions containing a soluble hypochlorite. The concentration of the solutions vary over a wide range. The solutions are alkaline and corrosive but are not flammable. If the hypochlorite solution contacts strong acids, a decomposition takes place to produce the noxious chlorine type gases. Contact with textile fibres, etc., will cause severe damage to fibres and to colours. The solutions are used for water-treatment, bleaching, etc. IATA App. A

REFERENCES

GOCT, HCC, KOE, MH14

Carbon

Activated carbon ■ Activated charcoal ■ Carbon, activated, 4.2 ■ Carbon, animal or vegetable origin, 4.2 ■ Carbon black (animal or vegetable origin) ■ Carbon, non-activated, mineral origin ■ Charcoal ■ Charcoal, activated ■ Charcoal briquettes, shell, screenings, wood, etc , 4.2 ■ Charcoal, non-activated ■ Charcoal screenings, wet ■ Charcoal, wet ■ Lamp black ■ Non-activated carbon ■ Non-activated charcoal

1361 ■ 1362

Naturally occurring *carbon* exists as a solid in its pure crystalline forms (diamond and graphite); in its amorphous form (e.g., coal, coke, peat) along with many impurities; and combined with other elements to form the molecular basis of all animal and vegetable life. Graphite and high carbon-concentration amorphous forms combust readily; the smaller the particle size the greater the surface area exposed to the air and the greater the fire risk, even to the point of spontaneous combustion.

Destructive distillation (slow heating in an atmosphere deficient of oxygen) of animal, vegetable, or carbonaceous mineral matter (e.g., wood, animal carcasses, bones, nut shells) yields a high-carbon residue called *charcoal* which can be *activated* by the action of heat (700 to 900°C) in the presence of superheated steam or carbon dioxide to remove the hydrocarbons adhering to the carbon. *Activated charcoal syn. activated carbon* (16291-96-6) is amorphous and highly porous. In fact, the internal surface area can reach 420 to $1700m^2/g$ which is around 2000 times that of nonactivated charcoal, greatly enhancing its filtration and absorption properties.

Charcoal may also be screened (filtration of contaminants and separation into size classes), crushed into powder, or made into *briquettes* wherein the charcoal is compressed into compact, high carbon-density masses with or without a binder. Wet charcoal presents a particular combustion hazard because moisture catalyzes oxidation leading to combustion and self-heating.

Burning carbonaceous material in a closed system with insufficient oxygen for complete combustion generates fine particles of high-carbon soot: *lamp black* results from burning solids or liquids (e.g., resins, oils, coal tar); *carbon black* (1333-86-4) results from burning gases and vaporized liquids. Lamp black generally contains considerable quantities of mineral matter and other organic substances while carbon black can approach elemental levels of purity. Blacks from the combustion of many different types of materials have long been used as pigments, on carbon paper, and in rubber reinforcement.

<u>RELATED TERMS</u>

Activated, see *Terminology, Activated*, p.231

Carbon paper, see *Fibres and Fibrous Products*, p.90

Charcoal, see also *Solid Bulk Materials*,

p.221

Non-activated, see *Terminology, Non-Activated*, p.244

<u>REGULATORY DEFINITIONS</u>

None

<u>REFERENCES</u>

CBI, CSS, GOCT, HCC, KOC, MEOS, STRAC, STRCB

Castor Beans

Castor beans, 9 ■ Castor flake, 9 ■ Castor meal, 9 ■ Castor pomace, 9

2969

The castor plant (*Ricinus communis*) is a coarse herb of the spurge family *Euphorbiaceae*. When processed, the seed *syn.* **castor bean** yields half its weight in a triglyceride oil that is, uniquely, up to 85% ricinoleic acid. The oil is used as an ingredient or precursor in the manufacture of paints, varnishes, resins, synthetic fibres, lubricants, polishes, cosmetics, and many other materials.

The cake remaining after mashing, crushing, or solvent extraction of the oil is known variously as **castor pomace, castor flake, castor meal**, or poonac. It contains ricin, a highly toxic protein; ricinine, a toxic alkaloid; and CB-1A, an extremely potent allergen. The cake with its high potassium, fibre, and protein concentration is used as a fertilizer and as animal feed after detoxification.

RELATED TERMS
Meal, see *Fibres and Fibrous Products*, p.90

REGULATORY DEFINITIONS
Castor Beans; Castor Meal; Castor Pomace; Castor Flake Whole beans or meal. The latter is the residue remaining after the oil has been extracted from the seeds. IMO 9026

Castor Beans or Castor Meal or Castor Pomace, or Castor Flake. The residue from extraction of oil from the castor seed. IATA App. A

REFERENCES
CSA, KOE

Catalysts

Cobalt catalyst ■ Metal catalyst ■ Metal catalyst, dry, 4.2 ■ Metal catalyst, wetted with a visible excess of liquid, 4.2 ■ Metal catalyst, wetted without a visible excess of liquid ■ Nickel catalyst

1378 ■ 2881

Catalysts change the rate of a chemical reaction without being chemically altered, although they may undergo physical change. They are used extensively in almost all industrial processes (e.g., chemical synthesis, petroleum refining, emission control) to increase reaction speed and efficiency. *Metal catalysts* are the most important commercially although catalysts may be organic, inorganic, or organometallic. They may have any physical form, although gases are rare. A typical catalyzed reaction involves platinum and rhodium in catalytic converters which speed up the rate at which carbon monoxide (emitted from the combustion of gasoline in internal combustion engines) is oxidized to carbon dioxide:

$$2CO + O_2 \rightarrow 2CO_2$$

Without the catalyst, this reaction would require sufficient energy to bring the gaseous molecules close together and to dissociate (split) the O-O bonds. The catalytic metals provide sites on which the oxygen is absorbed and dissociated whereupon it reacts with carbon monoxide which is absorbed intact close by.

Although the exact nature of most catalyzed reactions is not well understood, those involving solid catalysts depend upon the catalytic surface area exposed to reactants. Catalysts, then, are often manufactured as fine powders or granules or coated onto highly porous substrates or other substrates. Either way, reactive metals so prepared will be at particular risk to unwanted reactions (principally oxidation) and are often handled wetted or otherwise protected from the atmosphere. Among the enormous number of metals and compounds used as catalysts are *nickel catalysts* (used, for example, in hydrocarbon reforming, ammonia, and diamond synthesis) and *cobalt catalysts* (used, for example, in hydroformulaton and gasoline-to-coal synthesis).

RELATED TERMS

Dry, see *Terminology, Dry*, p.235
Liquid, see *Terminology, Liquid*, p.241

Metal, see *Terminology, Metals*, p.242
Wetted, see *Terminology, Wetted*, p.253

REGULATORY DEFINITIONS

Metal Catalyst, Finely Divided, Activated, or Spent. Is metal in an extremely finely divided form. It must be shipped wet in moisture-tight and air-proof packagings. If exposed to air, the metal may become hot and may even ignite. IATA App. A

REFERENCES

HCC, KOE, MH14, TPD

Chemical Kits and Samples

Chemical kit, 8, 9 ■ Chemical sample, liquid, 6.1 ■ Chemical sample, solid, 6.1 ■ Chemical sample, toxic, liquid or solid, 6.1 ■ First aid kit, 9 ■ Gas identification set, 2.3 ■ Gas sample, non-pressurised, n.o.s., not refrigerated liquid, 2.1, 2.3 ■ Permeation devices, containing dangerous goods, for calibrating air quality monitoring ■ Samples, explosive other than initiating explosives, 1

0190 ■ 1760 ■ 3167 ■ 3168 ■ 3169 ■ 3315 ■ 3316

CHEMICAL KITS

Chemical kits describe any container housing small quantities of different chemicals used for medical, analytical, or testing purposes. The individual chemicals may exhibit one or more hazards, although they must be compatible with the other contents. *First aid kits* are convenient packages or boxes of medical supplies to provide immediate and sometimes makeshift medical attention. They may include multiple different materials, some of which will be hazardous (e.g., drugs, alcohols, mercury in thermometers):

- Topical or internal products: drugs such as skin antiseptics, anaesthetics, topical antibiotics, wound cleansers, topical hydrocortisones, adsorption dressings, internal analgesics.
- Dressing products: gauze bandages and dressings, adhesive bandages, adhesive tape, film dressings, hydrocolloids, gels.
- Other products: cotton balls, swabs, sponges, thermometers, scissors, etc.

SAMPLES

Chemical samples are collected for environmental reasons, research, quality control, industrial hygiene, law enforcement, and other purposes. The materials being sampled may range from solids, sludges, contaminated soil and debris, blood, explosives, wastewaters, and gases and exhibit any of the hazard classes. A wide range of sample handling techniques are designed to inhibit reactions that alter the chemical and physical nature of the sample after it has been taken. *Explosive samples* may be the subject of quality control analyses or of explosives identification at crime scenes. The transportation term *chemical sample, toxic* is reserved for those samples taken to demonstrate conformance to the *Chemical Weapons Convention* (COTC) which prohibits their manufacture or use.

Gas samples are taken by equipment that draws the gas stream into empty vials or through tubes containing some absorbent that concentrates the sample. Many systems analyse the sample *in situ* using colour indicators, gas chromatographs, or solid state analysers. Other gas samples are transported in sealed containers for analysis, but they are rarely under pressure.

A gas identification set includes containers of small quantities of various gases of known concentration that are used as standards against which to calibrate analytical equipment. *Permeation devices* contain small containers of compressed or liquefied gases that, on opening, pass through a gas permeable membrane at a known rate and constant temperature. They are also used for calibration.

RELATED TERMS

Gas, see *Terminology, Gas*, p.239

Liquid, see *Terminology, Liquid*, p.241

Refrigerated liquid, see *Gases and Class 2*, p.104

Solid, see *Terminology, Solid*, p.247

Toxic, see *Toxic Substances and Division 6.1*, p.255 or *Ammunition, Toxic*, p.19

REGULATORY DEFINITIONS

Chemical Kit or First Aid Kit The entry Chemical Kit or First Aid Kit is intended to apply to boxes, cases, etc. containing small quantities of various dangerous goods which are used for medical, analytical or testing purposes. [Text continues.] IMO 9026-1

Chemical Kits. Are boxes, cases, etc., containing small amounts of various dangerous goods used for analytical or other purposes. IATA App. A

Chemical Sample, Toxic This entry may only be used for chemical samples taken for analysis in connection with the implementation of the Convention on the Prohibition of the Development, Production, Stockpiling and Use of Chemical Weapons and on Their Destruction. The transport of substances under this entry should be in accordance with the chain of custody and security procedures specified by the Organization for the Prohibition of Chemical Weapons. IMO 6096-1

Chemical Sample, Toxic. This entry may only be used for chemical samples taken for analysis in connection with the implementation of the Chemical Weapons Convention. IATA App. A

First Aid Kits Are boxes, cases, etc., containing small amounts of various dangerous goods used for medical purposes IATA App. A

REFERENCES

KOE

Cleaning Liquids

Cleaning fluid or liquid ■ Compound, cleaning liquid, 3, 8

1760 ■ 1993

Cleaning liquids, *cleaning compounds*, and *cleaning fluids* remove unwanted materials (soil such as oil, sugars, proteins, dust, lubricants, etc.) from a solid substrate (fabrics and fibrous materials and hard surfaces like glass, metal, and painted surfaces) by bringing the soil into mechanical suspension or colloidal, molecular, or ionic solution in some solvent. The nature of the solvent provides a means of classification.

AQUEOUS CLEANERS

Aqueous cleaners are called detergents. They are based on water (although some other inorganic solvents are possible) and active ingredients that encourage dissolution of the soil by altering the interface between it and the solvent. Diluents, fillers, abrasives, foaming agents, bleaches, brighteners, antiseptics, and emollients may also be added. There are two types of active ingredients:

- Surfactants (short for *surface-active agent*), which are based on organic compounds; e.g., soaps, sulphates, sulphonates, acid esters, glycol ethers, copolymers, and others. These are not usually hazardous.
- Inorganic compounds; e.g., alkalis, phosphates, silicates, salts, acids, and others. These often include corrosives like sodium hydroxide, alkali silicates, phosphoric acid, hydrochloric acid, and sulphuric acid.

For example, water by itself will not remove some oils from skin, but addition of a sodium salt of stearic acid (a common hand soap) lowers the surface tension of water allowing the oil to be emulsified. Soap is less dense than water and the action of washing aerates the mixture so that the soap scum (soil and soap products) float to the surface.

NONAQUEOUS CLEANERS

Nonaqueous cleaners include a wide range of hydrocarbons and other organic solvents in which soil is dissolved directly by the solvent; there is no other active ingredient, although solvent mixtures are common. Many of these nonaqueous cleaners are flammable and toxic.

RELATED TERMS

Compound, see *Terminology, Compound*, p.234

Corrosive; Corrosive liquid, n.o.s., see *Corrosives and Class 8*, p.47

Flammable; Flammable liquid, n.o.s., see *Flammable Liquids and Class 3*, p.96

Fluid, see *Terminology, Fluid*, p.238

Liquid, see *Terminology, Liquid*, p.241

REGULATORY DEFINITIONS
None

REFERENCES
HCC, KOE

Coal

Blau gas ■ Coal ■ Coal briquettes, hot ■ Coal gas, 2.3 ■ Coal gas, compressed, 2.1, 2.3 ■ Coal tar, crude and solvent ■ Coal tar distillates, flammable, 3, 3.2, 3.3 ■ Coal tar naphtha ■ Coal tar oil ■ Coke, hot ■ Creosote ■ Creosote (coal tar or wood tar) ■ Creosote salts ■ Cresols (o-, m-, p-), 6.1, 8 ■ Cresols (ortho-; meta-; para-), liquid or solid, 6.1 ■ Dead oil ■ Fischer Tropsch gas ■ Fischer-Tropsch gas compressed, 2.2 ■ Iron oxide, spent (obtained from coal gas purification), 4.2 ■ Iron sponge, spent, 4.2 ■ Iron sponge, spent (obtained from coal gas purification), 4.2 ■ Prilled coal tar ■ Synthesis gas ■ Synthesis gas, compressed ■ Water gas ■ Water gas, compressed

1023 ■ 1044 ■ 1136 ■ 1376 ■ 2076

Coal is a black or brown, solid, combustible mineral formed by the alteration of prehistoric plant life by bacterial decomposition, with subsequent chemical changes caused by temperature and pressure. These processes result in a range of carbonaceous materials, the first of which is peat and the last of which is graphite (pure carbon). The coals lie between these two extremes:

- Lignite *syn.* brown coal, less than 75% carbon.
- Lignitous, 75 to 84% carbon.
- Bituminous, 84 to 91% carbon.
- Carbonaceous, 91 to 93% carbon.
- Anthracite, 93% carbon or higher.

The balance of coal is water (up to 40% for brown coal), a complex range of hydrocarbons, other organic compounds, and inorganic compounds. Coal is an important source of heat and energy, often used unprocessed or manipulated into *coal briquettes*: cylindrical, oval, or otherwise shaped blocks composed of granulated particles compressed and usually embedded in a binder made, for example, of starches, heavy hydrocarbons or asphalts, or inorganic materials. Briquettes are used to ease handling, produce smokeless solid fuel, and to convert cheap waste coal or low grade coal (often brown coal) into useful lumps.

COAL DERIVATIVES

The destructive distillation (heating in the absence of oxygen) of coal (mostly bituminous coal) at temperatures ranging from 500 to 1200°C, generates a number of derivatives:

- *Coke* (65996-77-2), the solid remains, about 70% of the coal weight. Coke is primarily used for the reduction of iron and other metal ores in blast furnaces. Coke may be generated from other sources including petroleum residuum (petroleum coke).
- *Coal gas syn.* town gas, a complex mixture of flammable and toxic gases based on hydrogen, methane, and ammonia.
- *Crude coal tar*, about 5% by weight, a generic name for the resultant toxic, black, oily, viscous distillation liquid or semisolid.

Coal tar contains an estimated 10,000 compounds, many of which are important organic chemicals. The use of coal tar as a source of these compounds has been largely relegated to a position below numerous synthetic processes, primarily based on petroleum. Fractionation of coal tar yields the following (approximate temperatures and yields given):

RANGE	YIELD	PRODUCT
up to 200°C	5%	Light oil *syn. coal tar distillates, coal tar solvents*, or *coal tar oil*: a highly flammable mixture of toluene, xylene, cumenes, etc.
200 to 250°C	17%	Middle oil, e.g., heavy naphtha, phenol.
250 to 300°C	7%	Heavy oil, e.g., naphthalene, cresols.
300 to 375°C	9%	Anthracene oil, e.g., phenanthrene, anthracene, quinolines.
over 375°C	62%	Pitch, a dark brown or black residue of toxic aromatic hydrocarbons and other compounds often used as road and roofing tar. *Prilled coal tar* is coal tar pitch manufactured into small globules or granules.

Coal tar naphtha (71-43-2) is an indefinite term generally applied to the flammable coal tar distillates extracted at around 160 to 220°C. The term *naphtha*, however, is mostly applicable to petroleum products. *Cresol* (1319-77-3) is a mixture of methyl phenyl isomers extracted from the middle- to heavy-oil fractions (it is also derived from petroleum). Cresol is a toxic irritant and corrosive to skin and mucous membranes, which explains its common use as a disinfectant.

Depending on the source, *creosote syn. dead oil* (8001-58-9) collectively describes the distillation range from 200 to 400°C, or just the narrower cut of anthracene oils (creosote also derives from the distillation of wood). It is generally agreed, though, to contain significant proportions of anthracene and naphthalene. Its combined properties as a moderately toxic compound and its insolubility in water make it an ideal wood preservative, particularly in marine environments. *Creosote salts* are those generated by converting a variety of acidic or basic compounds present in creosote (e.g., phenol and pyridine) to salts so they can be more easily extracted.

OTHER TERMS

Blau gas syn. Fischer-Tropsch gas, synthesis gas, water gas is a mixture of carbon monoxide and hydrogen gas generated from passing steam over hot coal (the word *blau*, an obsolete word for *blow*, is derived from this action). The mixture is used to generate liquid or gaseous hydrocarbons and was developed by German chemists, Fischer and Tropsch.

Iron sponge is the finely divided and spontaneously combustible porous form of iron oxide (sesquioxide iron, 1332-37-2) made by heating iron ore and carbon. It is used, among other applications, to purify coal gas generated from coke ovens by removing sulphur compounds and carbonic acid. As a consequence, *spent iron sponge* syn. *spent iron oxide* evolves the toxic gases hydrogen sulphide, sulphur dioxide, and hydrogen cyanide.

RELATED TERMS

Coal, see also *Solid Bulk Materials*, p.221

Coal tar dye, corrosive, liquid, n.o.s., see *Dyes and Pigments*, p.63

Compressed, see *Gases and Class 2*, p.104

Dye intermediate, liquid, corrosive, n.o.s., see *Dyes and Pigments*, p.63

Dye, liquid, corrosive, n.o.s., see *Dyes and Pigments*, p.63

Flammable, see *Flammable Liquids and Class 3*, p.96

Gas, see *Terminology, Gas*, p.239

Liquid, see *Terminology, Liquid*, p.241

o-, m-, p-, see *Terminology, Structural Notation*, p.248

ortho-; meta-; para-, see *Terminology, Structural Notation*, p.248

Petroleum distillates, n.o.s., see *Petroleum*, p.183

Petroleum products, n.o.s., see *Petroleum*, p.183

Prilled coal tar, see also *Solid Bulk Materials*, p.221

Salts, see *Terminology, Salts*, p.247

Solid, see *Terminology, Solid*, p.247

Solvent, see *Solvents*, p.224

Sponge, see *Terminology, Sponge*, p.248

Tars, liquid, see *Bituminous Products*, p.32

REGULATORY DEFINITIONS

Coal Gas Compressed. The gas obtained by the destructive distillation of bituminous coal. ICAO A2

Coal Gas Compressed. Is the gas obtained by the destructive distillation of bituminous coal. It is shipped in steel cylinders and classed as a toxic flammable compressed gas. IATA App. A

Iron Mass and Iron Sponge. Consist of a mixture of wood shavings with iron oxide and possibly lime or other material. If properly made and all the iron is properly oxidised, the material is free from the hazard of spontaneous heating or ignition. If there is an undue amount of metallic unoxidised iron, further oxidisation is liable to occur, producing sufficient heat in closely packed material to cause fire. This material may be ignited by external sparks. Iron mass is used for gas purification. IATA App. A

Iron Oxide, Spent or Iron Sponge, Spent. A mixture of wood shavings with iron oxide and possibly lime or other material, which has been obtained from coal gas purification after saturation with sulphur. This spent material is very liable to spontaneous heating and ignition. ICAO A2

Iron Oxide, Spent or Iron Sponge, Spent. Consists of iron mass or iron sponge after saturation with sulphur in gas purification. This spent material is very liable to spontaneous heating and ignition. IATA App. A

REFERENCES

ADM, CPS, CTA, DOCO, FAF, HCC, KOE, MEOS, MH14, KOC, SSD, STRCC, TCAOC, TIP, TIR, VNS

Corrosives and Class 8

Acid liquid, n.o.s. ■ Acid mixture, nitrating acid ■ Acid mixture, spent, nitrating acid ■ Alkaline caustic liquid, n.o.s. ■ Alkaline corrosive liquid, n.o.s. ■ Alkaline corrosive solid, n.o.s. ■ Alkylsulphonic acids, liquid, 8 ■ Alkylsulphonic acids, solid, 8 ■ Alkylsulphuric acids, 8 ■ Caustic alkali liquid, n.o.s., 8 ■ Corrosive liquid, acidic, inorganic, n.o.s., 8 ■ Corrosive liquid, acidic, organic, n.o.s., 8 ■ Corrosive liquid, basic, inorganic, n.o.s., 8 ■ Corrosive liquid, basic, organic, n.o.s., 8 ■ Corrosive liquid, n.o.s., 3, 4.2, 4.3, 5.1, 6.1, 8 ■ Corrosive solid, acidic, inorganic, n.o.s., 8 ■ Corrosive solid, acidic, organic, n.o.s., 8 ■ Corrosive solid, basic, inorganic, n.o.s., 8 ■ Corrosive solid, basic, organic, n.o.s., 8 ■ Corrosive solid, n.o.s., 4.1, 4.2, 4.3, 5.1, 6.1, 8 ■ Etching acid, liquid, n.o.s. ■ Matting acid ■ Mixed acid ■ Mixed acid, spent ■ Nitrating acid mixture, 5.1, 8 ■ Nitrating acid mixture, spent, 5.1, 8 ■ Nitrating acid, mixture, spent, all concentrations, unstable ■ Solids containing corrosive liquids, n.o.s., 8

1719 ■ 1759 ■ 1760 ■ 1796 ■ 1826 ■ 2571 ■ 2583 ■ 2584 ■ 2585 ■ 2586 ■ 2920 ■ 2921 ■ 2922 ■ 2923 ■ 3084 ■ 3093 ■ 3094 ■ 3095 ■ 3096 ■ 3244 ■ 3260 ■ 3261 ■ 3262 ■ 3263 ■ 3264 ■ 3265 ■ 3266 ■ 3267 ■ 3301

Corrosive materials are those chemicals which, on contact, damage living tissue (skin, eyes, mucous membranes, plant life) and inorganic materials (metals, textiles, ceramics, glass, etc.). They include acids and bases.

ACIDS

Acids describe a vast range of organic and inorganic chemicals that characteristically contain hydrogen and react with metals and alkalis to form salts (*acidic* describes the property of an acid). In water, acids ionize so that some concentration of hydrogen forms the hydronium ion (H_3O^+), abbreviated to the hydrogen ion (H^+), in solution. The hydrogen ion is a powerfully reactive oxidizer in that it works to complete the following reaction:

$$2H^+ + 2e^- \rightarrow H_2 \text{ (flammable gas) where } e^- \text{ is an electron}$$

Acids can also exist in non water-based solvents such as ammonia (NH_3), where the ammonium ion NH_4^+ is generated by the acid. This concept is generally extended to define acids as those substances that ionize to generate the positive ion of the solvent in which they reside.

Common *inorganic acids* may ionize completely and include the extremely corrosive sulphuric, nitric, and hydrochloric acids. Often weaker, *organic acids* principally contain the -COOH group such as acetic acid, which becomes CH_3COO^- in solution with H_3O^+, but many other organic acid groupings are possible including the sulphonic group (-SO_2OH) and the phenols (substances based on C_6H_5OH).

BASES

In contrast, bases *syn. alkalis* are those substances which in aqueous solution form the hydroxyl ion (OH^-) that acts powerfully to be oxidized by gaining a positive ion (often a hydrogen ion) to form a new compound. Alkalis work by the reducing power of the hydroxyl ion. In this way, bases can react violently with acids to form a salt and water:

$$AH + BOH \rightarrow AB + H_2O, \text{ where } AH \text{ is an acid and } BOH \text{ is a base.}$$

In nonaqueous solutions, acids accept electrons and bases donate electron pairs; for example, ammonia and boron trifluoride react as follows:

$$NH_3 \text{ (base)} + BF_3 \text{ (acid)} \rightarrow F_3B\text{-}NH_3$$

In general usage, the terms *alkali* and *caustic* are synonymous with the term *base*, and just as *basic* describes the properties of a base, *alkaline* describes the properties of an alkali. However, *caustic* is often used to specify a substance corrosive to metals, particularly aluminium, zinc, and tin; for example, the *inorganic base* caustic soda *syn.* sodium hydroxide and caustic potash *syn.* potassium hydroxide. *Organic bases* include the amines.

pH

The strength of an acid or alkali in aqueous solutions is measured by its ability to dissociate (divide into its pair of ions). The unit of measurement (pH) indicates the acidic or alkaline strength compared to that of water. Mathematically, pH is the logarithm of the reciprocal hydrogen ion concentration, or $\log_{10} 1/[H^+]$. In pure water (H_2O) according to the equation

$$H_2O \leftrightarrows H^+ + OH^-$$

there is an extremely small concentration of hydrogen ions and hydroxyl ions: 1×10^{-7} moles/l. The reciprocal logarithm of this number sets the pH of water at 7, which is regarded as neutral. Acids are those with a lower pH and, therefore, a higher hydrogen ion concentration. Alkalis are those with a higher pH and a lower hydrogen ion concentration. For example, a 0.1 molar solution of hydrochloric acid (HCl) has a hydrogen ion concentration of 0.083 moles/l and a pH of 1.08; a 0.01 molar solution of sodium hydroxide (NaOH) has a hydrogen ion concentration of 1×10^{-12} moles/l and a pH of 12.

CORROSIVE FORMS

Corrosives may act directly on contact as *corrosive liquids* or solutions. They may also be generated by

- The evolution of corrosive vapours through evaporation, heat, decomposition, or fire.
- Contact with water. This is commonly the case with the *corrosive solids*, such as the inorganic base potassium hydroxide and the organic crotonic acid. As solids, these do not have any available hydrogen or hydroxyl ions but easily form solutions in contact with water or moisture from the air or from skin, eyes, lung tissues, etc.
- Reaction with other substances.

CORROSION

The action of corrosion depends on the chemical and the substrate attacked, but in general there are three possible types of corrosion:

1. Direct contact with an acid or base in which oxidation or reduction occurs according to electrochemical principles.
2. The oxidizing effects of oxygen in the atmosphere.
3. Solvents in which the substrate is dissolved, such as the dissolution of aluminium by mercury.

Corrosive materials in transportation are regulated when the circumstances of concentration, time, temperature, substrate composition, radiation, inhibitors, environment, and other factors are such that the reaction will be particularly violent, aggressive, destructive, and rapid enough to cause a hazard (although longer-term corrosion of packaging is also considered). The corrosive effects of an acid may be represented by the action of sulphuric acid on zinc in which the acid exists in equilibrium in solution as

$$H_2SO_4 \leftrightarrows 2H^+ + SO_4^{2-}$$

which reacts with zinc:

$$2H^+ + SO_4^{2-} + Zn \leftrightarrows Zn^{2+} + SO_4^{2-} + H_2$$

The practical effect is that if sulphuric acid is placed in contact with zinc, some of the zinc will be oxidized and dissolved into solution, thus weakening the substrate, and the flammable gas, hydrogen, is generated. Acids work similarly to destroy organic materials such as skin, an effect humans experience as acid burns.

SPECIFIC CORROSIVES

A number of specific corrosive terms are used in the transportation regulations:

□ *Alkylsulphonic acids*, extremely reactive acids with an alkyl group attached to the sulphonic group ($-SO_3H$), such as methyl sulphonic acid (CH_3SO_3H).

□ *Alkylsulphuric acids,* similarly corrosive acids based on an alkyl group attached to the sulphuric group ($-SO_4H$), such as methyl sulphuric acid (CH_3OSO_3H).

□ *Etching acid* usually refers to a solution of hydrofluoric acid which will attack any silicon-containing substrate like glass.

□ *Matting acid*, sulphuric acid when used as a matting agent. Matting is the process by which metals and other substrates have their surface sheen removed to leave a dull, lustreless finish. It can also be used to describe the use of sulphuric acid as an etchant.

□ *Solids containing corrosive liquids* include sand, soil, and debris contaminated with corrosive liquids.

NITRATING ACIDS

Nitration, the addition of the nitro group (-NO$_2$) to a compound, is one of the most important commercial reactions used during the manufacture of dyes, fertilizers, explosives (e.g., nitrocellulose), plastics, etc. The principal nitrating agent is nitric acid (7697-37-2) which acts in a two stage process to generate the nitronium ion (NO$_2^+$):

$$2HNO_3 \leftrightarrows NO_2^+ + NO_3^- + H_2O$$

However, the second step of the reaction is relatively slow and the concentration of the nitronium ion is low. Sulphuric acid (7664-93-9), which speeds the reaction and maximizes the nitronium ion concentration, is mixed with nitric acid (nitric acid:sulphuric acid, 36:61%) and the two are called *mixed acid syn. nitrating acid mixture*. Spent nitrating acids mixtures have a lower nitronium ion concentration, a higher water concentration, and some organic matter.

RELATED TERMS

Acid sludge, see *Petroleum*, p.183

Beverage extract (concentrate), see *Extracts*, p.86

Flammable liquid, see *Flammable Liquids and Class 3*, p.96

Flammable solid, see *Flammable Solids and Division 4.1*, p.99

Inorganic, see *Terminology, Inorganic*, p.241

Liquid, see *Terminology, Liquid*, p.241

Mixture, see *Terminology, Mixture*, p.243

Organic, see *Terminology, Organic*, p.244

Oxidizing, see *Oxidizers and Class 5*, p.170

Self-heating, see *Spontaneously Combustible Materials and Division 4.2*, p.226

Sludge, acid, see *Petroleum*, p.183

Solid, see *Terminology, Solid*, p.247

Toxic, see *Toxic Substances and Division 6.1*, p.255

Water-reactive, see *Dangerous When Wet Materials and Division 4.3*, p.58

REGULATORY DEFINITIONS

Class 8

Class 8 substances (corrosive substances) are substances which, by chemical action, will cause severe damage when in contact with living tissue, or, in the case of leakage, will materially damage, or even destroy, other goods or the means of transport; they may also cause other hazards. UN 2.8.1

Substances which, in the event of leakage, can cause severe damage by chemical action when in contact with living tissue or can materially damage other freight or the means of transport are included in Class 8. ICAO 2-8.1

The substances in this class are solids or liquids possessing, in their original state, the common property of being able, more or less severely, to damage living tissue. The escape of such a substance from its packaging may also cause damage to other cargo or to the ship. IMO Class 8, 1.1

All substances in this class have a more or less destructive effect on materials such as metals and textiles. IMO Class 8, 1.3

A few substances in this class can corrode glass, earthenware and other siliceous materials. [Text continues.] IMO Class 8, 1.3.3

Many substances in this class only become corrosive after having reacted with water, or with moisture in the air. [Text continues.] IMO Class 8, 1.4

A few substances in this class generate heat in reaction with water or organic materials, including wood, paper, fibres, some cushioning materials and certain fats and oils. [Text continues.] IMO Class 8, 1.5

For the purpose of this subchapter, *"corrosive material"* (Class 8) means a liquid or solid that causes full thickness destruction of human skin at the site of contact within a specified period of time. A liquid that has a severe corrosion rate on steel or aluminum based on the criteria in Sec. 173.137(c)(2) is also a corrosive material. US 173.136(a)

Substances which, in the event of leakage, can cause severe damage by chemical action when in contact with living tissue or can materially damage other freight or the means of transport. IATA 3.8.1

Other Definitions

Acidic. In general, an acidic substance is one which contains hydrogen and which dissolves in water to produce one or more hydrogen ions. Such solutions turn litmus dye red and cause other indicator dyes to change to characteristic colours. They also react with certain metals and bases/alkalis to form salts. Acidity is commonly measured using the pH scale; on this scale water has a "neutral" pH, i.e. neither acidic nor basic, or 7 and acids have a pH lower than 7. Some examples of acidic substances are hydrochloric acid, sulphuric acid, hydrogen sulphide (inorganic acids) and acetic acid, e.g. vinegar, and citric acid (organic acids). IATA App. A

Basic (Alkali). In general, a basic substance dissolves in water to produce one or more hydroxyl ions. Such substances have the ability to turn litmus blue and to cause other indicators to take on characteristic colours. They also react with (neutralise) acids to form salts. Basicity is commonly measured using the pH scale; on this scale water has a "neutral" pH (neither acidic nor basic) of 7 and bases have a pH higher than 7. Some examples of basic substances are sodium hydroxide, (e.g. caustic soda or lye); calcium hydroxide (e.g. lime); potassium hydroxide and ammonium hydroxide. [Text continues.] IATA App. A

Nitrating Acid Mixture; Acid Mixture, Nitrating Acid; Mixed Acid Mixture of concentrated nitric and sulphuric acids. Oxidant, may cause fire in contact with organic materials such as wood, cotton or straw, developing highly toxic gas (brown fumes). Highly corrosive to most metals. IMO 8194

Nitrating Acid Mixture. A mixture of nitric acid and sulphuric acids used for the nitration of glycerin, cellulose or other organic substances. This acid mixture coming in contact with organic matter commonly causes fire, unless the mixture contains much water. ICAO A2

Nitrating Acid Mixture. Mixture of nitric and sulphuric used for the nitration of glucose, cellulose or other organic substances. This acid mixture coming in contact with organic matter commonly causes fire, unless the mixture contains much water. IATA App. A

REFERENCES

E2, HCC, KOE, TPD

Cosmetics and Perfumes

Cologne spirits ■ Cosmetics ■ Cosmetics, corrosive, liquid, n.o.s. ■ Cosmetics, corrosive, solid, n.o.s. ■ Cosmetics, flammable, liquid, n.o.s. ■ Cosmetics, flammable, solid, n.o.s. ■ Cosmetics, n.o.s. ■ Cosmetics, oxidizing material, liquid, n.o.s. ■ Cosmetics, oxidizing material, solid, n.o.s. ■ Perfumery products in small inner packagings ■ Perfumery products, with flammable liquid, 3.2, 3.3 ■ Perfumery products with flammable solvents, 3

1266

With an enormous variety of creams, lotions, deodorants, antiperspirants, sunscreens, makeup, hair preparations, bath salts, soaps, and shaving products, *cosmetics* are applied to alter, preserve, or beautify the outer surface of the body (skin, hair, nails, lips, eyes, and teeth) by cleaning, colouring, conditioning, or protecting. *Perfumery products* are applied to our skin and hair to emit pleasant odours. The constituents of cosmetics and perfumes can be grouped as follows:

□ Preservatives which prevent infection of the skin and decomposition of the product. An ideal preservative is nontoxic to humans, non-irritating, and nonsensitizing, however, they may include flammable alcohols or mercury-based anti-infectives for the eye areas.

□ Acids, alkalis, buffers, and neutralizers which maintain an acidity level that prevents skin irritation and maintains product formulation. As raw materials, including citric acid, ammonium carbonate, calcium carbonate, and tartaric acid, these are potential sources of corrosivity. Other cosmetics, like depilatories (cuticle and hair removers), are corrosive by function.

□ Moisture content controls which prevent powders from becoming damp and creams and lotions from drying out; rarely do these raw materials, including glycerine and propylene glycol, present hazards in transportation.

□ Colouring agents which disguise colour, decolourize, or impart colour. These include hundreds of different dyes and pigments, some of which are flammable, corrosive, or toxic although the debate about toxicity continues; some "safe" colours have caused tumours in animals and others have been banned. Hair bleaches (e.g., hydrogen peroxide) can be oxidizing, and products such as nail polish contain flammable solvents and compounds, including pyroxylin. Mercury compounds have been used as skin-bleaching agents.

□ Flavourings and fragrances which include thousands of substances, sometimes blended in complex proportions by experts to produce prized odours. Some of these products may be hazardous in their own right like mirbane oil *syn.* nitrobenzene which has an almond smell, but they are often associated with solvents to aid handling and use.

□   Processing aids and carriers which include surfactants to emulsify, stabilize, aid dissolution, and texturize, as well as solvents, clarifiers and chelators, opacifiers, foaming agents, and aerosol systems. Some of these, particularly alcohols or acids, may be flammable, corrosive, or otherwise hazardous.

Compared to normal perfumes, *cologne spirits* contain a relatively high alcohol-fragrance ratio (up to 80% for solid deodorants) to impart a pleasant cooling effect on application as the alcohol draws heat from the skin to evaporate. As liquids, gels, pastes, or solids, colognes may be flammable liquids or flammable solids.

RELATED TERMS

Aerosol, see *Aerosols*, p.3

Consumer commodity, see *Miscellaneous Dangerous Goods (including Class 9)*, p.158

Corrosive; Corrosive, liquid, n.o.s.; Corrosive, solid, n.o.s., see *Corrosives and Class 8*, p.47

Flammable; Flammable liquid; Flammable liquid, n.o.s., see *Flammable Liquids and Class 3*, p.96

Flammable; Flammable solid organic, n.o.s.;

Flammable solid inorganic, n.o.s., see *Flammable Solids and Division 4.1*, p.99

Flammable solvents, *Flammable Liquids and Class 3*, p.96

Liquid, see *Terminology, Liquid*, p.241

Oxidizing; Oxidizing liquid, n.o.s.; Oxidizing solid, n.o.s., see *Oxidizers and Class 5*, p.170

Solid, see *Terminology, Solid*, p.247

Solvent, see *Solvents*, p.224

REGULATORY DEFINITIONS
None

REFERENCES
ACD, ICI, KOC, MH14

Dangerous Goods and Hazardous Materials

Dangerous goods are capable of posing a significant risk to health, safety, or property when transported. *Hazardous materials* is the term applied to dangerous goods in the United States and some other countries. In 1956, the United Nations published its first recommendations (*Recommendations on the Transport of Dangerous Goods* (ROT)) designed to provide a uniform and basic scheme for dangerous goods classification, identification, packaging, marking, labelling, placarding, documentation, and handling procedures. Addressed to the international community, these recommendations allow governments and organizations to incorporate requirements unique to different shipping modes and to cover special concerns. Over the years, many national bodies and international organizations have adopted the UN's recommendation in whole or in part. Other bodies have progressively aligned their standards to be more compatible with those of the UN.

At the core of the UN's scheme lies nine physical, chemical, and physiological properties that present hazards in transportation:[2]

- Class 1: Explosives
  - Division 1.1: Substances and articles which have a mass explosion hazard
  - Division 1.2: Substances and articles which have a projection hazard but not a mass explosion hazard
  - Division 1.3: Substances and articles which have a fire hazard and either a minor blast hazard or a minor projection or both, but not a mass explosion hazard
  - Division 1.4: Substances and articles which present no significant hazard
  - Division 1.5: Very insensitive substances which have a mass explosion hazard
  - Division 1.6: Extremely insensitive articles which do not have a mass explosion hazard
- Class 2: Gases
  - Division 2.1: Flammable gases
  - Division 2.2: Nonflammable, nontoxic gases
  - Division 2.3: Toxic gases
- Class 3: Flammable liquids

---

[2] ROT 2.0.1.1.

- Class 4: Flammable solids; substances liable to spontaneous combustion; substances which, on contact with water, emit flammable gases
  - Division 4.1: Flammable solids, self-reactive and related substances, and desensitized explosives
  - Division 4.2: Substances liable to spontaneous combustion
  - Division 4.3: Substances which in contact with water emit flammable gases
- Class 5: Oxidizing substances and organic peroxides
  - Division 5.1: Oxidizing substances
  - Division 5.2: Organic peroxides
- Class 6: Toxic and infectious substances
  - Division 6.1: Toxic substances
  - Division 6.2: Infectious substances
- Class 7: Radioactive material
- Class 8: Corrosive substances
- Class 9: Miscellaneous dangerous substances and articles

RADIOACTIVE SHIPMENTS

Regulations applicable to the shipment of radioactive materials provide an acceptable level of control of radiation, criticality, and thermal hazards to persons, property, and the environment during routine, normal, and accidental conditions by all modes, as governed by the International Atomic Energy Agency. These rules are promulgated in the *Regulations for the Safe Transport of Radioactive Materials* (RFT).

TRANSPORT BY SEA

Transport by sea of dangerous goods takes into account possible damage to the ship and the marine environment as required by the *International Convention for the Safety of Life at Sea*, 1974 (SOLAS), as amended; the *International Convention for Prevention of Pollution from Ships*, 1973, as modified by the Protocol of 1978 (MARPOL); and preceding regulations. These rules are implemented by the International Maritime Organization (IMO) and promulgated in the *International Maritime Dangerous Goods Code* (IMDG).

Supplement 3 to the IMDG Code is the *Code of Safe Practice for Solid Bulk Cargoes* which recognizes that certain large volume solid materials, dangerous goods or not, present hazards when transported without intermediate containment in the cargo space of ships. These hazards include possible structural damage to the ship, loss or reduction of stability due to cargo shifting or liquefaction, adverse chemical reactions, and the release of hazardous concentrations of dust.

TRANSPORT BY AIR

The principles associated with the transport of dangerous goods by air are contained in Annex 18 to the Convention of International Civil Aviation—

The Safe Transport of Dangerous Goods by Air. This Annex, which is based on the UN recommendations, sets out the rules covering materials that can be transported on passenger and cargo aircraft and those whose risks warrant shipment only by cargo aircraft. The International Civil Aviation Organization (ICAO) amplifies the rules of Annex 18 in the *Technical Instructions for the Safe Transport of Dangerous Goods by Air* (TI).

In 1956, the member airlines of the International Air Transport Association (IATA) used their specialized knowledge of air transport to respond to the growing volume of dangerous goods shipments by publishing the *Restricted Articles Regulations*. Today, these have been superseded by the *Dangerous Goods Regulations* (DGR) which contain all the requirements of ICAO's Technical Instructions and include more restrictive requirements to reflect industry practices and operational considerations.

U.S. DEPARTMENT OF TRANSPORTATION
In the United States, the Associate Administrator for Hazardous Materials Safety, Research and Special Programs Administration, U.S. Department of Transportation (DOT), is the national authority for the control and regulation of hazardous materials shipments. Under the Hazardous Materials Transportation Act, 49 USC 1801 *et seq.*, the DOT promulgates its multimodal hazardous materials transportation regulations in Title 49 of the *Code of Federal Regulations*, Subtitle B—Other Regulations Relating to Transportation, Parts 100 to 199 (49CFR).

FORBIDDEN DANGEROUS GOODS
Dangerous goods regulations make transport feasible while eliminating or reducing risk to a minimum. Some materials, however, present an unacceptable risk under any circumstances or by certain modes. Others may be transported only with special permission and arrangements. The majority of *forbidden dangerous goods* are extremely shock-sensitive explosives or explosives that become unstable when subjected to conditions normally incident to transportation. Other forbidden material may include

- Electrical devices that generate sparks or excessive heat.
- Excessively magnetic materials.
- Incompatible materials transported in the same container.
- Materials that decompose or polymerize violently under normal conditions.
- Materials that give off flammable vapour concentrations.
- Materials that detonate in a fire.
- Specifically forbidden materials that have been involved in transportation incidents in the past.

It should be noted that the majority of otherwise forbidden materials may be diluted, protected, refrigerated, mixed, or stabilized to enable their transportation by one or more modes.

RELATED TERMS
None

REGULATORY DEFINITIONS

Dangerous goods. Articles or substances which are capable of posing a significant risk to health, safety or property when transported by air and which are classified according to Part 2, Chapters 1 to 10. ICAO 1-3.1

Dangerous goods are articles or substances which are capable of posing a significant risk to health, safety or to property when transported by air and which are classified according to Section 3. IATA 1.0

Dangerous goods are defined as those goods which meet the criteria of one or more of nine UN hazard classes and, where applicable, to one of three UN packing groups according to the provisions of this section. The nine classes relate to the type of hazard whereas the packing groups relate to the applicable degree of danger within the class. IATA 3.0.1.1

Hazardous material means a substance or material, which has been determined by the Secretary of Transportation to be capable of posing an unreasonable risk to health, safety, and property when transported in commerce, and which has been so designated. The term includes hazardous substances, hazardous wastes, marine pollutants, and elevated temperature materials as defined in this section, materials designated as hazardous under the provisions of Sec. 172.101 of this subchapter, and materials that meet the defining criteria for hazard classes and divisions in part 173 of this subchapter. US 171.8

Hazardous Material. The U.S. Department of Transportation term for dangerous goods. IATA 2.9.2

REFERENCES
49CFR, COS, DGR, IMDG, RFT, ROT, TI

Dangerous When Wet Materials and Division 4.3

Substances which in contact with water emit flammable gases ■ Water-reactive liquid, n.o.s., 4.3, 6.1, 8 ■ Water-reactive solid, n.o.s., 4.1, 4.2, 4.3, 5.1, 6.1, 8

2813 ■ 3129 ■ 3130 ■ 3131 ■ 3132 ■ 3133 ■ 3134 ■ 3135 ■ 3148

Water (7732-18-5) is a simple molecule of two hydrogens attached to an oxygen atom, H-O-H. However, the few types of reaction it undergoes are universally fundamental to all forms of life.

HYDRATION

Water is an extremely polar substance meaning that its electrical charge is spread unevenly over the molecule as the oxygen pulls the electron bonds towards itself, leaving the hydrogens strongly electropositive. This property makes it an ideal solvent for a vast number of ionic and polar molecules like salts and alcohols in which the solute is surrounded by a tight film of water molecules bound by electrostatic forces. As sodium chloride (NaCl) is hydrated in water, for example, the sodium ions (Na^+) are attracted to the electronegative oxygen in water while the chloride ions (Cl^-) are attracted to the hydrogens:

$$NaCl \rightarrow Na(OH_2)_{4\text{-}6}^+ + Cl(H_2O)_{4\text{-}6}^-$$

abbreviated to

$$NaCl \rightarrow Na^+ \text{ (aqueous)} + Cl^- \text{ (aqueous)}$$

Certain compounds contain crystalline water as part of their structure and form hydrates. This water can be driven off by heating. One of the hydrates of sodium sulphate (Na_2SO_4), for example, is its decahydrate, $Na_2SO_4 \cdot 10HOH$.

HYDROLYSIS

In hydrolysis, the H-OH bond in the water molecule splits and the reactions take place in the presence of the H^+ and OH^- ions. For example, calcium carbide in water reacts to produce acetylene and calcium hydroxide in solution:

$$CaC_2 + 2H_2O \rightarrow Ca^{2+} + 2OH^- + C_2H_2 \text{ (gas)}$$

The excess of OH^- ions makes the solution alkaline. Or, in the case of many metals (M), hydrogen gas is given off leaving a hydroxide solution:

$$M + H_2O \rightarrow M^+ + OH^- + H_2 \text{ (gas)}$$

Hydrolysis can also result in the addition of a water molecule to another chemical, e.g.,

$$C_3H_5CN + H_2O \rightarrow C_3H_5CONH_2$$

and displacement reactions in which a radical in a molecule is replaced by the OH⁻ ion, e.g.,

$$CH_3CH_2Cl + H_2O \rightarrow CH_3CH_2OH + HCl$$

CATALYZATION

Water catalyzes many reactions including the oxidation of metals and in saponification (the conversion of an ester to an alcohol and the sodium salt of the acid corresponding to the ester). Saponification is the basis for soap-making.

HAZARDS

Many reactions with water occur without incident. Other *water-reactive* chemicals can present a transportation risk because the reactions are violent, evolve high heat, or generate toxic, flammable, corrosive, or otherwise dangerous reaction products:

- Hydrides, like aluminium hydride, react violently with water to produce hydrogen gas.
- Phosphides, like sodium phosphide, react violently to produce phosphine, a toxic and flammable gas.
- Alkali and alkaline earth metals react with varying degrees of violence to produce hydrogen and the metal hydroxide.
- Strongly electropositive metal carbides react to form flammable hydrocarbon gases, e.g., aluminium carbide gives methane.
- Chlorosilanes react to generate corrosive gases.

In some instances flammable gases can build up to explosive concentrations. In others and when the mixture is exposed to the air, the heat of reaction may be sufficient to ignite the flammable gases as they evolve. Like many reactions with solids, the rate of reaction increases with decreasing particle size.

RELATED TERMS

Corrosive, see *Corrosives and Class 8*, p.47
Flammable, see *Flammable Solids and Division 4.1*, p.99
Flammable gas, see *Gases and Class 2*, p.104
Liquid, see *Terminology, Liquid*, p.241
Oxidizing, see *Oxidizers and Class 5*, p.170

Self-heating, see *Spontaneously Combustible Materials and Division 4.2*, p.226
Solid, see *Terminology, Solid*, p.247
Toxic, see *Toxic Substances and Division 6.1*, p.255

REGULATORY DEFINITIONS

Certain substances in contact with water emit flammable gases that can form explosive mixtures with air. Such mixtures are easily ignited by all ordinary sources of ignition, for example naked lights, sparking handtools or unprotected light bulbs. The resulting blast wave and flames may endanger people and the environment. [Text continues.] UN 2.4.4.1.2

Substances which in contact with water emit flammable gases Substances which, by interaction with water, are liable to become spontaneously flammable or to give off flammable gases in dangerous quantities. UN 2.4.1.1(c); ICAO 2-4

The substances in this class are either liquids or solids which, by interaction with water, are liable to become spontaneously flammable or to give off flammable gases in dangerous quantities. IMO Class 4.3, 1.1

Certain substances on contact with water emit flammable gases in such quantities that explosive mixtures with air will be formed. Such mixtures are easily ignited by ordinary sources of ignition, for example naked lights, sparking hand tools, unprotected light bulbs, or by the heat of the reaction. IMO Class 4.3, 1.2

Some of these substances also evolve toxic gases when in contact with moisture, water or acids. IMO Class 4.3, 1.3

Certain substances in contact with water emit flammable gases which can form explosive mixtures with air. Such mixtures are easily ignited by all ordinary sources of ignition, for example, naked lights, sparking handtools or unprotected light bulbs. The resulting blast wave and flames may endanger people and the environment. [Text continues.] ICAO 2-4.3.2, IATA 3.4.3.2

(Dangerous when wet material). For the purposes of this chapter, dangerous when wet material (Division 4.3) means a material that, by contact with water, is liable to become spontaneously flammable or to give off flammable or toxic gas at a rate greater than 1 liter per kilogram of the material, per hour [text continues]. US 173.124(c)

Substances which, in contact with water, emit flammable gases (Dangerous when wet). Substances which, by interaction with water, are liable to become spontaneously flammable or to give off flammable gases in dangerous quantities. IATA 3.4.3.1

REFERENCES
EOI, HCC, KOE, TPD

Drugs and Medicines

Anaesthetic ether ■ Disinfectants, liquid, n.o.s., 6.1, 8 ■ Disinfectants, solid, n.o.s., 6.1 ■ Drugs, corrosive, liquid, n.o.s ■ Drugs, corrosive, solid, n.o.s. ■ Drugs, flammable, liquid, n.o.s ■ Drugs, flammable, solid, n.o.s. ■ Drugs, n.o.s. ■ Drugs, oxidizing, liquid, n.o.s. ■ Drugs, oxidizing, solid, n.o.s. ■ Drugs, toxic, liquid, n.o.s. ■ Drugs, toxic, solid, n.o.s. ■ Medicinal preparation ■ Medicine, liquid, n.o.s., 3, 3.2, 3.3, 6.1, 8 ■ Medicine, n.o.s. ■ Medicine, solid, n.o.s., 4.1, 5.1, 6.1, 8 ■ Tinctures, medicinal, 3, 3.2, 3.3

1293 ■ 1325 ■ 1479 ■ 1601 ■ 1759 ■ 1760 ■ 1851 ■ 1903 ■ 1993 ■ 3142 ■ 3248 ■ 3249

A *drug* is any natural, semisynthetic, or synthetic substance taken as a narcotic or to help diagnose, treat, cure, or prevent disease or other abnormal condition; for the relief of pain or suffering; or to control any physiologic or pathologic condition. *Medicine* is the collective term often used by patients to describe prescribed drugs.

Most drugs are toxic when the useful dose is exceeded; even "safe" doses may produce unwanted effects: they can affect unintended organs, promote allergies, act singly or in concert with other drugs to cause foreseen and unforeseen side effects, and form addictions. The range of drugs is enormous:

alkaloids	astringents	expectorants
anaesthetics (drugs	bitters and tonics	glandular products
used to depress the	chemotherapeutic	hallucinogens
nervous system)	agents	hormones
antacids	depressants and	mydriatics and
antibiotics	stimulants	myotics
anticoagulants	counterirritants	narcotics
antihistamines	demulcents	parasympathetic
analgesics	diuretics	nerve drugs
antiseptics	emetics	psychotropic agents
antiserums	emmenagogues	purgatives
antitoxins	enzymes	vitamins

SPECIFIC TYPES

Medicinal tinctures are solutions involving an animal, vegetable, or chemical substance or extract dissolved in alcohol or alcohol and water. For example, tincture of iodine is a preparation of iodine and sodium iodide in diluted ethanol used as a topical anti-infective.

Antiseptics are substances that inhibit the growth or development of microbes; *disinfectants* do so by killing or destroying the growing form, but not necessarily resistant spore forms. A wide variety of toxic, flammable, and corrosive compounds are used as disinfectants, including phenols, cresols, alcohols, mercury, silver, and other metal compounds.

HAZARDS

Drugs exhibit a number of possible hazards in addition to toxicity. *Anaesthetic ether* (diethyl ether, 60-29-7) forms extremely explosive and poisonous substances on exposure to atmospheric oxygen; sodium hypochlorite and hydrogen peroxide used as antiseptics are oxidizers; multiple acids, metals, metal salts, and bases are used. In addition, drugs are rarely administered pure. Instead, they are coupled with solvents, diluents, solutions, and administered by aerosol sprays, etc., which may represent hazards.

RELATED TERMS

Aerosol, see *Aerosols*, p.3

Corrosive; Corrosive liquid, see *Corrosives and Class 8*, p.47

Corrosive solid, n.o.s., see *Corrosives and Class 8*, p.47

Flammable; Flammable liquid; Flammable liquid, n.o.s.; see *Flammable Liquids and Class 3*, p.96

Flammable; Flammable solid, inorganic, n.o.s.; Flammable solid, organic, n.o.s., see *Flammable Solids and Division 4.1*, p.99

Liquid, see *Terminology, Liquid*, p.241

Oxidizing; Oxidizing liquid; Oxidizing solid Oxidizing liquid, n.o.s.; Oxidizing solid, n.o.s., see *Oxidizers and Class 5*, p.170

Solid, see *Terminology, Solid*, p.247

Toxic; Toxic liquid; Toxic solid; Toxic liquid, organic, n.o.s.; Toxic solid, organic, n.o.s., see *Toxic Substances and Division 6.1*, p.255

REGULATORY DEFINITIONS

None

REFERENCES

BMS, DIM, KOC, MH14, P, TPD, VNS

Dyes and Pigments

Coal tar dye, corrosive, liquid, n.o.s. ■ Dye and dye intermediate, n.o.s. ■ Dye intermediate, liquid, n.o.s., 6.1, 8 ■ Dye intermediate, solid, n.o.s., 6.1, 8 ■ Dye, liquid, n.o.s.,
6.1, 8 ■ Dye, solid, n.o.s., 6.1, 8 ■ Nitrates of diazonium compounds ■ Organic pigments, self-heating, 4.2

1602 ■ 2801 ■ 3143 ■ 3147 ■ 3313

Dyes, a term which often includes *pigments*, are chemicals that are intensely
coloured or impart colour to natural products, textiles, food and drugs, consumer products, and synthetic products. They work by being adsorbed onto
or held mechanically by the substrate; by the formation of ionic, covalent,
and complex bonds; or by forming solutions. Dyes are distinguishable from
pigments by the manner in which they are applied: dyes often lose their
physical form by dissolution, often being water-soluble, although they may
regain a crystalline form after application; pigments retain their physical
form, often being held in suspension like the pigments in paint. Tens of
thousands of metric tons of dyes and pigments, inorganic and organic, are
produced worldwide each year.

When light falls on a dye or pigment, the electrons making intramolecular
bonds absorb discrete and characteristic portions of the light spectrum and
move to a different configuration. The remaining light (that which is not
absorbed) no longer covers the full spectrum and is reflected into our eyes as
colour.

There are over 8000 chemicals that exhibit commercially significant optical
properties. Some of these are natural products such as indigo, chlorophyll,
and cochineal, but the majority are synthesized in a series of steps involving
dye intermediates (any of the 3000 or more organic and inorganic chemicals
used as raw material precursors to manufactured dyes, the most important of
which are benzene and naphthalene). *Coal tar dyes* refers to the dyestuffs
originating from the complex mix of hydrocarbons (benzene, toluene, xylene,
pyrene, naphthalene, anthracene, etc.) present in coal tar (8007-45-2). Petroleum has succeeded coal as the dominant source of dye intermediates. Dyes,
pigments, and dye intermediates include

□ Azo compounds, those with one or more -N=N- chromophores in
   which at least one nitrogen is attached to an aromatic group; e.g., ben-
   zeneazophenol. These are the most important group of dyestuffs.
□ Anthraquinone-based dyes.
□ Other benzene intermediates and derivatives.
□ Phthalocyanine-based dyes.
□ Nitro and nitroso compounds.
□ Sulphur compounds.
□ Metal compounds.

Diazonium salts are formed by the diazotization reaction, the action of nitrous acid on primary aromatic amines in the presence of a mineral acid. As salts, these compounds are able to stabilize the reactive azo group in preparation for making dyes and other products. ***Nitrates of diazonium compounds*** are diazonium nitrate salts.

HAZARDS

Dyes and pigments and the many chemicals associated with them (e.g., intermediates, dispersants, carriers, etc.) may exhibit any number of hazardous properties.

RELATED TERMS

Coal tar, see *Coal*, p.44

Corrosive, see *Corrosives and Class 8*, p.47

Flammable liquid, n.o.s., see *Flammable Liquids and Class 3*, p.96

Liquid, see *Terminology, Liquid*, p.241

Organic, see *Terminology, Organic*, p.244

Self-heating, see *Spontaneously Combustible Materials and Division 4.2*, p.226

Solid, see *Terminology, Solid*, p.247

Toxic, see *Toxic Substances and Division 6.1*, p.255

REGULATORY DEFINITIONS

Dye Intermediate, n.o.s. A cyclic compound, containing an amino, hydroxy, sulfonic acid, or quinone group or a combination of these groups used in the manufacture of dyes. ICAO A2

Dye, n.o.s., or Dye Intermediate, n.o.s. Are cyclic or ring compounds, containing an amino, hydroxy, sulphonic acid, or quinone group or a combination of these groups used in the manufacture of dyes. IATA App. A

REFERENCES

KOE, MEOS, MH14, TFD

Elevated Temperature Materials

Elevated temperature liquid, n.o.s., 3, 3.3, 9 ■ Elevated temperature solid, n.o.s., 9

3256 ■ 3257 ■ 3258

Certain materials are transported at *elevated temperatures*. They may have been intentionally heated or have undergone self-heating exothermic reactions (e.g., oxidation, radiation, bacterial decay). Primarily, though, applied heat controls viscosity as the molecules absorb the heat as kinetic energy and move more actively in relation to one another or as the intramolecular and intermolecular bonds vibrate more. Heat turns certain solids into liquids, viscous liquids into less viscous liquids, and liquids into gases. Examples of materials that are transported at elevated temperatures include the following:

- Sulphur, a solid at normal temperatures, is shipped as a molten liquid (119°C melting point) to improve the ease of handling, loading, and unloading. Many other solids and otherwise viscous liquids are heated for the same reason.
- Road oils, asphalts, roofing, and other bituminous materials are heated to aid their application on arrival.
- Other materials, such as metals, molten aluminium, molten glass, and molten metal salts are transported hot from the manufacturing process because extensive amounts of energy would be required to reheat them to be suitably malleable or viscous on delivery for subsequent processing.

Other materials are shipped hot because unwanted chemical or physical changes would take place if they were allowed to cool or solidify.

HAZARDS
Elevated temperature materials present a number of hazards in transportation:

- Hot liquids and solids present an immediate physical burn hazard, regardless of their other properties.
- Heated hazardous materials evolve more vapour which, if flammable, increases the chance of ignitable or explosive concentrations or higher concentrations of other hazards.
- Flammable liquids heated to above their flash points present a particular risk of ignition.
- Heat may accelerate or cause self-decomposition, polymerization, or reaction between components of mixtures, possibly with the accompaniment of hazardous reaction products or an evolution of energy as heat through fire.
- Heating systems that fail threaten vehicle damage and adverse reactions as chemicals cool.

Elevated Temperature Materials

RELATED TERMS

Flammable, see *Flammable Liquids and Class*
   3, p.96
Flash point, see *Terminology, Flash Point,*

p.237
Liquid, see *Terminology, Liquid*, p.241
Solid, see *Terminology, Solid*, p.247

REGULATORY DEFINITIONS

This class also includes: (a) Liquids offered for transport at temperatures at or above their flash point; and (b) Substances that are transported or offered for transport at elevated temperatures in a liquid state and which give off a flammable vapour at a temperature at or below the maximum transport temperature. UN 2.3.1.1

Class 3 also includes substances transported or offered for transport at elevated temperatures in a liquid state, which give off a flammable vapour at temperatures equal to or below the maximum transport temperature. IMO Class 3, 1.1.1

Notwithstanding 3.1.1 and 3.1.2 above, liquids offered for transport at temperatures at or above their flash point are considered as flammable liquids. ICAO 2-3.1.3

Substances that are transported or offered for transport at elevated temperatures in a liquid state and which give off a flammable vapour at a temperature at or below the maximum transport temperature (i.e. the maximum temperature likely to be encountered by the substance in transport) are also considered to be flammable liquids. ICAO 2-3.1.4; IATA 3.3.1.4

Substances that are transported or offered for transport in a liquid state at temperatures equal to or exceeding 100°C and below their flash point, or in a solid state at temperatures equal to or exceeding 240°C. ICAO 2-9.1

Liquid phase means a material that meets the definition of liquid when evaluated at the higher of the temperature at which it is offered for transportation or at which it is transported, not at the 37.8°C (100°F) temperature specified in ASTM D 4359-84. US 171.8

Elevated temperature material means a material which, when offered for transportation or transported in a bulk packaging: (1) Is in a liquid phase and at a temperature at or above 100°C (212°F); (2) Is in a liquid phase with a flash point at or above 37.8°C (100°F) that is intentionally heated and offered for transportation or transported at or above its flash point; or (3) Is in a solid phase and at a temperature at or above 240°C (464°F). US 171.8

Notwithstanding 3.3.1.1 and 3.3.1.2, liquids offered for transport at temperatures at or above their flash point are considered as flammable liquids. IATA 3.3.1.3

Substances that are transported or offered for transport in a liquid state at temperatures equal to or exceeding 100°C (212°F) and below their flash point, or in a solid state at temperatures equal to or exceeding 240°C (464°F). [Text continues.] IATA 3.9.1.3

REFERENCES
49CFR, DGR, HCC

Environmentally Hazardous Substances

Environmentally hazardous substance, liquid, n.o.s., 9 ▪ Environmentally hazardous substance, solid, n.o.s., 9 ▪ Hazardous substances, liquid or solid, n.o.s.

3077 ▪ 3082

Many substances of all hazard classes can affect the environment by polluting air, water, and land and causing physiological damage to human, animal, and plant life. The extraction of raw materials and use of their derivatives modifies the balance of the earth's ecosystems.

The United States' Comprehensive Environmental Response, Compensation, and Liability Act of 1980, CERCLA (Superfund) as amended (42 USC 9601 et seq) requires that *hazardous substances* be listed and regulated as hazardous materials in transportation.[3] The responsibility for identifying hazardous substances lies with the U.S. Environmental Protection Agency (EPA) which lists over 2000 hazardous substances and radionuclides in 40CFR302.4.[4] As the EPA conducts its regulatory activities, new substances are added while others are removed.

The U.S. Department of Transportation (DOT) reproduces the EPA's list in 49CFR172.101.[5] The list includes *reportable quantities* (RQs) which indicate the amount of a hazardous substance which if released to the environment triggers the reporting, spill cleanup, and financial responsibility elements of CERCLA.

Most, but not all, hazardous substances are already covered by chemical name or generic entry on the lists of regulated dangerous goods in transportation, for example, methanol, sodium cyanide, and methyl chloroform each appear under their chemical names in both the DOT's *List of Hazardous Substances and Reportable Quantities* and on the United Nations' *Dangerous Goods List*. If no entry is appropriate, and if they are not hazardous wastes, the DOT requires that hazardous substances be shipped under the entry *environmentally hazardous substances*.[6]

RELATED TERMS
Liquid, see *Terminology, Liquid*, p.241
Marine pollutants, liquid or solid, n.o.s., see

Marine Pollutants, p.141
Solid, see *Terminology, Solid*, p.247

---

[3] 42 USC 9656(a).

[4] *Table 302.4-List of Hazardous Substances and Reportable Quantities.*

[5] Appendix A (*List of Hazardous Substances and Reportable Quantities*).

[6] 49CFR172.101(c)(8)(ii).

Environmentally hazardous substances Many of the substances listed in Classes 1 to 9 are deemed, without additional labelling, as being environmentally hazardous. ICAO 2-(1)

Hazardous substance for the purposes of this subchapter, means a material, including its mixtures and solutions, that--(1) Is listed in the Appendix A to Sec. 172.101 of this subchapter; (2) Is in a quantity, in one package, which equals or exceeds the reportable quantity (RQ) listed in the Appendix A to Sec. 172.101 of this subchapter; and (3) When in a mixture or solution--(i) For radionuclides, conforms to paragraph 6 of the Appendix A to Sec. 172.101. (ii) For other than radionuclides, is in a concentration by weight which equals or exceeds the concentration corresponding to the RQ of the material, as shown in the following table:

RQ pounds (kilograms)	Concentration by weight	
	Percent	PPM
5000 (2270)	10	100,000
1000 (454)	2	20,000
100 (45.4)	0.2	2,000
10 (4.54)	0.02	200
1 (0.454)	0.002	20

The term does not include petroleum, including crude oil or any fraction thereof which is not otherwise specifically listed or designated as a hazardous substance in Appendix A to Sec. 172.101 of this subchapter, and the term does not include natural gas, natural gas liquids, liquefied natural gas, or synthetic gas usable for fuel (or mixtures of natural gas and such synthetic gas). US 171.8

Hazardous Substance. Any substance which, if spilled, would adversely affect the environment. IATA 2.9.2

Reportable quantity (RQ) for the purposes of this subchapter means the quantity specified in column 2 of the appendix to Sec. 172.101 for any material identified in Column 1 of the appendix. US 171.8

Reportable Quantity. For a given substance, the minimum quantity that would adversely affect the environment significantly enough to warrant reporting. The requirement to report a spillage is indicated on the documentation and the package. IATA 2.9.2

REFERENCES
49CFR, CERCLA, ISO, MEOE

Explosive Articles

Actuating cartridge, explosive ▪ Ammunition, industrial ▪ Articles, EEI, 1.6N ▪ Articles, explosive, extremely insensitive, 1.6N ▪ Articles, explosive, n.o.s., 1.1C, 1.1D, 1.1E, 1.1F, 1.1L, 1.2C, 1.2D, 1.2E, 1.2F, 1.2L, 1.3C, 1.3L, 1.4B, 1.4C, 1.4D, 1.4E, 1.4F, 1.4G, 1.4S ▪ Cable cutters, explosive ▪ Cartridge cases ▪ Cartridge cases, empty, primed ▪ Cartridges, actuating, for aircraft ejector seat catapult, fire extinguisher, canopy removal or apparatus ▪ Cartridges, actuating, for fire extinguisher or apparatus or apparatus valve ▪ Cartridges, explosive ▪ Cartridges, oil well, 1.3C, 1.4C ▪ Cartridges, power device, 1.2C, 1.3C, 1.4C, 1.4S ▪ Cartridges, starter, jet engine ▪ Cases, cartridge, empty, with primer, 1.4C, 1.4S ▪ Cases, combustible, empty, without primer, 1.3C, 1.4C ▪ Charges, demolition, 1.1D ▪ Charges, expelling, explosives, for fire extinguishers ▪ Charges, explosives, commercial, without detonator, 1.1D, 1.2D, 1.4D, 1.4S ▪ Charges, shaped, flexible, linear, 1.1D, 1.4D ▪ Charges, shaped, without detonator, 1.1D, 1.2D, 1.4D, 1.4S ▪ Contrivances, water-activated with burster, expelling charge or propelling charge, 1.2L, 1.3L ▪ Cutters, cable, explosive, 1.4S ▪ Explosive articles ▪ Fire extinguisher charges, expelling, explosive ▪ Fracturing devices, explosive, for oil wells, without detonators, 1.1D ▪ Jet perforating guns, charged, oil well, with detonator, 1.1D, 1.4D ▪ Jet perforating guns, charged, oil well, without detonator, 1.1D, 1.4D ▪ Jet perforators ▪ Jet tappers, without detonator ▪ New explosive device ▪ Power device, explosive ▪ Release devices, explosive, 1.4S ▪ Rivets, explosive, 1.4S ▪ Shaped charges ▪ Shaped charges, commercial ▪ Sounding devices, explosive, 1.1D, 1.1F, 1.2D, 1.2F ▪ Water-activated contrivances

0048 ▪ 0055 ▪ 0059 ▪ 0070 ▪ 0099 ▪ 0124 ▪ 0173 ▪ 0174 ▪ 0204 ▪ 0237 ▪ 0248 ▪ 0249 ▪ 0275 ▪ 0276 ▪ 0277 ▪ 0278 ▪ 0288 ▪ 0296 ▪ 0323 ▪ 0349 ▪ 0350 ▪ 0351 ▪ 0352 ▪ 0353 ▪ 0354 ▪ 0355 ▪ 0356 ▪ 0374 ▪ 0375 ▪ 0379 ▪ 0381 ▪ 0439 ▪ 0440 ▪ 0441 ▪ 0442 ▪ 0443 ▪ 0444 ▪ 0445 ▪ 0462 ▪ 0463 ▪ 0464 ▪ 0465 ▪ 0466 ▪ 0467 ▪ 0468 ▪ 0469 ▪ 0470 ▪ 0471 ▪ 0472 ▪ 0486 ▪ 0494

Explosive articles include a wide variety of devices or tools that employ explosives which on initiation produce a sudden, powerful expansion of gases used to some mechanical advantage: drive turbines, move pistons, shear bolts and wires, operate pumps, and start engines. Often, these articles contain an *explosive cartridge*, an explosive charge encased in a *cartridge case* made of metal, fibre, paper, plastic, or other material.

POWER DEVICES

Power devices syn. power device cartridges or *industrial ammunition* are explosives that

- Activate valves, switches, or diaphragms. For example, *actuating cartridges* trigger the catapults that project a pilot's canopy and seat away from an aircraft in emergencies or open fire extinguisher valves allowing them to dispel their contents.
- Inflate air bags in the event of a vehicle crash.
- Function as *release devices*, such as rod or bolt cutters that sever a rod or link between two items of hardware. A *cable cutter* functions by forcing a blade into a receiving anvil, severing connectors interposed

between them. Release devices are used in remote, inhospitable, or inaccessible locations to release solar array panels from the side of a satellite, drop fuel rods into nuclear reactors, or separate the stages of a space launch vehicle, etc.

□ Project fastening devices like *explosive rivets* (used in inaccessible places) which contain small explosive charges and which are initiated when the head of the rivet is touched by a heat source.

□ Provide rotary motion like *jet engine starter cartridges* used to initiate the rotation of jet engine turbine rotors.

EXPLOSIVE CHARGES

Other explosive articles are used directly for their explosive power used in ammunition, blasting, and as

□ *Demolition charges* for commercial or military earth-moving or levelling buildings.

□ *Commercial charges* used in metallurgy to join, weld, or form metals: metal parts may be shaped in a confined die cavity by the force of an explosive blast; sheets and charges of explosive may be placed on metals to weld or fuse them.

□ Explosives in seismic thumpers which generate regular, highly controlled signals for petroleum prospecting.

□ *Explosive fracturing devices* used to fracture rock formations at the base or shaft walls of oil wells to increase the flow of gas or oil. These devices may use shaped charges or *oil well cartridges* (*syn.* industrial ammunition) which propel steel or hardened projectiles.

SHAPED CHARGES AND JETS

Shaped charges are high explosives in one of two forms:

□ Cone- or V-shaped hollows formed in the surface of solid explosive charges. On detonation, the explosive force is multiplied along the axis of the cone as the shock waves reinforce themselves, thus producing a powerful penetrating effect several times that otherwise possible. This focused force is sometimes called a *jet*.

□ Hollows lined with metal such as copper which acts as an energy carrier as it collapses and converges along the axis of symmetry of the charge to create a penetrating projectile that punches the target.

Shaped charges find wide use in armour-piercing ammunition; in mining to break boulders and cut deep well linings; in metallurgy as *jet tappers*, in which insulated charges detonate to create a tapping channel in open-hearth metal furnaces or break up blast furnace hang-ups. Some oil well fracturing devices use shaped charges in *jet perforator guns* which consist of canisters supported by a metal framework lowered down an oil well and detonated electronically.

WATER-ACTIVATED CONTRIVANCES

When immersed in water, the explosives in *water-activated contrivances* are initiated by electric current as water (acting as an electrolyte) immerses the electrodes of specially designed batteries; by chemical reaction with water; and by pressure sensors triggered at certain depths. These contrivances include ammunition, signal flares and other pyrotechnics, *sounding devices* (which are dropped by ships to determine depth), and actuating cartridges for gas cylinders that automatically inflate life rafts and jackets.

COMBUSTIBLE CASES

Some cartridge cases are made of combustible materials (e.g., paper, plastic) that leave no residue or spent case to be removed from the device, tool, or weapon. In transportation terms, however, true *combustible cases* are those made of nitrocellulose which, being an explosive, adds force to the overall power of the explosive charge as well as leaving no residue.

RELATED TERMS

Ammunition, smoke (water-activated contrivances), white phosphorus, with burster, expelling charge or propelling charge, see *Ammunition*, p.8

Ammunition, smoke (water-activated contrivances), with or without white phosphorus or phosphides, with burster, expelling charge or propelling charge, see *Ammunition*, p.8

Ammunition, toxic (water-activated contrivances), see *Ammunition*, p.8

Articles, pyrotechnic, see *Pyrotechnics and Signals*, p.194

Detonator, see *Initiating Explosives*, p.119

EEI, see *Explosives and Class 1*, p.74

Explosive, extremely insensitive, see *Explosives and Class 1*, p.74

Explosive, see *Explosives and Class 1*, p.74

Fire extinguisher, see *Fire Extinguishers*, p.94

Flares, water-activated, see *Pyrotechnics and Signals*, p.194

Gas cartridges, see *Gases and Class 2*, p.104

REGULATORY DEFINITIONS

An explosive article is an article containing one or more explosive substances. IMO Class 1, 1.4.3

Explosive article is an article containing one or more explosive substances. UN 2.1.1.3(c); ICAO 2-1.5; IATA App. A

Articles, Explosive, Extremely Insensitive (Articles, EEI) Articles that contain only extremely insensitive detonating substances and which demonstrate a negligible probability of accidental initiation or propagation (under normal conditions of transport) and which have passed Test Series 7. UN App. B, US 173.59

Articles, Explosive, Extremely Insensitive (Articles, EEI). Articles that contain only extremely insensitive detonating substances and which demonstrate a negligible probability of accidental initiation or propagation (under normal conditions of transport). ICAO A2, IATA App. A

Cartridges. Generic term, applied to any explosive article designed to deliver combustion gases, under pressure, with a view to performing a given mechanical action, for example to propel a projectile. In particular, it applies to assembled ammunition consisting of a case fitted with a primer, filled with propellant powder with or without projectile. The term cartridge is used also to indicate a unit charge of blasting explosive, wrapped in a thin paper, plastic or other envelop, the shape of which is ordinarily a cylinder. However, cartridge blasting explosives are not considered as articles but as substances. IATA App. A

Cartridges, Actuating for Fire Extinguisher. Contrivances containing a small explosive charge with a primer, the functioning of which ruptures a metal piece (for example, a bursting disc) and thereby actuates a fire extinguisher. ICAO A2

Cartridges, Actuating for Fire Extinguisher or Apparatus Valve, Explosive. Contrivances containing a small explosive charge with a primer, the functioning of which ruptures a metal piece (for example, a bursting disc) and thereby actuates a fire extinguisher or either opens or closes a valve. See "Cartridges, power device". IATA App. A

Cartridges, Oil Well Articles consisting of a casing of thin fibre, metal or other material containing only propellant which projects a hardened projectile. The term excludes the following articles which are listed separately: Charges, Shaped. UN App. B, ICAO A2

Cartridges, oil well. Articles consisting of a casing of thin fiber, metal or other material containing only propellant explosive. The term excludes charges, shaped, commercial. US 173.59

Cartridges, Oil Well. Articles consisting of a casing of thin fibre, metal or other material containing only propellant which projects a hardened projectile. The term excludes Charges, shaped, which are listed separately. IATA App. A

Cartridges, Power Device Articles designed to accomplish mechanical actions. They consist of a casing with a charge of deflagrating explosive and a means of ignition. The gaseous products of the deflagration produce inflation, or linear or rotary motion, or activate diaphragms, valves or switches or project fastening devices or extinguishing agents. UN App. B, ICAO A2, IATA App. A

Cartridges, power device. Articles designed to accomplish mechanical actions. They consist of a casing with a charge of deflagrating explosive and a means of ignition. The gaseous products of the deflagration produce inflation, linear or rotary motion; activate diaphragms, valves or switches, or project fastening devices or extinguishing agents. US 173.59

Cases, Cartridge, Empty, With Primer Articles consisting of a cartridge case made from metal, plastics or other non-flammable material, in which the only explosive component is the primer. UN App. B, ICAO A2, US 173.59, IATA App. A

Cases, Combustible, Empty Without Primer Articles consisting of cartridge cases made partly or entirely from nitrocellulose. UN App. B, ICAO A2, US 173.59, IATA App. A

Charges, Demolition Articles containing a charge of a detonating explosive in a casing of fibreboard, plastics, metal or other material. The term excludes the following articles which are listed separately: bombs, mines, etc. UN App. B, ICAO A2

Charges, demolition. Articles consisting of a charge of detonating explosive in a casing of fiberboard, plastics, metal or other material. The term excludes articles identified as bombs, mines, etc. US 173.59

Charges, Demolition. Articles containing a charge of a detonating explosive in a casing of fibreboard, plastics, metal or other material. The term excludes Bombs, Mines, etc. which are listed separately. IATA App. A

Charges, Explosive, Commercial without detonator Articles consisting of a charge of detonating explosive without means of initiation, used for explosive welding, jointing, forming and other metallurgical processes. UN App. B, ICAO A2, IATA App. A

Charges, explosive, commercial without detonator. Articles consisting of a charge of detonating explosive without means of initiation, used for explosive welding, joining, forming, and other commercial processes. US 173.59

Charges, shaped commercial, without detonator. Articles consisting of a casing containing a charge of detonating explosive with a cavity lined with rigid material, without means of initiation. They are designed to produce a powerful, penetrating jet effect. US 173.59

Charges, Shaped, Flexible, Linear Articles consisting of a V-shaped core of a detonating explosive clad by a flexible metal sheath. UN App. B, ICAO A2, US 173.59, IATA App. A

Charges, Shaped, without detonator Articles consisting of a casing containing a charge of detonating explosive with a cavity lined with rigid material, without means of initiation. They are designed to produce a powerful, penetrating jet effect. UN App. B, ICAO A2, IATA App. A

Contrivances, Water Activated with burster, expelling charge or propelling charge Articles whose functioning depends upon physico-chemical reaction of their contents with water. UN App. B, ICAO A2, IATA App. A

Contrivance, water-activated with burster, expelling charge or propelling charge. Articles whose functioning depends of [on] physico-chemical reaction of their contents with water. US 173.59

Cutters, Cable, Explosive Articles consisting of a knife-edged device which is driven by a small charge of deflagrating explosive into an anvil. UN App. B, ICAO A2, US 173.59, IATA App. A

Fracturing Devices, Explosive for oil wells, without detonator Articles consisting of a charge of detonating explosive contained in a casing without means of initiation. They are used to fracture the rock around a drill shaft to assist the flow of crude oil from the rock. UN App. B, ICAO A2, US 173.59

Fracturing Devices, Explosive, for Oil Wells, without Detonators. Articles consisting of a charge of detonating explosive without the means of initiation. They are used to fracture the rock around a drill shaft to assist the flow of crude oil from the rock. IATA App. A

Jet Perforating Guns, oil well, without detonator Articles consisting of a steel tube or metallic strip, into which are inserted shaped charges connected by detonating cord, without means of initiation. UN App. B, ICAO A2, US 173.59, IATA App. A

Release Devices, Explosive Articles consisting of a small charge of explosive with means of initiation. They sever rods or links to release equipment quickly. UN App. B, ICAO A2, US 173.59, IATA App. A

Sounding Devices, Explosive Articles consisting of a charge of detonating explosive. They are dropped from ships and function when they reach a predetermined depth or the sea-bed. UN App. B, ICAO A2, US 173.59, IATA App. A

REFERENCES
29CFR1910, ADM, BH, DOSATT, E3, HOO, KOE, PII, POG, RDA, TIP

Explosives and Class 1

Agents, blasting type B ■ Agents, blasting type, E ■ Amatols ■ Ammonium nitrate explosives ■ Ballistite ■ Black powder, compressed, 1.1D ■ Black powder for small arms, 4.1 ■ Black powder granular or as a meal, 1.1D ■ Black powder in pellets, 1.1D ■ Blasting agent, n.o.s. ■ Cordite ■ Deflagrating metal salts of aromatic nitro-derivatives, n.o.s., 1.3C ■ Dynamite ■ Explosive, blasting, type A, 1.1D ■ Explosive, blasting, type B, 1.1D, 1.5D ■ Explosive, blasting, type C, 1.1D ■ Explosive, blasting, type D, 1.1D ■ Explosive, blasting, type E, 1.1D, 1.5D ■ Explosive, emulsion ■ Explosive, seismic ■ Explosive, slurry ■ Explosive substances ■ Explosive, water gel ■ Gelatin, blasting ■ Gelatin dynamite ■ Gelatine, blasting ■ Gelatine dynamite ■ Gunpowder, compressed, 1.1D ■ Gunpowder granular or as a meal, 1.1D ■ Gunpowder in pellets, 1.1D ■ High explosives ■ New explosive ■ Nitroglycerin mixture, desensitized, liquid, n.o.s., 3 ■ Nitroglycerin mixture, desensitized, solid, n.o.s., 4.1 ■ Pentaerythrite tetranitrate mixture, desensitized, solid, n.o.s., 4.1 ■ Plastic explosives ■ Powder cake, wetted, 1.1C, 1.3C ■ Powder paste, wetted, 1.1C, 1.3C ■ Powder, smokeless, 1.1C, 1.3C ■ Propellant, single, double or triple base ■ Rifle powder ■ Slurry explosives ■ Smokeless powder ■ Smokeless powder for small arms, 4.1 ■ Substances, EVI, n.o.s., 1.5D ■ Substances, explosive, n.o.s., 1.1A, 1.1C, 1.1D, 1.1G, 1.1L, 1.2L, 1.3C, 1.3G, 1.3L, 1.4C, 1.4D, 1.4G, 1.4S ■ Substances, explosive, very insensitive, n.o.s., 1.5D ■ Water gels

0027 ■ 0028 ■ 0081 ■ 0082 ■ 0083 ■ 0084 ■ 0132 ■ 0159 ■ 0160 ■ 0161 ■ 0241 ■ 0331 ■ 0332 ■ 0357 ■ 0358 ■ 0359 ■ 0433 ■ 0473 ■ 0474 ■ 0475 ■ 0476 ■ 0477 ■ 0478 ■ 0479 ■ 0480 ■ 0481 ■ 0482 ■ 0485 ■ 3178 ■ 3319 ■ 3343 ■ 3344

Explosive substances are pure chemicals or mixtures, usually liquid or solid, which once initiated react suddenly to yield stable products (principally gaseous) and energy in the form of heat, light, sound, and the potential energy of gases under great pressure. (When suspended or mixed with air or oxygen, aerosols of flammable solids, vapours of flammable liquids, and flammable gases can form explosive concentrations.) Generally, two exothermic reactions are involved, one following the other closely: (1) molecular nitrogen groups decompose to generate nitrogen gas, and (2) molecular carbon and hydrogen combust with oxygen to generate, in an ideal explosion, carbon dioxide gas and water (gunpowder yields around 300 times its original volume in gas). Explosives combust without external oxygen from the atmosphere.

The sudden conversion of explosives to gases at high temperature and high pressure causes immediate expansion of the gas to do mechanical work as the potential energy is converted into kinetic energy (e.g., nitroglycerine, RDX, and gunpowder after explosion but before expansion reach 4500°C, 4200°C, and 2000°C, respectively; confined, PETN produces 120 metric ton/cm^2). Explosive reactions progress in one of two ways:

□ Deflagrating explosives *syn.* low explosives or *propellants* react subsonically as the reaction propagates by conduction and diffusion from the initiation point layer-by-layer through the substance; e.g., gun-powder burns at a rate of around 1m/s.

□ Detonating explosives *syn.* **high explosives** are unstable chemicals or mixtures that react almost instantly as a shock or detonating wave propagates through the entire mass from the initiation point at super-sonic speeds that can approach 8,000 or 9,000 m/s in the case of ni-troglycerine. Reaction occurs at the wave front.

Individual chemicals may be so prepared or mixed to be suitable for use as either deflagrating or detonating explosives. Explosives find wide use as mining and construction to blast ore, coal, and rock; in petroleum prospecting and oil well production; in manufacturing to bond metals and manufacture diamonds; as pyrotechnics; and in the aerospace industry. Military explo-sives are used in demolition, ammunition, pyrotechnics, signals, and the like.

CLASSIFICATION

In transportation terms, explosives are divided into six divisions (1.1 to 1.6) based on the: (1) speed with which they react; (2) sensitivity and modes of initiation; (3) explosive power; and (4) effects of the packaging or article to contain the explosive.

Explosives are further assigned to one of thirteen compatibility groups (A through H, J, K, L, N, S) which identify the explosive article or substance and allow identification of other compatible explosives.

EXPLOSIVE CHEMICALS

A large number of chemicals, usually organic compounds of carbon, hydro-gen, nitrogen, and oxygen, exhibit explosive properties:

□ Amino guanidine derivatives, e.g., tetrazene, a common initiating ex-plosive, or DINGU.

□ Azides, e.g., lead azide widely used as a detonator.

□ Diazo compounds, e.g., diazodinitrophenol.

□ Fulminates, e.g,. mercury fulminate, a common initiating chemical.

□ Nitramines, with the -C-N-NO$_2$ group, eg., the military explosives cy-clotrimethylenetrinitramine *syn.* Cyclonite, Hexogen, or RDX (*Research Department Explosive*); cyclotetraethylene-tetranitramine *syn.* Octogen or HMX (*High Melting Explosive*); nitroguanidine; and tetryl.

□ Nitrate esters (among the most powerful explosives) which contain the -C-ONO$_2$ group including nitrocellulose which is also used to provide mechanical strength to other explosives; **nitroglycerine** (55-63-0); **pentaerythrite tetranitrate** (78-11-5) *syn.* PETN or pentaerythritol tetranitrate widely used in detonators; nitrosugars; and nitrostarch.

□ Nitro compounds, with the -C-NO$_2$ group like Hexyl and TNT *syn.* trinitrotoluene (118-96-7), a widely used explosive and explosive sensitizer. This group includes the **_aromatic nitro-derivatives_** (chemicals derived from the nitrobenzene ring usually with three -NO_2 groups), like picric acid, ammonium picrate, sodium picramate, and other explosive metal salts.

- Nitrophenol salts, e.g., lead styphnate.
- Organic peroxides, e.g., hexamethylenediamine peroxide.
- Salts of nitric, chloric, or perchloric acid, e.g., **_ammonium nitrate explosives_** (6484-52-2), guanidine nitrate, ammonium chlorate.

INITIATION

Initiation of explosives can occur by mechanical means (friction or impact), application of heat, or shock. Deflagrating explosives are normally ignited (the source of initiation often being a flame or electric spark). Detonating explosives are initiated by the application of shock to produce intense, momentary pressure. If ignited, detonating explosives will normally just burn, sometimes sluggishly, although sufficient volumes will detonate if combustion induces sufficiently high heat and pressure.

EXPLOSIVE COMPOSITIONS

An ideal explosive reacts completely to yield stable, gaseous products. However, some chemicals explode less efficiently and yield free carbon (e.g., TNT) or oxygen (e.g., ammonium nitrate), as well as some incompletely reacted gases (e.g., carbon monoxide, nitrous oxide, nitric oxide, hydrogen, methane). To increase efficiency, explosives are intimately mixed with oxidizers to react with excess carbon or fuels to react with excess oxygen. The resultant mixtures may have slightly slower reaction speeds but higher reaction temperatures. They also have fewer solid reaction products (gunpowder alone may yield half its weight as smoke—solid carbon—as a reaction product). Explosives are also mixed with other explosives or chemicals to modify their mechanical, thermal, sensitivity, or other properties.

Gunpowder and Smokeless Powder

The first explosive, a solid low explosive, was **_gunpowder syn. black powder_**, a mixture of potassium nitrate (7757-79-1), charcoal (16291-96-6), and sulphur (7704-34-9) typically in 75:15:10% ratio by weight. These chemicals are ground separately into powders, mixed wet, pressed, crushed, formed, dried, finished, and sorted into grain size. During the wet stage they may be described as **_powder paste_**, following which they are consolidated in presses into **_powder cakes_**, although any mixture of black powder with water or other liquid can be described as a paste, and if compressed, as a cake.

Black powder meal, the finest grade, burns the fastest; larger black powder granules range from around 0.43 to 1.75mm across. By varying the size and shape of individual grains, the rate of burning and the explosive effect can be modified. Thus black powder may be compressed into various formations such as pellets (cylinders around 5cm long and half as wide) used extensively as blasting agents. Although still used in pyrotechnics, larger calibre ammunition, and in initiating devices (fuses and primers), the major drawback of gunpowder is that it generates large volumes of smoke and other solid combustion products. It has been largely replaced by smokeless powders and other propellants.

Smokeless powders, solid low explosives, are virtually smokeless in comparison to black powder. They are also less susceptible to damp, store better, are more powerful, and burn at a more easily controlled rate. These benefits come with the disadvantage that they burn hotter and cause greater damage to the barrels of weapons in which they are used extensively as ammunition propellants. The length of a weapon's barrel and other ballistic requirements result in smokeless powder for pistol ammunition being in flakes, which burns quickly, while slower burning balls, cylinders, or tubes are used for rifle ammunition (*rifle powder*). There are three types of smokeless powder:

1. *Single-base propellants*, those composed chiefly of nitrocellulose (NC, 9004-70-0) with a nitrogen content above 13%.
2. *Double-base propellants* of NC and nitroglycerine (NG, 55-63-0) although nitrate esters may replace NG. More powerful than single-base propellants, these also generate higher temperatures and more wear on the barrel of the weapon.
3. *Triple-base propellants* of NC, NG, and nitroguanidine (556-88-7). These marry the higher power of double-bases with the cooling affect of nitroguanidine to minimize barrel wear.

Smokeless powders contain stabilizers like diphenylamine to remove nitrogen oxides and other products of the decomposition of nitro compounds, and other additives such as potassium acid salts (to smother the muzzle flash) and camphor and dimethyl phthalate which are manufacturing processing aids.

Cordite is a double-base powder (NG:NC typically 55:37% by weight) with mineral jelly gelatinized by the action of acetone. The term comes from *cords* of wood, the source of cellulose used to manufacture nitrocellulose. *Ballistite* is the United States term for much the same material (NG:NC around 40:60% by weight).

Other Mixtures
Other explosive mixtures are common:

- *Amatol*: TNT and ammonium nitrate.
- ANFO: ammonium nitrate, a widely used blasting agent, with around 6% fuel oil (No.2 diesel fuel), sometimes modified with aluminium powder to increase brisance.
- Hexolite *syn.* Composition B (RDX:TNT, 60-40:40-60% by weight); Hexotol (RDX:TNT, 80:20% by weight).
- Octol *syn.* Octolite: commonly HMX:TNT, 76:24% by weight.

- Pentolite: TNT:PETN, 50:50% by weight.
- Tritonal: TNT and aluminium powder.
- *Plastic explosives* syn. plastics-bonded explosives, sensitive crystalline explosives coated with a polymer, like polystyrene or polybutadiene, so that they may be pressed and handled.

BLASTING AGENTS

The term *blasting agent* originated in the United States and was applied to explosives, often ammonium nitrate-based, that could not be detonated with blasting caps but required stronger detonators. This degree of insensitivity afforded relatively safe handling. Worldwide, the term has become associated with a wider number of explosives used in mining and demolition applications in which blasting explosives and initiating devices are placed into boreholes drilled in rock, ore, or coal and fired from remote locations.

Contemporary explosive blasting agents consist of an oxidizer intimately mixed with a fuel so that they are oxygen-balanced to generate as little toxic or flammable reaction products (e.g., carbon monoxide) as possible. In this way they present minimum risk to miners and other operators. They are likely to contain many other ingredients, including materials to prevent stiffening and separation of the components and to provide plasticity, storage stability, and resistance to the desensitizing effects of additives, temperature, and pressure. Fluid blasting agents are pumped or poured into boreholes to fill any irregularities within; solid blasting agents are often pelletized and blown pneumatically into position.

In a transportation context, blasting agents are divided into five categories:

- Type A, based on liquid organic nitrates.
- Type B, based on inorganic nitrates.
- Type C, based on chlorates and perchlorates.
- Type D, based on nitrated organics and combustibles.
- Type E, with water as the principal ingredient.

Dynamites

Although finding wide use in nonblasting applications, *dynamite* (the first trade name for a commercial explosive) has historically been a major explosive used in mining and demolition. The original dynamite was nitroglycerine desensitized in an inert substance such as kieselguhr, a diatomaceous earth. Later dynamites used desensitizing materials that were also fuels like sawdust, charcoal, wood pulp, and flour. Contemporary dynamites encompass a range of explosives rated on their concentration of nitroglycerine (around 25 to 60% by weight) with other chemicals (e.g., sodium nitrate, ammonium nitrate, nitrocellulose, nitrostarch) bringing the total to around 80% oxidizer and 10 to 15% fuel. The remainder includes preservatives, sensitizers, and other additives (preservatives are required to inhibit the separation of nitroglycerine, particularly in contact with water).

Gelatine dynamite includes those dynamites in which nitroglycerine is gelatinized with nitrocellulose which constitutes up to 8% of the mixture. Ammonium gelatines are those with some portion of the nitroglycerine replaced by ammonium nitrate. They are water resistant, plastic (making them suitable for loading in underground drill holes), and well suited for underwater work and as *seismic explosives*, those which generate shock waves used to identify oil- and gas-bearing formations in petroleum prospecting and other geological investigations. *Blasting gelatine* is composed almost wholly of plasticized nitroglycerine and nitrocellulose (NG:NC around 91:7.9%, respectively) and a minute quantity of an alkali, such as magnesium carbonate or zinc oxide. Blasting gelatines are dense, expensive, and particularly well adapted for underwater use.

Other Dry Blasting Agents

Increasingly, ammonium nitrate has replaced nitroglycerine in commercial explosives. ANFO, the most extensively used blasting agent, began replacing dynamites, gelatines, and other dry agents in the 1950s. It is inexpensive and works well in dry conditions.

Aqueous Blasting Agents

In the early 1960s, aqueous explosives *syn. water gels* or *slurry explosives*, designed for wet conditions but often replacing those used in dry conditions, were introduced. They are thickened, emulsified, or gelled mixtures of high proportions of inorganic nitrates or perchlorates suspended in substantial proportions of water, a fuel (e.g., coal, starch, wax, oil, aluminium, or sugar), and sensitized by the addition of another explosive: TNT or a smokeless powder; others may be sensitized by non-explosives, such as aluminium or fuels. They are easily pumped or poured into boreholes.

Explosive emulsions are minute droplets of ammonium nitrate solution emulsified to the texture of margarine in a fuel (often diesel). Because the ammonium nitrate remains in solution, it is not an explosive and maintains an inherent high degree of safety during transportation. Emulsions are sensitized just prior to use by the introduction of gas bubbles or glass microballoons which create voids around which ammonium nitrate solidifies (the explosive form). Initiation is caused by the shockwave of a high explosive detonator.

RELATED TERMS

Aromatic, see *Terminology, Organic*, p.244
By Mass, see *Terminology, By Mass*, p.232
Components, explosive train, n.o.s., see *Initiating Explosives*, p.119
Desensitized, see *Terminology, Desensitized*,

p.234
Flammable, see *Flammable Solids and Division 4.1*, p.99
Gelatin, see *Terminology, Gel*, p.239
Liquid, see *Terminology, Liquid*, p.241

Metal, see *Terminology, Metals*, p.242
Mixture, see *Terminology, Mixture*, p.243
Propellant, liquid, see *Ammunition*, p.8
Propellant, solid, see *Ammunition*, p.8
Salts, see *Terminology, Salts*, p.247
Samples, explosive other than initiating

explosives, see *Chemical Kits and Samples*, p.40
Small arms, see *Ammunition*, p.8
Solid, see *Terminology, Solid*, p.247
Wetted, see *Terminology, Wetted*, p.253

REGULATORY DEFINITIONS

Class 1

Class 1 comprises: (a) Explosive substances (a substance which is not itself an explosive but which can form an explosive atmosphere of gas, vapour or dust is not included in Class 1), except those that are too dangerous to transport or those where the predominant hazard is appropriate to another class; (b) Explosive articles, except devices containing explosive substances in such quantity or of such a character that their inadvertent or accidental ignition or initiation during transport shall not cause any effect external to the device either by projection, fire, smoke, heat or loud noise; and (c) Substances and articles not mentioned under (a) and (b) which are manufactured with a view to producing a practical, explosive or pyrotechnic effect. UN 2.1.1.1, ICAO 2-1.1.1

Class 1 comprises: .1 explosive substances, except those which are too dangerous to transport or those where the predominant hazard is one appropriate to another class; .2 explosive articles, except devices containing explosive substances in such quantity or of such a character that their inadvertent or accidental ignition or initiation during transport shall not cause any effect external to the device either by projection, fire, smoke, heat or loud noise; and .3 substances and articles not mentioned under 1.1.1 and 1.1.2 which are manufactured with a view to producing a practical explosive or pyrotechnic effect. IMO Class 1, 1.1

Explosive. For the purpose of this subchapter, an explosive means any substance or article, including a device, which is designed to function by explosion (i.e., an extremely rapid release of gas and heat) or which, by chemical reaction within itself, is able to function in a similar manner even if not designed to function by explosion, unless the substance or article is otherwise classed under the provision of this subchapter. US 173.50(a)

Class 1 comprises: (a) explosive substances (a substance which is not itself an explosive but which can form an explosive atmosphere or gas, vapour or dust is not included in Class 1), except those that are too dangerous to transport or those where the predominant hazard is appropriate to another class; (b) explosive articles, except devices containing explosive substances in such a quantity or of such a character that their inadvertent or accidental ignition or initiation, during transport, will not cause any effect external to the device either by projection, fire, smoke, heat or loud noise; and (c) articles and substances not mentioned under (a) and (b) above which are manufactured with a view to producing a practical, explosive or pyrotechnic effect. IATA 3.1.1

Divisions

Division 1.1 Substances and articles which have a mass explosion hazard (a mass explosion is one which affects almost the entire load virtually instantaneously). UN 2.1.1.4(a); ICAO 2-1.2

Division 1.1 Substances and articles which have a mass explosion hazard. IMO Class 1, 1.5.2

Division 1.1 consists of explosives that have a mass explosion hazard. A mass explosion is one which affects almost the entire load instantaneously. US 173.50(b)(1)

Division 1.1 Articles and substances having a mass explosion hazard (a mass explosion is one which affects almost the entire load virtually instantaneously). IATA 3.1.3.1

Division 1.2 Substances and articles which have a projection hazard but not a mass explosion hazard. UN 2.1.1.4(b); IMO Class 1, 1.5.2; ICAO 2-1.2

Division 1.2 consists of explosives that have a projection hazard but not a mass explosion hazard. US 173.50(b)(2)

Division 1.2 Articles and substances having a projection hazard but not a mass explosion hazard. IATA 3.1.3.2

Division 1.3 Substances and articles which have a fire hazard and either a minor blast hazard or a minor projection hazard or both, but not a mass explosion hazard. This division comprises substances and articles: (i) which give rise to considerable radiant heat; or (ii) which burn one after another, producing minor blast or projection effects or both. UN 2.1.1.4(c)

Division 1.3 Substances and articles which have a fire hazard and either a minor blast hazard or a minor projection hazard or both, but not a mass explosion hazard. This division comprises substances and articles: (a) which give rise to considerable radiant heat; or (b) which burn one after another, producing minor blast or projection effects or both. IMO Class 1, 1.5.2

Division 1.3 - Substances and articles which have a fire hazard and either a minor blast hazard or a minor projection hazard or both, but not a mass explosion hazard. This division comprises substances and articles which: a) give rise to considerable radiant heat, or b) burn one after another, producing minor blast or projection effects or both. ICAO 2-1.2

Division 1.3 consists of explosives that have a fire hazard and either a minor blast hazard or a minor projection hazard or both, but not a mass explosion hazard. US 173.50(b)(3)

Division 1.3 Articles and substances having a fire hazard and either a minor blast hazard or a minor projection hazard or both, but not a mass explosion hazard. This division comprises articles and substances that: (a) give rise to considerable radiant heat; or (b) burn one after another, producing minor blast and/or projection effects. IATA 3.1.3.3

Division 1.4 Substances and articles which present no significant hazard. This division comprises substances and articles which present only a small hazard in the event of ignition or initiation during transport. The effects are largely confined to the package and no projection of fragments of appreciable size or range is to be expected. An external fire must not cause virtually instantaneous explosion of almost the entire contents of the package. UN 2.1.1.4(d); IMO Class 1, 1.5.2; ICAO 2-1.2

Division 1.4 consists of explosives that present a minor explosion hazard. The explosive effects are largely confined to the package and no projection of fragments of appreciable size or range is to be expected. An external fire must not cause virtually instantaneous explosion of almost the entire contents of the package. US 173.50(b)(4)

Division 1.4 Articles and substances that present no significant hazard. This division comprises articles and substances which present only a small hazard in the event of ignition or initiation during transport. The effects are largely confined to the package and no projection of fragments of appreciable size or range is to be expected. An external fire must not cause virtually instantaneous explosion of almost the entire contents of the package. IATA 3.1.3.4

Division 1.5 Very insensitive substances which have a mass explosion hazard. This division comprises substances which have a mass explosion hazard but are so insensitive that there is very little probability of initiation or of transition from burning to detonation under normal conditions of transport. UN 2.1.1.4(e); IMO Class 1, 1.5.2; ICAO 2-1.2

Division 1.5 Very insensitive substances, having a mass explosion hazard, which are so insensitive that there is very little probability of initiation or of transition from burning to detonation under normal conditions of transport. IATA 3.1.3.5

Division 1.5 consists of very insensitive explosives. This division is comprised of substances which have a mass explosion hazard but are so insensitive that there is very little probability of initiation or of transition from burning to detonation under normal conditions of transport. US 173.50(b)(5)

Division 1.6 Extremely insensitive articles which do not have a mass explosion hazard. This division comprises articles which contain only extremely insensitive detonating substances and which demonstrate a negligible probability of accidental initiation or propagation. UN 2.1.1.4(f); IMO Class 1, 1.5.2; ICAO 2-1.2; IATA 3.1.3.6

Division 1.6 consists of extremely insensitive articles which do not have a mass explosive hazard. This division is comprised of articles which contain only extremely insensitive detonating substances and which demonstrate a negligible probability of accidental initiation or propagation. US 173.50(b)(6)

<u>Blasting Agents</u>

Explosive, blasting Detonating explosive substances used in mining, construction and similar tasks. Blasting explosives are assigned to one of five types. In addition to the ingredients listed, blasting explosives may also contain inert components such as kieselguhr, and minor ingredients such as colouring agents and stabilizers. UN App. B, ICAO A2, IATA App. A

Explosive, blasting. Detonating explosive substances used in mining, construction, and similar tasks. Blasting explosives are assigned to one of five types. In addition to the ingredients listed below for each type, blasting explosives may also contain inert components, such as kieselguhr, and other minor ingredients, such as coloring agents and stabilizers. US 173.59

Explosive, Blasting Type A Substances consisting of liquid organic nitrates such as nitroglycerin or a mixture of such ingredients with one or more of the following: nitrocellulose; ammonium nitrate or other inorganic nitrates; aromatic nitro-derivatives, or combustible materials, such as wood-meal and aluminium powder. Such explosives shall be in powdery, gelatinous or elastic form. UN App. B

Explosive, Blasting, Type A. Substances consisting of liquid organic nitrates such as nitroglycerin or a mixture of such ingredients with one or more of the following: nitrocellulose, ammonium nitrate or other inorganic nitrates, aromatic nitro derivatives or combustible materials such as wood-meal and aluminium powder. Such explosives must be in powdery, gelatinous or elastic form. The term includes dynamite, gelatine, blasting and gelatine dynamites. ICAO A2, US 173.59, IATA App. A

Explosive, Blasting Type B Substances consisting of (a) a mixture of ammonium nitrate or other inorganic nitrates with an explosive such as trinitrotoluene, with or without other substances such as wood-meal and aluminium powder, or (b) a mixture of ammonium nitrate or other inorganic nitrates with other combustible substances which are not explosive ingredients. Such explosives shall not contain nitroglycerin, similar liquid organic nitrates, or chlorates. UN App. B, ICAO A2, IATA App. A

Explosive, blasting, type B. Substances consisting of a mixture of ammonium nitrate or other inorganic nitrates with an explosive, such as trinitrotoluene, with or without other substances, such as wood-meal or aluminum powder, or a mixture of ammonium nitrate or other inorganic nitrates with other combustible substances which are not explosive ingredients. Such explosives may not contain nitroglycerin, similar liquid organic nitrates, or chlorates. US 173.59

Explosive, Blasting Type C Substances consisting of a mixture of either potassium or sodium chlorate or potassium, sodium or ammonium perchlorate with organic nitro-derivatives or combustible materials such as wood-meal or aluminium powder or a hydrocarbon. Such explosives shall not contain nitroglycerin or similar liquid organic nitrates. UN App. B, ICAO A2, US 173.59, IATA App. A

Explosive, Blasting Type D Substances consisting of a mixture of organic nitrated compounds and combustible materials such as hydrocarbons and aluminium powder. Such explosives shall not contain nitroglycerin, similar liquid organic nitrate, chlorate or ammonium nitrate. The term generally includes plastic explosives. UN App. B, ICAO A2, IATA App. A

Explosive, blasting, type D. Substances consisting of a mixture of organic nitrate compounds and combustible materials, such as hydrocarbons and aluminum powder. Such explosives must not contain nitroglycerin, any similar liquid organic nitrate, chlorate or ammonium-nitrate. The term generally includes plastic explosives. US 173.59

Explosive, Blasting Type E Substances consisting of water as an essential ingredient and high proportions of ammonium nitrate or other oxidizers, some or all of which are in solution. The other constituents may include nitro-derivatives such as trinitrotoluene, hydrocarbons or aluminium powder. The term includes explosives, emulsions; explosives slurry and explosives, watergel. UN App. B; ICAO A2, US 173.59, IATA App. A

Other Definitions

Ammonium-nitrate–fuel oil mixture (ANFO). A blasting explosive containing no essential ingredients other than prilled ammonium nitrate and fuel oil. US 173.59

Black Powder (Gunpowder) Substance consisting of an intimate mixture of charcoal or other carbon and either potassium nitrate or sodium nitrate, with or without sulphur. It may be meal, granular, compressed or pelletized. UN App. B, ICAO A2, IATA App. A

Black powder (gunpowder). Substance consisting of an intimate mixture of charcoal or other carbon and either potassium or sodium nitrate, and sulphur. It may be meal, granular, compressed, or pelletized. US 173.59

Dynamite. A detonating explosive containing a liquid explosive ingredient (generally nitroglycerin, similar organic nitrate esters, or both) that is uniformly mixed with an absorbent material, such as wood pulp, and usually contains materials such as nitrocellulose, sodium and ammonium nitrate. US 173.59

Explode The verb used to indicate those explosive effects capable of endangering life and property through blast, heat and projection of missiles. It encompasses both deflagration and detonation. UN App. B; ICAO A2, IATA App. A

Explode. The term indicates those explosive effects capable of endangering life and property through blast, heat, and projection of missiles. It encompasses both deflagration and detonation. US 173.59

Explosive article means an article or device which contains one or more explosive substances. [Text continues.] US 176.2

Explosive, deflagrating A substance, e.g. propellant, which reacts by deflagration rather than detonation when ignited and used in its normal manner. UN App. B, ICAO A2, US 173.59, IATA App. A

Explosive, detonating A substance which reacts by detonation rather than deflagration when initiated and used in its normal manner. UN App. B, ICAO A2, US 173.59, IATA App. A

Explosive, Extremely Insensitive Detonating Substance (EIDS) A substance which, although capable of sustaining a detonation, has demonstrated through tests that it is so insensitive that there is very little probability of accidental initiation. UN App. B, ICAO A2, US 173.59, IATA App. A

Explosive, primary Explosive substance manufactured with a view to producing a practical effect by explosion which is very sensitive to heat, impact or friction and which, even in very small quantities, either detonates or burns very rapidly. It is able to transmit detonation (in the case of initiating explosive) or deflagration to secondary explosives close to it. The main primary explosives are mercury fulminate, lead azide and lead styphnate. UN App. B, ICAO A2, IATA App. A

Explosive, primary. Explosive substance which is manufactured with a view to producing a practical effect by explosion, is very sensitive to heat, impact, or friction, and even in very small quantities, detonates. The major primary explosives are mercury fulminate, lead azide, and lead styphnate. US 173.59

Explosive, secondary Explosive substance which is relatively insensitive (when compared to primary explosives), which is usually initiated by primary explosives with or without the aid of boosters or supplementary charges. Such an explosive may react as a deflagrating or as a detonating explosive. UN App. B, ICAO A2

Explosive, secondary. An explosive substance which is relatively insensitive (when compared to primary explosives) and is usually initiated by primary explosives with or without the aid of boosters or supplementary charges. Such an explosive may react as a deflagrating or as a detonating explosive. US 173.59

Explosive, Secondary An explosive substance which is relatively insensitive (when compared to primary explosives), which is usually initiated by primary explosives with or without the aid of boosters or supplementary charges. Such an explosive may react as a deflagrating or as a detonating explosive. IATA App. A

Explosive substance is a solid or liquid substance (or a mixture of substances) which is in itself capable by chemical reaction of producing gas at such a temperature and pressure and at such speed as to cause damage to the surroundings. Pyrotechnic substances are included even when they do not evolve gases. UN 2.1.1.3(a)

An explosive substance is a solid or liquid substance, or a mixture of substances, which is in itself capable by chemical reaction of producing gas at such a temperature and pressure and at such a speed as to cause damage to the surroundings. Pyrotechnic substances are included even when they do not evolve gases. IMO Class 1, 1.4.1

Explosive substance. A solid or liquid substance (or a mixture of substances) which is in itself capable by chemical reaction of producing gas at such a temperature and pressure and at such a speed as to cause damage to the surroundings. Pyrotechnic substances are included even when they do not evolve gases. ICAO 2-1.5

Explosive substance means a solid or liquid material, or a mixture of materials, which is in itself capable by chemical reaction of producing gas at such a temperature and pressure and at such a speed as to cause damage to its surroundings. [Text continues.] US 176.2

Explosive Substance. A solid or liquid substance (or a mixture of substances) which is in itself capable by chemical reaction of producing gas at such a temperature and pressure and at such a speed as to cause damage to the surroundings. Included are pyrotechnic substances even when they do not evolve gases. A substance which is not itself an explosive but which can form an explosive atmosphere of gas, vapour or dust is not included. IATA App. A

Powder Cake (Powder Paste), Wetted Substance consisting of nitrocellulose impregnated with not more than 60% of nitroglycerin or other liquid organic nitrates or a mixture of these. UN App. B, ICAO A2

Powder cake (powder paste). Substance consisting of nitrocellulose impregnated with not more than 60 percent of nitroglycerin or other liquid organic nitrates or a mixture of these. US 173.59

Power Cake, Wetted. Substances consisting of nitrocellulose impregnated with not more than 60% of nitroglycerin or other liquid organic nitrates or a mixture of these. IATA App. A

Powder, Smokeless Substance based on nitrocellulose used as propellant. The term includes propellants with a single base (nitrocellulose (NC) alone), those with a double base (such as NC and nitroglycerin (NG)) and those with a triple base (such as NC/NG/nitroguanidine). Cast, pressed or bag-charges of smokeless powder are listed under "Charges, Propelling" or "Charges, Propelling for Cannon". UN App. B, ICAO A2, IATA App. A

Powder, smokeless. Substance based on nitrocellulose used as propellant. The term includes propellants with a single base (nitrocellulose (NC) alone), those with a double base (such as NC and nitroglycerin (NG)) and those with a triple base (such as NC/NG/nitroguanidine). Cast pressed or bag-charges of smokeless powder are listed under charges, propelling and charges, propelling for cannon. US 173.59

Substances, Explosive, Very Insensitive (Substances, EVI), N.O.S. Substances which present a mass explosion hazard but which are so insensitive that there is very little probability of initiation, or of transition from burning to detonation (under normal conditions of transport) and which have passed Test Series 5. UN App. B, US 173.59

Substances, Explosive, Very Insensitive (Substances, EVI), n.o.s. Substances that present a mass explosion hazard but are so insensitive that there is very little probability of initiation or of transition from burning to detonation (under normal conditions of transport) and that have passed Test Series 5. ICAO A2, IATA App. A

REFERENCES
ADM, BH, D, DOW, E1, E2, E3, EOE, GAE, GOCT, HG, ICT, ITT, JDOMS, KOC, MEOS, MH14, POG, RBA, REI, TAOE, TEA, TIE

Extracts

Aromatic liquids ■ Beverage extract (concentrate) ■ Camphor oil, 3.3 ■ Extract, aromatic or flavouring ■ Extracts, aromatic, liquid, 3, 3.2, 3.3 ■ Extracts, flavouring, liquid, 3, 3.2, 3.3 ■ Flavouring liquids ■ Other regulated substance, aromatic extract or flavouring ■ Terpene hydrocarbons, n.o.s., 3.3 ■ Terpenes, n.o.s.

1130 ■ 1169 ■ 1197 ■ 2319

Around 3000 additives are used to flavour or add aroma to food, cosmetics, and drugs. Of these, around 500 derive from animal (e.g., musk, civet, and ambergris) and vegetable sources such as spices, herbs, fruits, and roots; the remainder are synthetic. Natural sources may be used directly, but often the desirable compounds, *essential oils*, are separated by pressing, steam distillation, or solvent extraction (which avoids the destructive heat of distillation) resulting in **aromatic extracts** (*syn. aromatic liquids*) and *flavouring extracts* (*syn. flavouring liquids*). Essential oils also contain waxes, fats, other oils, acids, and pigments. Essential oils are often associated with the solvents (often a flammable alcohol or ether) used in extraction or added to aid handling and use.

Many extracts are based chemically on the benzene ring, a class of often fragrant compounds which have come to be called *aromatics* as a consequence. They include benzaldehyde, which smells like almonds, and vanillin, the basis of vanilla. **Camphor oil** (8008-51-3) is the essential oil removed from the camphor tree. It contains camphor, as well as many other compounds. Camphor may also be extracted from other natural sources, but usually in smaller quantities. Vegetable extracts usually contain **terpenes** ($C_{10}H_{16}$), a class of flammable hydrocarbons based on the isoprene unit (C_5H_8).

Carbonated beverages are often sold as concentrated syrups *syn.* **beverage extracts** that are subsequently mixed with water, sweeteners, and other components at bottling facilities. In addition to the flavour, colours, preservatives, and other constituents, these syrups contain acidulants: acids that give a tart flavour, maintain pH levels, disinfect, and perform other functions. Acidulants include phosphoric, citric, ascorbic, tartaric, adipic, and malic acids.

HAZARDS

Individual extracts may be flammable, toxic, or otherwise hazardous (e.g., benzaldehyde is a combustible liquid with an irritating odour in high concentrations, vanillin is a flammable solid, and beverage acidulants are corrosive) but it is the extensive use of solvents, like alcohols and ethers, to carry the extracts that presents the usual hazard in transportation. However, even if the extracts, carriers, or additives are not directly hazardous, they may be so irritating and obnoxious that they present a hazard in transportation to passengers and crew when released into confined spaces like aircraft cabins (i.e., *other regulated substances*).

Aromatic, see *Terminology, Organic*, p.244

Aviation regulated liquid, see *Miscellaneous Dangerous Goods (including Class 9)*, p.158

Aviation regulated solid, see *Miscellaneous Dangerous Goods (including Class 9)*, p.158

Corrosive liquid, acidic, inorganic, n.o.s., see

Corrosives and Class 8, p.47

Hydrocarbons, see *Terminology, Hydrocarbons*, p.240

Liquid, see *Terminology, Liquid*, p.241

Oil, see *Terminology, Oil*, p.244

Other regulated substance, see *Miscellaneous Dangerous Goods (including Class 9)*, p.158

REGULATORY DEFINITIONS

Extracts, Aromatic or Extracts, Flavouring. Substances used for fragrances or for flavouring foods or beverages. Where they contain a solvent or other liquid with a sufficiently low flash point they are classified as flammable liquids. However, where they contain a liquid which has corrosive or toxic properties they must be classified according to that criteria. They may have obnoxious properties such that in the event of a leakage from the package they may cause extreme discomfort to the crew or passengers. ICAO A2

Extracts, Aromatic or Flavouring. Consist of substances used for flavouring/odorising/ aromatising food, beverages, cosmetics, etc. They may have obnoxious properties such as an overpowering odour which, in the case of a leak, may cause extreme discomfort to passengers and crew. Some may contain flammable solvents and hence have a flash point sufficiently low to require classification as flammable liquids. Others may have corrosive or toxic properties and will require appropriate classification. Note that although there is a technical difference between "extracts" and "flavouring", for the purpose of these Regulations they are treated alike under the term "extracts". IATA App. A

REFERENCES

21CFR182, ACD, GOCT, HCC, ICI, KOC, KOE

Fertilizers

Ammonium nitrate fertilizer, 5.1, 1.1D, 9 ■ Ammonium nitrate fertilizer, n.o.s., 5.1 ■ Ammonium nitrate mixed fertilizers, 5.1 ■ Fertilizer ammoniating solution with free ammonia, 2.2 ■ Fertilizers containing ammonium nitrate, n.o.s. ■ Fertilizer with ammonium nitrate, n.o.s. ■ Tankage

0223 ■ 1043 ■ 2067 ■ 2068 ■ 2069 ■ 2070 ■ 2071 ■ 2072

There are 22 nutrients considered essential for plant growth:

MACRONUTRIENTS	MICRONUTRIENTS	
calcium	boron	manganese
carbon	chlorine	molybdenum
hydrogen	cobalt	silicon
magnesium	copper	sodium
nitrogen	gallium	vanadium
oxygen	iodine	zinc
potassium	iron	
phosphorus		
sulphur		

Carbon is supplied to plant cells as atmospheric carbon dioxide is converted via photosynthetic action; hydrogen and oxygen enter in the form of water. The remaining elements are taken up from the soil where they exist naturally or applied in the form of *fertilizers*.

Fertilizers supply nutrients directly, make available nutrients already present, regulate, and condition the soil. They supply nutrients from a wide variety of chemical compounds and mixtures:

ammonia (7664-41-7)	kieserite
ammonium nitrate (6484-52-2)	limestone
ammonium sulphate	marble dust
anhydrous ammonia	other rocks and minerals
brine	potash
chalk	potassium ores
dolomite	sodium phosphate
epsom salts	tankage
fluorspar	urea solutions
gypsum	

Nitrogen (a constituent of plant proteins) is by far the most heavily applied nutrient, in part because nitrogen fertilizers are generally water soluble and about half are leached by rain and runoff out of the soil. Nitrogen is applied directly as ammoniating solutions, including those with free ammonia (normally a gas, ammonia is highly soluble in water), or as other ammonia-based compounds, most significantly as *ammonium nitrate fertilizers*, crystals, granules, or prills, wholly or partially soluble in water.

Although not easily detonated, ammonium nitrate will explode under confinement at high temperatures. When mixed with other nutrient-bearing compounds hazards vary but when combined with combustible materials (e.g., fuel oil), its strong oxidizing properties increase the likelihood of explosion. At lower concentrations and in the absence of combustibles the mixture presents only a limited hazard.

Tankage comes from two sources: the dried solid product of boiling the bones, skin, meat scraps, and other animal by-products from abattoirs; and garbage treated with high-pressure steam and subsequently pressed. Both products are high in nitrogen, in the form of ammonia, and phosphorus as phosphoric acid, potash, and phosphates. Tankage can be flammable and presents a spontaneous combustion hazard.

RELATED TERMS

Combustible, see *Terminology, Combustion*, p.233

Inert, see *Terminology, Inert*, p.240

Inorganic, see *Terminology, Inorganic*, p.241

Mixture, see *Terminology, Mixture*, p.243

Organic, see *Terminology, Organic*, p.244

Solution, see *Terminology, Solutions*, p.247

Tankage, see also *Solid Bulk Materials*, p.221

REGULATORY DEFINITIONS

None

REFERENCES

EOA, GOCT, HCC, KOC, KOE, MH14

Fibres and Fibrous Products

Animal fabrics, oily ∎ Animal fibres, burnt, wet or damp ∎ Animal fibres, oily ∎ Bhusa, 4.1 ∎ Carbon paper ∎ Coir ∎ Copra, 4.2 ∎ Cotton, 9 ∎ Cotton, dry ∎ Cotton seed, cut linters, hull fibres, pulp, waste, and shavings, with animal or vegetable oil ∎ Cotton waste, oily, 4.2 ∎ Cotton, wet, 4.2 ∎ Fabrics, animal, n.o.s. with oil, 4.2 ∎ Fabrics, synthetic, n.o.s. with oil, 4.2 ∎ Fabrics, vegetable, n.o.s. with oil, 4.2 ∎ Fibres, animal, burnt, wet, or damp, 4.2 ∎ Fibres, animal, n.o.s. with oil, 4.2 ∎ Fibres, synthetic, n.o.s. with oil, 4.2 ∎ Fibres, vegetable, burnt, wet or damp, 4.2 ∎ Fibres, vegetable, dry, 4.1 ∎ Fibres, vegetable, n.o.s. with oil, 4.2 ∎ Flax, dry ∎ Hair, wet ∎ Hay, 4.1 ∎ Hemp, dry ∎ Jute ∎ Jute, dry ∎ Kapok ∎ Kapok, dry ∎ Meal, oily, 4.2 ∎ Oil cake ∎ Paper stock, wet ∎ Paper, unsaturated oil treated, incompletely dried, 4.2 ∎ Paper, unsaturated oil treated, incompletely dried (includes carbon paper), 4.2 ∎ Paper waste, wet ∎ Rags, oily, 4.2 ∎ Rags, wet ∎ Sawdust ∎ Seed cake, 4.2 ∎ Seed cake, containing vegetable oil mechanically expelled seeds, 4.2 ∎ Seed cake containing vegetable oil, solvent extractions, 4.2 ∎ Seed cake, containing vegetable oil, solvent extractions and expelled seeds, 4.2 ∎ Seed expellers ∎ Seed expellers, oily ∎ Sisal ∎ Sisal, dry ∎ Straw, 4.1 ∎ Synthetic fabrics, oily ∎ Synthetic fibres, oily ∎ Textile waste, wet, 4.2 ∎ Vegetable fabrics, oily ∎ Vegetable fibres, burnt, wet or damp ∎ Vegetable fibres, dry ∎ Vegetable fibres, oily ∎ Wood chips ∎ Wood pulp, pellets ∎ Wool waste, wet, 4.2

1327 ∎ 1363 ∎ 1364 ∎ 1365 ∎ 1373 ∎ 1379 ∎ 1386 ∎ 1856 ∎ 2217

Fibre is the fundamental component of textiles and paper goods. It is woven, pressed, or bonded directly into various products including *fabrics* or undergoes the interim step of being spun into yarn. Fibres are commonly characterized by having a length at least 100 times their diameter. There are four distinct fibre categories:

- *Animal fibres* and filaments, proteinaceous substances taken from *hair* or fur (e.g., wool), down or feathers, and cocoon materials.
- Mineral fibre, of which asbestos is the principal and only true natural example. Synthetic mineral fibres include fibreglass and steel wool.
- *Synthetic fibres*; manufactured fibres can be divided into those derived from natural polymers (such as regenerated protein fibres: rayon, cellulose acetates, or alginates) and those derived from synthetic polymers including nylons, polyesters, acrylics, and polyolefins.
- *Vegetable fibre*, those derived from the seed, stem, wood, bark (bast fibres), leaf, or fruit of plants. They are composed chiefly of cellulose (up to 90%), the remainder being primarily lignin, hemicellulose, and pectins.

VEGETABLE FIBRES

Seeds are one source of vegetable fibres, oils, and protein. Oil-bearing seeds are mechanically squeezed in a *seed expeller* or the oil is extracted with a solvent leaving behind the pulp and fibre *syn. seed cake*, *oil cake*, or *meal*. Seed cake retains some oils as does the seed expeller. Common seed cakes include cottonseed, peanut, linseed, maize, palm, rape seed, rice bran, soy beans, and sunflower.

Certain vegetable fibres are mentioned directly in the transportation regulations:

- *Coir*, a strong and elastic fibre from the outer husk of coconuts (*Cocos nicifera*).
- *Copra* (8001-31-8), the dried meat of the coconut. It is a potentially combustible material due to the high residual oil content and has an acrid odour.
- *Cotton* (*Malvaceae*), an almost pure cellulose fibre from the *cotton seed*. *Cut linters* is the short, fibrous material remaining in the cotton seed after the spinnable lint has been removed by ginning; *hull fibre* refers to the shortest linters. The term *cotton waste* is applied to fibrous waste *syn. textile waste* of any vegetable origin rejected from textile manufacturing and removed from the processing equipment. It is often recovered and reworked or used as a general absorbent material for equipment and hands, paper manufacturing, furniture padding, etc.
- *Flax,* fibre from *Linum usitatissimum* (the source of linseed oil) and numerous other plants.
- *Hemp*, fibre principally from annual herbaceous *Cannalissativa* but also from many other plants.
- *Jute*, bast fibre from the inner bark of the round or long pod jute plant (*Tiliaceae*) and from other closely related plants such as kenaf.
- *Kapok*, light fibre from the seed pods of the kapok tree (*Ceiba pentandra*).
- *Sisal* or *sizal*, hard leaf fibre from the sisal plant (*Agave sisalana*).
- *Hay*, cut and dried grass used for animal feed.
- *Straw*, stems and stalks from cereals (e.g., wheat, barley, oats, and rye) that is the product of threshing to remove the seeds. *Bhusa*, or more accurately *bhoosa*, is the husks and broken straw from threshing operations. Straw and bhoosa are used as fodder, while straw is also used as thatch, fillings, and weaving.
- Wood, the hard, fibrous substance consisting of holocellulose and lignin from trees and other plants. *Sawdust* and *wood chips* are by-products from a number of wood fabrication techniques, or they may be produced purposely to provide feedstocks for paper manufacture, compressed wood products, mulch, etc.

OILS AND FIBRES

There are many reasons why animal, vegetable, or mineral oils are associated with fibre products:

- Vegetable and animal fibres contain significant fractions of oil even after pressing or processing. Kapok fibre, for example, contains up to 25% of a nondrying oil similar to cottonseed oil. The term *dry*, when applied to fibres, may just mean that they are free of water but still retain a concentration of oils.

□ Oils may be added to soften fibres and ease combing.

□ Oils are added to stop certain fibres from drying out; jute, for example, becomes brittle on exposure to moisture and is protected by oil.

□ Pine and other oils are added as preservatives.

□ Finished fabrics are coated with oils to produce waterproofing and insulating materials.

□ The manufacture of certain fabrics requires the use of oil. Crepe is produced by treatment with oils: sulphonated, lauric acid ester oils, or other soluble oils.

□ Fibrous materials are often used as absorbents and become contaminated with oils on use, hence *oily rags*.

PAPER

Paper is a product manufactured from cellulose fibres derived from any number of vegetable sources, although softwood trees are the most common. During the processing of wood, wood chips are mechanically and chemically disintegrated into *wood pulp syn. paper stock*, a liquid consisting of cellulose fibres and water. Resins and pigments are added to the stock to impart desired properties before it is formed into sheets, dried, and possibly coated. During the process *paper waste* is generated containing high proportions of cellulose and which, if wet, constitutes a significant fire risk. Paper stock presents a similar hazard. Cellulose in the presence of moisture undergoes oxidation and generates sufficient heat to cause ignition. Oxidation can also be initiated through the heat generated by microbial action of bacteria on paper, although properly aged and dried paper should not present a spontaneous combustion hazard.

The term *carbon paper* may apply to

□ A lightweight paper coated on one or both sides with carbon black or other colouring and wax and used to make duplicate copies.

□ Certain photographic papers coated with gelatine and a pigment.

□ Thin sheets of carbon produced in the same fashion as paper from a carbon fibre or graphite and binder slurry. Its conductivity and resistance to corrosion have led to its use as electrodes in electrostatic precipitators.

HAZARDS

Under appropriate circumstances the oils associated with fibres undergo oxidation and other reactions in the presence of air or through self-oxidation as they dry. These reactions can generate sufficient heat to cause the fibres (many of which are made of combustible materials and all of which have high relative surface area on which reactions can proceed) to reach the point of spontaneous combustion (linters present a particular risk of spontaneous combustion). In certain fibres, moisture catalyzes oxidation to generate heat, making wet materials such as wet hair, cotton, rags, wool waste, and other textile wastes particularly hazardous.

RELATED TERMS

Animal oil, see *Terminology, Oil*, p.244

Carbon, see *Carbon*, p.36

Fabrics impregnated with weakly nitrated nitrocellulose, n.o.s. (including toe puffs, nitrocellulose base), see *Nitrocellulose Products*, p.161

Fibres impregnated with weakly nitrated nitrocellulose, n.o.s. (including toe puffs, nitrocellulose base), see *Nitrocellulose Products*, p.161

Fibres impregnated with weakly nitrated nitrocellulose, n.o.s., see *Nitrocellulose*

Products, p.161

Nitrated paper, see *Nitrocellulose Products*, p.161

Oily, see *Terminology, Oil*, p.244

Sawdust, see also *Solid Bulk Materials*, p.221

Synthetic, see *Terminology, Synthetic*, p.252

Unsaturated oil, see *Terminology, Oil*, p.244

Vegetable oil, see *Terminology, Oil*, p.244

Woodchips, see also *Solid Bulk Materials*, p.221

Wood pulp pellets, see also *Solid Bulk Materials*, p.221

REGULATORY DEFINITIONS

Copra. The dried meat of coconuts used to produce coconut oil. Copra contains up to 67% oil and may be subject to spontaneous combustion. ICAO A2

Copra. The dried meat of coconuts used to produce coconut oil. Copra contains up to 67% oil and is subject to spontaneous combustion. IATA App. A

Cotton Waste, Oily Fibres of vegetable origin. IMO 4228

Seed Cake; Meal, Oily; Oil Cake; Seed Expellers, Oily Residue remaining after oil has been expelled mechanically from oil-bearing seeds. Used mainly as animal feed or fertilizer. The most common seed cakes include those derived from coconut (copra), cottonseed, groundnut (peanut), linseed, maize (hominy chop), niger seed, palm kernel, rape seed, rice bran, soya bean and sunflower seed and they may be shipped in the form of cake, flakes, pellets, meal etc. May self-heat slowly and, if wet or containing an excessive proportion of unoxidized oil, ignite spontaneously. IMO 4257

REFERENCES

CAC, CON, EOTF, FDO, GOCT, HCC, IMDG, KOC, KOE, MH14, PF, STRF, TEA, TOE

Fire Extinguishers

Fire extinguisher charges, corrosive liquid, 8 ■ Fire extinguishers, containing compressed or liquefied gas ■ Fire extinguishers with compressed or liquefied gas, 2.2

1044 ■ 1774

Portable *fire extinguishers* and fixed extinguishing systems put out fires by the ejection of a substance (extinguishant) that inhibits fire by cooling, quenching, oxygen isolation, or the interruption of chemical chain reactions. Depending on the extinguisher design, both the extinguishant and the ejection system can cause a hazard in transportation.

FIRE EXTINGUISHER TYPES

aqueous film-forming	halogenated agent
carbon dioxide	loaded stream
cartridge-operated water	pressurized water
combustible metal	pump tank
dry chemical	soda and acid
foam	vaporizing liquid

Common extinguishants include

- Gases: compressed air, nitrogen, argon, carbon dioxide, and halocarbons.
- Liquids: halocarbons, synthetics, and aqueous (water with calcium chloride, antifreeze, detergents, thickening and foaming agents, slurries, salts, proteins).
- Solids: dry chemicals such as sodium chloride, lithium bromide, sodium bicarbonate, potassium bicarbonate, ammonium phosphate, potassium chloride, potassium bicarbonate, and graphite.

Some ejection systems rely on manual or electric pumps while others employ gases to force and carry extinguishants out of the extinguisher as they expand. Compressed gases may be released from cartridges, held under pressure above the extinguishant, or dissolved under pressure into the extinguishant itself. In other instances, gases are generated by explosive cartridges or other chemical reactions. Soda-acid extinguishers contain a charge (*fire extinguisher charge*) of dry sodium bicarbonate (144-55-8) which is mixed on inversion of the extinguisher with a sulphuric acid solution (7664-93-9). The carbon dioxide generated by the reaction expels the now neutral solution.

RELATED TERMS

Cartridges, actuating, for aircraft ejector seat catapult, fire extinguisher, canopy removal or apparatus, see *Explosive Articles*, p.69
Cartridges, actuating, for fire extinguisher or apparatus or apparatus valve, see *Explo-*
sive Articles, p.69
Charges, expelling, explosive, for fire extinguishers, see *Explosive Articles*, p.69
Compressed gas, see *Gases and Class 2*, p.104

Corrosive liquid, see *Corrosives and Class 8*, p.47

Fire extinguisher charges, expelling, explo

sive, see *Explosive Articles*, p.69

Liquefied gas, see *Gases and Class 2*, p.104

Liquid, see *Terminology, Liquid*, p.241

REGULATORY DEFINITIONS

Fire Extinguishers. Are devices containing one of more non-flammable gases under pressure. They have a mechanism to spray the contained gas, or expel a liquid or powder through some type of nozzle. IATA App. A

Fire Extinguisher Charges. These commonly consist of packages containing sodium bicarbonate (a dry powder) which is non-hazardous, and bottles containing concentrated sulphuric acid, a corrosive liquid. ICAO A2

Fire Extinguisher Charges. Commonly consist of packages containing sodium bicarbonate (a dry powder) which is non-hazardous, and bottles containing concentrated sulphuric acid, a corrosive liquid. IATA App. A

REFERENCES

DOSATT, FSA, FOM, KOC, KOE, MH14

Flammable Liquids and Class 3

Combustible liquid, n.o.s. ■ Flammable liquid, n.o.s., 3, 3.1, 3.2, 3.3, 6.1, 8 ■ Flammable liquid preparations, n.o.s. ■ Hydrocarbons, liquid, n.o.s., 3.1, 3.2, 3.3

1992 ■ 1993 ■ 2924 ■ 3286 ■ 3295

According to regulatory definitions, *flammable liquids* and **combustible liquids** are those which can undergo *combustion* under conditions incident to transportation. A *flammable liquid preparation* is any liquid mixture or compound readied for some kind of use, possibly medicinal, that remains flammable. A material's tendency to be flammable is characterized using the following factors:

□ Flash point: the lowest temperature at which vapours above a liquid will ignite when exposed to an ignition source. For example, the flash point of a No.1 fuel oil is 54°C (open cup) and 59°C (closed cup). Materials transported above their flash points present a significant risk of fire if they come into contact with electrical equipment, electrostatic sparks, automotive ignition systems, open flames, and other ignition systems.

□ Fire point: the lowest temperature at which the vapours evolve above a liquid at a sufficient rate to support ongoing combustion; e.g., the same No.1 fuel oil cited above may have a fire point of 63°C.

□ The presence of other substances in the liquid; e.g., high concentrations of a nonflammable material will reduce the liquid's ability to sustain combustion.

□ The presence of water if the material is water-miscible; the dissolution of a miscible flammable liquid in water affects the flammability. For example, pure ethanol will flash at 12.7°C; a 70% mixture by weight in water will flash at 23°C; and a 24% mixture in water at 60.5°C.

□ Transportation temperature: materials with high flash points may be considered flammable if transported at temperatures above their flash points.

RELATED TERMS

Combustible, see *Terminology, Combustion*, p.233

Corrosive, see *Corrosives and Class 8*, p.47

Elevated temperature materials, see *Elevated Temperature Materials*, p.65

Hydrocarbons, see *Terminology, Hydrocarbons*, p.240

Liquid, see *Terminology, Liquid*, p.241

Toxic, see *Toxic Substances and Division 6.1*, p.255

REGULATORY DEFINITIONS

Flammable liquids are liquids, or mixtures of liquids, or liquids containing solids in solution or suspension (for example paints, varnishes, lacquers, etc., but not including substances otherwise classified on account of their dangerous characteristics) which give off a flammable vapour at temperatures of not more than 60.5°C, closed-cup test, or not more than 65.6°C, open-cup test, normally referred to as the flash point. [Text continues.] UN 2.3.1.1

Liquids meeting the definition in 2.3.1.1 with a flash point of more than 35°C which do not sustain combustion need not be considered as flammable liquids for the purposes of these Regulations. Liquids are considered to be unable to sustain combustion for the purposes of these Regulations (i.e. they do not sustain combustion under defined test conditions) if: (a) They have passed a suitable combustibility test [text continues]; (b) Their fire point according to ISO 2592:1973 is greater than 100°C; or (c) They are water miscible solutions with a water content of more than 90% by mass. UN 2.3.1.2

Class 3 deals with liquids, or mixtures of liquids, or liquids containing solids in solution or suspension (e.g. paints, varnishes, lacquers, etc., but not including substances which, on account of their other dangerous characteristics, have been included in other classes) which give off a flammable vapour at or below 61°C (141°F) closed cup test (corresponding to 65.6°C (150°F) open cup test), normally referred to as the flashpoint. IMO Class 3, 1.1

However, the provisions of this Code need not apply to such liquids with a flashpoint of more than 35°C (95°F) which do not sustain combustion. Liquids offered for transport at temperatures equal to or above their flashpoint are, in any case, considered as flammable liquids. IMO Class 3, 1.1.2

For the purpose of subsection 1.1.2, liquids are not considered to sustain combustion if: .1 they have passed the combustibility test (see Chapter 5 of the United Nations Recommendations on the Transport of Dangerous Goods); or .2 their fire point according to ISO 2592 is greater than 100°C; or .3 they are miscible solutions with a water content of more than 90% by mass. IMO Class 3, 1.1.3

Because of the presence of impurities, the flashpoint may be lower or higher than the reference temperature indicated in the schedule for the substance. IMO Class 3, 1.1.4

Flammable liquids are liquids or mixtures of liquids or liquids containing solids in solution or suspension (for example paints, varnishes, lacquers, etc., but not including substances otherwise classified on account of their dangerous characteristics) which give off a flammable vapour at temperatures of not more than 60.5°C closed-cup test or not more than 65.6°C open-cup test, normally referred to as the flash point. ICAO 3.1.1

Liquids described in 3.1.1 above with a flash point of more than 35°C need not be considered as flammable liquids for the purposes of these Instructions; if: a) they do not sustain combustion when subjected to the method of testing for combustibility given in Part 8, Chapter 3; or b) their fire point according to ISO 2592 is greater than 100°C; or c) they are miscible solutions with a water content of more than 90 per cent by mass. ICAO 2-3.1.2

Flammable liquid. For the purpose of this subchapter, a flammable liquid (Class 3) means a liquid having a flash point of not more than 60.5°C (141°F), or any material in a liquid phase with a flash point at or above 37.8°C (100°F) that is intentionally heated and offered for transportation or transported at or above its flash point in a bulk packaging, with the following exceptions: (1) Any liquid meeting one of the definitions specified in Sec. 173.115. (2) Any mixture having one or more components with a flash point of 60.5°C (141°F) or higher, that make up at least 99 percent of the total volume of the mixture, if the mixture is not offered for transportation or transported at or above its flash point. (3) Any liquid with a flash point greater than 35°C (95°F) which does not sustain combustion according to ASTM 4206 or the procedure in Appendix H of this part. (4) Any liquid with a flash point greater than 35°C (95°F) and with a fire point greater than 100°C (212°F) according to ISO 2592. (5) Any liquid with a flash point greater than 35°C (95°F) which is in a water-miscible solution with a water content of more than 90 percent by mass. US 173.120(a)

Combustible liquid. (1) For the purpose of this subchapter, a combustible liquid means any liquid that does not meet the definition of any other hazard class specified in this subchapter and has a flash point above 60.5°C (141°F) and below 93°C (200°F). US 173.120(b)

A flammable liquid with a flash point at or above 38°C (100°F) that does not meet the definition of any other hazard class may be reclassed as a combustible liquid. This provision does not apply to transportation by vessel or aircraft, except where other means of transportation is impracticable. An elevated temperature material that meets the definition of a Class 3 material because it is intentionally heated and offered for transportation or transported at or above its flash point may not be reclassed as a combustible liquid. US 173.120(b)(2)

A combustible liquid which does not sustain combustion is not subject to the requirements of this subchapter as a combustible liquid. [Text continues.] US 173.120(b)(3)

This class has no subdivisions. It comprises liquids or mixtures of liquids or liquids containing solids in solution or in suspension (for example paints, varnishes, lacquers, etc., but not including substances otherwise classified on account of their dangerous characteristics) which give off a flammable vapour at temperatures of not more than 60.5°C (141°F) closed-cup test or not more than 65.6°C (150°F) open-cup test normally referred to as the flash point. IATA 3.3.1.1

Liquids described in 3.3.1.1 with a flash point exceeding 35°C (95°F) need not be considered as flammable liquids for the purposes of these Regulations, if: (a) they do not sustain combustion when subjected to the method of testing for combustibility given in 3.3.5; or (b) their fire point according to ISO 2592 is greater than 100°C (212°F); or (c) they are miscible solutions with a water content of more than 90% by weight. IATA 3.3.1.2

REFERENCES
27CFR21, HCC, HOC, ISH, THO

Flammable Solids and Division 4.1

Firelighters, solid with flammable liquid, 4.1 ■ Flammable solid, inorganic, n.o.s., 4.1, 6.1, 8 ■ Flammable solid, n.o.s., 4.1, 5.1 ■ Flammable solid, organic molten, n.o.s., 4.1 ■ Flammable solid, organic, n.o.s., 4.1, 6.1, 8 ■ Musk xylene, 4.1 ■ Self-reactive liquid type B, C, D, E, F, 4.1 ■ Self-reactive liquid type B, C, D, E, F, temperature controlled, 4.1 ■ Self-reactive solid type B, C, D, E, F, 4.1 ■ Self-reactive solid type B, C, D, E, F, temperature controlled, 4.1 ■ Solids containing flammable liquid, n.o.s., 4.1

1325 ■ 2623 ■ 2925 ■ 2926 ■ 2956 ■ 3097 ■ 3175 ■ 3176 ■ 3178 ■ 3179 ■ 3180 ■ 3221 ■ 3222 ■ 3223 ■ 3224 ■ 3225 ■ 3226 ■ 3227 ■ 3228 ■ 3229 ■ 3230 ■ 3231 ■ 3232 ■ 3233 ■ 3234 ■ 3235 ■ 3236 ■ 3237 ■ 3238 ■ 3239 ■ 3240

According to regulatory definitions, *flammable solids* include combustible, self-reactive, and desensitized explosive solids. In general, flammable solids decompose at elevated temperatures or, in fire, burn vigorously, decompose in contact with acids or alkalis, and can produce toxic reaction products.

FLAMMABLE SOLIDS

Many solids combust (oxidize with atmospheric oxygen) when exposed to an ignition source (spark, match, flame, or friction); examples include phosphorous trioxide, straw, and sulphur. The strength of the reaction is determined in part by the surface area exposed to the atmosphere. The more finely divided the substance, as in a powder or dust, the more surface is exposed and the quicker the reaction will proliferate. At some point the solid may be so finely divided that its reactivity increases to the point where it no longer needs an ignition source and ignites by itself (*spontaneous combustion*[7]) or even explodes. If moisture plays a major role in the reaction, the material may also be *dangerous when wet.*[8]

Included in the definition of flammable solids are materials such as soil, sand, production material contaminated with flammable liquids, and *firelighters*: combustible solids (e.g., wood, peat, cellular urea-formaldehyde resin, and compacted sawdust) that have been impregnated with a flammable liquid (usually kerosene or white spirit). When ignited, the liquid burns and propagates heat and flame to ignite, in time, the relatively less combustible solid. Firelighters are used as heat sources or to initiate the combustion of another material like coal.

SELF-REACTIVE AND RELATED SUBSTANCES

Self-reactive and related solids and liquids are those that will undergo an exothermic decomposition reaction without the need for oxygen, but without the force of an explosive. They may achieve self-accelerating decomposition

---

[7] See *Spontaneously Combustible Materials and Division 4.2*, p.226.

[8] See *Dangerous When Wet Materials and Division 4.3*, p.58.

at certain temperatures or be initiated through contact with other substances, friction, or impact. Consequently, certain self-reactive materials must be temperature-controlled or desensitized during transportation. They include

- Aliphatic azo compounds with the -CNNC- group.
- Organic azides with the -CN$_3$ group.
- Diazonium salts with the -CN$_2$Z group, where Z is an anion like the chloride ion, Cl$^-$.
- N-nitroso compounds with the -NNO group.
- Aromatic sulphohydrazides (-SO$_2$NHNH$_2$).

Musk xylene *syn.* 5-tert-butyl-2,4,6-trinitro-m-xylene (81-15-2), is self-reactive and particularly sensitive to shock.

Known self-reactive materials are sorted into five types (B, C, D, E, F) based on their UN Numbers. For example, 4-nitrosophenol (UN3236), which reacts violently with acids and bases, is assigned to Type D.

DENSENSITIZED EXPLOSIVES
Explosives may be desensitized by being wetted or diluted with other substances. Water is the most common desensitizer; for example, nitrocellulose is desensitized with water, although detonation can still occur with water concentrations up to 40%. Other desensitizers include oils, alcohols, waxes, castor oil, mineral jelly, and glycols. The mixtures may form pastes and cakes.

Desensitizers are used to reduce hazards in handling or to modify the explosive effect. Camphor is mixed with gunpowder to modify its burning rate. Easily detonated explosives may be mixed with less sensitive explosives resulting in a "desensitized" mixture. TNT, for example, is used to desensitize octol, even while it supplements its explosive power. These may not be sufficiently desensitized to meet the criteria of flammable solids.

RELATED TERMS

Corrosive, see *Corrosives and Class 8*, p.47
Flammable liquid, see *Flammable Liquids and Class 3*, p.96
Inorganic, see *Terminology, Inorganic*, p.241
Liquid, see *Terminology, Liquid*, p.241

Organic, see *Terminology, Organic*, p.244
Oxidizing, see *Oxidizers and Class 5*, p.170
Solid, see *Terminology, Solid*, p.247
Toxic, see *Toxic Substances and Division 6.1*, p.255

REGULATORY DEFINITIONS

Division 4.1
Division 4.1 includes the following types of substances: (a) Flammable solids (see 2.4.2.2); (b) Self-reactive and related substances (see 2.4.2.3); and (c) Desensitized explosives (see 2.4.2.4). UN 2.4.2.1

Flammable solids Solids which, under conditions encountered during transport, are readily combustible or may cause or contribute to fire through friction; self-reactive and related substances which are liable to undergo a strongly exothermic reaction; densensitized explosive which may explode if not diluted sufficiently. UN 2.4.1.1(a); ICAO 2-4

Flammable solids are readily combustible solids and solids which may cause fire through friction. UN 2.4.2.2.1.1; IMO Class 4.1, 1.6; ICAO 2-4.1.2.1; IATA 3.4.1.1.1

Readily combustible solids are powdered, granular, or pasty substances which are dangerous if they can be easily ignited by brief contact with an ignition source, such as a burning match, and if the flame spreads rapidly. The danger may not only come from the fire but also from toxic combustion products. Metal powders are especially dangerous because of the difficulty of extinguishing a fire since normal extinguishing agents such as carbon dioxide or water can increase the hazard. UN 2.4.2.2.1.2; IMO Class 4.1, 1.6; ICAO 2-4.1.2.1; IATA 3.4.1.1.1

The substances in this class are solids possessing the properties of being easily ignited by external sources, such as sparks and flames, and of being readily combustible, or being liable to cause fire through friction. This class also covers substances which are self-reactive (i.e. liable to undergo at normal or elevated temperatures a strong exothermic decomposition caused by excessively high transport temperatures or by contamination), and desensitized explosives, which may explode if not diluted sufficiently. IMO Class 4.1, 1.1

This class comprises: .1 readily combustible solids, and solids which may cause fire through friction; .2 self-reactive (solids and liquids) and related substances; .3 desensitized explosives. IMO Class 4.1, 1.2

Some substances, e.g. celluloid, may evolve toxic and flammable gases when heated or if involved in a fire. IMO Class 4.1, 1.3

The following types of substances are classified in Division 4.1: a) flammable solids; b) self-reactive and related substances; and c) desensitized explosives. ICAO 2-4.1.1

Division 4.1 (Flammable Solid). For the purposes of this subchapter, flammable solid (Division 4.1) means any of the following three types of materials: (1) Desensitized explosives that-- (i) When dry are Explosives of Class 1 other than those of compatibility group A, which are wetted with sufficient water, alcohol, or plasticizer to suppress explosive properties [text continues]. (2)(i) Self-reactive materials are materials that are thermally unstable and that can undergo a strongly exothermic decomposition even without participation of oxygen (air). [Text continues]. (3) Readily combustible solids are materials that--(i) Are solids which may cause a fire through friction, such as matches; (ii) Show a burning rate faster than 2.2 mm (0.087 inches) per second when tested in accordance with UN Manual of Tests and Criteria; or (iii) Any metal powders that can be ignited and react over the whole length of a sample in 10 minutes or less, when tested in accordance with UN Manual of Tests and Criteria. US 173.124(a)

Division 4.1 - Flammable Solids; Self-reactive and Related Substances; and Desensitised explosives Solids which, under conditions encountered in transport, are readily combustible or may cause or contribute to fire through friction; self-reactive and related substances which are liable to undergo a strongly exothermic reaction; desensitised explosives which may explode if not diluted sufficiently. Division 4.1 contains: flammable solids (3.4.1.1); self-reactive and related substances (3.4.1.2); desensitised explosives (3.4.1.3). IATA 3.4.1

Self-Reactive Substances

Self-reactive substances are thermally unstable substances liable to undergo a strongly exothermic decomposition even without participation of oxygen (air). Substances are not considered to be self-reactive substances of Division 4.1, if: (i) they are explosives according to the criteria of Class 1; (ii) they are oxidizing substances according to the assignment procedure of Division 5.1 (see 2.5.2.1.1); (iii) they are organic peroxides according to the criteria of Division 5.2; (iv) their heat of decomposition is less than 300J/g; or (v) their self-accelerating decomposition temperature (SADT) (see 2.4.2.3.4) is greater than 75°C for a 50kg package. UN 2.4.2.3.1.1(a)

The decomposition of self-reactive substances can be initiated by heat, contact with catalytic impurities (e.g. acids, heavy-metal compounds, bases), friction or impact. The rate of decomposition increases with temperature and varies with the substance. Decomposition, particularly if no ignition occurs, may result in the evolution of toxic gases or vapours. For certain self-reactive substances, the temperature shall be controlled. Some self-reactive substances may decompose explosively, particularly if confined. This characteristic may be modified by the addition of diluents or by the use of appropriate packagings. Some self-reactive substances burn vigorously. Self-reactive substances are, for example, some compounds of the types listed below: (a) Aliphatic azo compounds (-C-N=N-C-); (b) Organic azides (-C-N$_3$); (c) Diazonium salts (-CN$_2^+$Z$^-$); (d) N-nitroso compounds (-N-N=O); and (e) Aromatic sulphohydrazides (-SO$_2$-NH-NH$_2$). This list is not exhaustive and substances with other reactive groups and some mixtures of substances may have similar properties. UN 2.4.2.3.1.2

Self-reactive substances are thermally unstable substances liable to undergo a strongly exothermic decomposition even without the participation of oxygen (air). IMO Class 4.1, 1.7.1

The decomposition of self-reactive substances can be initiated by heat, contact with catalytic impurities (e.g. acids, heavy-metal compounds, bases), friction or impact. The rate of decomposition increases with temperature and varies with the substance. Decomposition, particularly if no ignition occurs, may result in the evolution of toxic gases or vapours. For certain self-reactive substances, the temperature should be controlled. IMO Class 4.1, 1.7.2

Some self-reactive substances may decompose explosively, particularly if confined. This characteristic may be modified by the addition of diluents or by the use of appropriate packagings. IMO Class 4.1, 1.7.3

Some self-reactive substances burn vigorously. IMO Class 4.1, 1.7.4

Self-reactive substances are, for example, some compounds of the types listed below: .1 aliphatic azo compounds (-C-N=N-C-); .2 organic azides (-C-N$_3$); .3 diazonium salts (-CN$_2^+$Z$^-$); .4 N-nitroso compounds (-N-N=O); and .5 aromatic sulphohydrazides (-SO$_2$-NH-NH$_2$). This list is not exhaustive. Substances with other reactive groups and some mixtures of substances may have similar properties. IMO Class 4.1, 1.7.5

Substances should not be considered to be self-reactive substances of class 4.1 if: .1 they are explosives according to the criteria of class 1; .2 they are oxidizing substances according to the assignment procedure of class 5.1; .3 they are organic peroxides according to the criteria of class 5.2; .4 their heat of decomposition is less than 300J/g; or .5 their self-accelerating decomposition temperature (SADT) (see section 21.2 of the General Introduction) is greater than 75°C for a 50kg package. IMO Class 4.1, 1.8

Self-reactive substances are thermally unstable substances liable to undergo a strongly exothermic decomposition even without the participation of oxygen (air). The following substances must not be considered to be self-reactive substances of Division 4.1: explosives according to the criteria of Class 1; oxidizing substances according to the assignment procedure in Part 2;5.2; organic peroxides according to the criteria of Division 5.2; substances where their heat of decomposition is less than 300J/g; or substances where their self-accelerating decomposition temperature is greater than 75°C or a 50kg package. ICAO 2-4.1.3.1.1

The decomposition of self-reactive substances can be initiated by heat, contact with catalytic impurities (e.g. acids, heavy-metal compounds, bases), friction or impact. The rate of decomposition increases with temperature and varies with the substance. Decomposition, particularly if no ignition occurs, may result in the evolution of toxic gases or vapours. For certain self-reactive substances, the temperature must be controlled. Some self-reactive substances may decompose explosively, particularly if confined; this characteristic may be modified by the addition of diluents or by the use of appropriate packagings. Some self-reactive substances burn vigorously. Self-reactive substances include some of the following types of compounds: aliphatic azo compounds (-C-N=N-C-); organic azides (-C-N$_3$); diazonium salts (-CN$_2^+$Z$^-$); N-nitroso compounds (-N-N=O); and aromatic sulphohydrazides (-SO$_2$-N-NH$_2$). ICAO 2-4.1.3.2, IATA 3.4.1.2.4

Self-Reactive substances of Division 4.1 are thermally unstable substances liable to undergo a strongly exothermic decomposition even without the participation of oxygen (air). [Text continues.] IATA 3.4.1.2.1

Related Substances

Substances related to self-reactive substances are distinguished from the latter by having a self-accelerating decomposition temperature greater than 75°C. They are liable to undergo, as are self-reactive substances, a strongly exothermic decomposition and are liable, in certain packagings, to meet the criteria for substances of Class 1. UN 2.4.2.3.1.1(b)

Substances related to self-reactive substances can be distinguished from the latter by a self-accelerating decomposition temperature greater than 75°C. They are liable to undergo, as are self-reactive substances, a strongly exothermic decomposition and are liable, in certain packagings, to meet the criteria for explosive substances of class 1. IMO Class 4.1, 1.9.1

"Related substances" are distinguished from self-reactive substances by having a self-accelerating decomposition temperature greater than 75°C. They are liable to undergo, as are self-reactive substances, a strongly exothermic decomposition. They are also liable, in certain packagings, to meet the criteria for explosive substances in Class 1. ICAO 2-4.1.3.1.2

Related substances These are distinguished from self-reactive substances by having a self-accelerating decomposition temperature greater than 75°C. They are liable to undergo, as are self-reactive substances, a strongly exothermic decomposition. They are also liable, in certain packagings, to meet the criteria for substances of Class 1. IATA 3.4.1.2.3

Desensitized Explosives

Desensitized explosives Desenstized explosives are substances which are wetted with water or alcohols or are diluted with other substances to suppress their explosive properties. UN 2.4.2.4; IMO Class 4.1, 1.10.1

Some of these substances, when in a dry state, are classified as explosives. Where reference is made to a substance which is wetted with water, or some other liquid, it should be permitted for transport as a class 4.1 substance only when in a wetted condition specified. IMO Class 4.1, 1.10.2

Desensitized explosives are substances which are wetted with water or alcohols or diluted with other substances to suppress their explosive properties. ICAO 2-4.1.4; IATA 3.4.1.3

Firelighters

Firelighters. These are usually made from peat, wood shavings, or sawdust and a flammable liquid. ICAO A2

Fire Lighters. Are usually made from peat, wood shavings, or sawdust and a flammable liquid. IATA App. A

Firelighters, Solid A porous solid, e.g. cellular urea-formaldehyde resin, compacted wood shavings, etc., impregnated with flammable liquid, usually white spirit or kerosene, and designed to burn in a controlled manner. When heated, evolves flammable vapours. IMO 4145

REFERENCES
49CFR, IMDG, KOE

Air, compressed, 2.2 ■ Air, refrigerated liquid, 2.2 ■ Air, refrigerated liquid, (cryogenic liquid), 2.2 ■ Air, refrigerated liquid, (cryogenic liquid), non-pressurized, 2.2 ■ Air, refrigerated liquid, low pressure or pressurised, 2.2, 5.1 ■ Air, refrigerated liquid, non-pressurised, 2.2, 5.1 ■ Calor gas ■ Camping gas ■ Compressed gas, n.o.s., 2.1, 2.2, 2.3, 5.1, 8 ■ Cryogenic liquid ■ Devices, small, hydrocarbon gas powered, 2.1 ■ Devices, small, hydrocarbon gas powered with release device, 2.1 ■ Dispersant gases, n.o.s. ■ Flammable gas ■ Flammable gas (small receptacles not fitted with a dispersion device, not refillable) ■ Gas candles, charged with flammable gas ■ Gas cartridges, 2.3, 5.1 ■ Gas cartridges, without a release device, non-refillable, 2.1, 2.2, 2.3, 5.1, 8 ■ Gas, compressed ■ Gas liquefied ■ Gas, refrigerated liquid, n.o.s., 2.1, 2.2, 5.1 ■ Hydrocarbon gases, compressed, n.o.s., 2.1 ■ Hydrocarbon gas, liquefied, n.o.s., 2.1 ■ Hydrocarbon gas mixture, compressed, n.o.s., 2.1 ■ Hydrocarbon gas mixture, liquefied, n.o.s., 2.1 ■ Hydrocarbon gas-powered small devices ■ Hydrocarbon gas refills for small devices with release device, 2.1 ■ Liquefied gases, non-flammable charged with nitrogen, carbon dioxide ■ Liquefied gas, n.o.s., 2.1, 2.2, 2.3, 5.1, 8 ■ Liquefied hydrocarbon gas ■ Liquids, other than those classified as flammable, corrosive, or toxic, charged with nitrogen, carbon dioxide, or air ■ Non-flammable gas, n.o.s. ■ Non-liquefied gas ■ Non-liquefied hydrocarbon gas ■ Poisonous gases, n.o.s. ■ Rare gases ■ Rare gases, mixture, compressed, 2.2 ■ Receptacles, small, containing gas without a release device, non-refillable, 2.1, 2.2, 2.3, 5.1, 8

1002 ■ 1003 ■ 1058 ■ 1845 ■ 1953 ■ 1954 ■ 1955 ■ 1956 ■ 1964 ■ 1965 ■ 1979 ■ 1980 ■ 1981 ■ 2037 ■ 2600 ■ 3150 ■ 3156 ■ 3157 ■ 3158 ■ 3160 ■ 3161 ■ 3162 ■ 3163 ■ 3303 ■ 3304 ■ 3305 ■ 3306 ■ 3307 ■ 3308 ■ 3309 ■ 3310 ■ 3311 ■ 3312 ■ 9035

Gases expand uniformly to fill the volume of the container that holds them. The relationship between the pressure (P), volume (V), and temperature (T) of a gas is expressed in the Boyle-Charles law:

$$PV = kT$$

where k is a constant based on the amount of gas and the units used. This equation shows that gas volume can be reduced by increasing pressure, decreasing temperature, or some combination of the two. These characteristics are exploited in the manner in which gases are transported:

- *Compressed gas*, gases under pressure; e.g., methane can be compressed to 1/600 of its original volume before liquefaction. *Non-liquefied gas* is synonymous with *compressed gas*.

- *Liquefied gas*, gases under sufficient pressure to be fully or partially liquefied; e.g., chlorine gas which normally liquefies at -34.04°C at atmospheric pressure (101 kPa), will liquefy at 340 kPa, a little over three times atmospheric pressure, at 20°C.

- *Refrigerated liquids* and gases, gases cooled to the point of liquefaction or solidification, such as methane which liquefies at -151°C. Refrigerated gases exert less pressure on their containers than do compressed gases.

    ❑   *Cryogenic liquids*, a subset of refrigerated gases, those with extremely low boiling points.

Gases may be mixtures with other gases or liquid vapours. Certain gases form solutions in which a liquid (the solvent) is able to dissolve many times its own volume of gas (the solute), a tendency which is increased with pressure; for example, acetone dissolves 25 times its volume of acetylene at atmospheric pressure and 250 times its volume at 10 atmospheres. Acetylene, due to its extreme flammability, is often shipped in this way. Many other liquids are used as solvents; water, for example, is used to dissolve hydrogen bromide. Liquefied gases are also *charged*, that is, made to dissolve other gases such as nitrogen (7727-37-9) and carbon dioxide (124-38-9).

SPECIFIC GAS TERMS

Air describes the atmospheric gases. By volume, its composition approximates as follows:

GAS	PERCENTAGE
nitrogen	78
oxygen	21
argon	1
carbon dioxide	0.035
neon	0.0018
helium	0.0005
methane	0.0002
krypton	0.0001
nitrous oxide	trace
hydrogen	trace
xenon	trace
ozone	trace
other gases	trace

The presence of significant concentrations of oxygen, a nonflammable gas that actively supports combustion, can introduce an oxidizing hazard into compressed or refrigerated air products.

Hydrocarbon gas is any gaseous compound containing only carbon and hydrogen. At normal temperatures and pressures the simplest in this series is methane (CH_4). With an increasing number of carbons, however, boiling point rises and between the isomers of butane (C_4H_{10}) and those of pentane (C_5H_{12}), hydrocarbons become liquid. (The transition for unsaturated hydrocarbons, those with double- or triple-carbon bonds, occurs from three to four carbons.) *Calor gas* is a trademark for liquefied petroleum gas (butane) sold in cylinders for commercial and domestic heating. LPG propane is also called *Calor propane*. *Camping gas* is mostly liquefied butane or propane used in camping stoves and lamps.

Dispersant gases are mixed with solids or liquids in pressurized containers. On release, the gases carry or disperse the contents to generate foams (blowing agents), aerosols, fogs, or manufacture such items as foamed plastics. Dispersant gases, like the widespread halocarbons, are generally nonflammable.

Rare gases are those elements in Group VIIIA of the periodic table:[9] helium, neon, argon, krypton, xenon, and radon. They are not necessarily rare, however, and are also called the *inert* or *noble* gases. Due to their electron configurations, helium, neon, and argon are extremely nonreactive, while others undergo some reactions.

GAS CONTAINERS

Conceptually, the minimum requirements for a fully operational gas container is a valve to permit filling and release or dispersion and some connection or manifold between the store of gas under pressure and the valve. These are common in the reusable metal cylinders, tank cars, and portable tanks used to transport bulk and nonbulk gas. Gases also come supplied in many devices and articles, including

- Nonrefillable metal containers, aerosols, and accumulators for engine starting fluids, foodstuffs, lighter fluids, soaps, shock absorbers, and carbonated beverage cans.
- Refillable metal containers, including LPG and compressed gas fuel tanks for vehicles, pressure vessels for refrigeration equipment, pump **tanks**, and fire extinguishers.
- **Sealed glass containers for gas samples.**
- **Gas-charged electron tubes.**
- Small plastic, metal, **and other** containers for cigarette lighters, mechanical limbs, and similar devices.

Gas-powered devices syn. gas candles, like cigarette lighters and hair curlers, burn hydrocarbon on use. These and other devices may be refilled by being coupled with a container holding a store of compressed or liquefied gas, or a replacement *gas cartridge* might be inserted into the device after the spent cartridge is removed.

HAZARDS

Gases present a number of risks in transportation:

- Explosion, if a container, pressure system, or refrigeration system fails the gases can rupture the container and expand with explosive force.
- Asphyxiation (suffocation), whether or not a gas is toxic, it can displace oxygen in a confined space making respiration impossible.

---

[9] See *Terminology, Elements*, p. 235.

- "Burning," as gases expand, particularly refrigerated gases and cyrogenic liquids, they can rapidly draw heat from the environment. If living tissue is in contact with the expanding gas, it will be used as a heat source and severe tissue damage may result.

- Chemical and physical hazards, individual chemicals may exhibit flammability, toxicity (*poisonous gas* is a synonym for toxic gas), corrosivity, oxidizing potential, etc.

- Self-reaction, other gases may self-react or polymerize along with the generation of heat.

Gases that are heavier than air accumulate closer to the ground, thus increasing the localized hazard of explosion, asphyxiation, etc., in valleys, basements, ship cargo areas, and other structures.

In the United States, the Department of Transportation assigns one of four *Hazard Zones* (A, B, C, and D; A being the most toxic) to toxic gases, which it terms *poisonous by inhalation* materials. The vapours of toxic liquids[10] can also share this designation which triggers special handling and communication precautions.

RELATED TERMS

Aerosol, see *Aerosols*, p.3

Blau gas, Fischer Tropsch gas, Synthesis gas, Water gas, see *Coal*, p.44

Candles, gas, see *Lighters*, p.137

Coal gas; Coal gas, compressed, see *Coal*, p.44

Corrosive, see *Corrosives and Class 8*, p.47

Flammable gas in lighters, see *Lighters*, p.137

Gas, see *Terminology, Gas*, p.239

Gas drips, hydrocarbon, see *Petroleum*, p.183

Gas generator assemblies, see *Lifesaving Equipment*, p.134

Gas identification set, see *Chemical Kits and Samples*, p.40

Gas sample, non-pressurized, flammable, n.o.s.; Gas sample, non-pressurized, toxic, flammable, n.o.s.; Gas sample, non-pressurized, toxic, n.o.s.; see *Chemical Kits and Samples*, p.40

Hydrocarbon, see *Terminology, Hydrocarbons*, p.240

Liquefied natural gas; LNG, see *Petroleum*, p.183

Liquefied petroleum gas; LPG, see *Petroleum*, p.183

Liquid, see *Terminology, Liquid*, p.241

Mixture, see *Terminology, Mixture*, p.243

Oil gas, see *Petroleum*, p.183

Oxidizing, see *Oxidizers and Class 5*, p.170

Rare gas, see *Terminology, Elements*, p.235

Refrigerant gas, n.o.s., see *Refrigerants and Halocarbons*, p.215

Tear gas candles; Tear gas cartridges; Tear gas devices containing tear gas substances; Tear gas grenades; Tear gas substance, liquid, n.o.s.; Tear gas substance, solid, n.o.s. , see *Ammunition, Toxic*, p.19

Toxic, see *Toxic Substances and Division 6.1*, p.255

REGULATORY DEFINITIONS

Class 2

A gas is a substance which: (a) At 50°C has a vapour pressure greater than 300 kPa; or (b) Is completely gaseous at 20°C at a standard pressure or 101.3 kPa. UN 2.2.1.1; IMO Class 2, 1.1; ICAO 2-2.1.1

---

[10] See *Toxic Substances and Division 6.1*, p.255.

The class comprises of compressed gases; liquefied gases; gases in solution; refrigerated lique-fied gases; mixtures of gases; mixtures of one or more gases with one or more vapours of sub-stances of other classes; articles charged with a gas; tellurium hexafluoride; aerosols. UN 2.2.1.3; IATA 3.2.1.3; ICAO 2-2.1.3

This class comprises: .1 compressed gases; .2 liquefied gases; .3 gases in solution; .4 refriger-ated liquefied gases; .5 mixtures of gases; .6 mixtures of one or more gases with one or more vapours of substances of other classes; .7 articles charged with a gas; .8 tellurium hexafluoride; .9 aerosols. IMO Class 2, 1.3

Some gases in this class, under conditions incident to transport, are liable to polymerize (combine or react with themselves) so as to cause dangerous liberation of heat or gas, possibly resulting in rupture of the receptacle. These gases should not be transported unless they are properly inhibited or stabilized; this is indicated in the proper shipping name. IMO Class 2, 1.10

Gas means a material which has a vapor pressure greater than 300 kPa (43.5 psi) at 50°C (122°F) or is completely gaseous at 20°C (68°F) at a standard pressure of 101.3 kPa (14.7 psi). US 171.8.

A gas is a substance which: (a) at 50°C (122°F) has a vapour pressure greater than 300 kPa (3.0 bar, 43.5 lb/in^2); or (b) is completely gaseous at 20°C (68°F) at a standard pressure of 101.3 kPa (1.01 bar, 14.7 lb/in2). IATA 3.2.1.1

Division 2.1

Flammable gases Gases which at 20°C and a standard pressure of 101.3 kPa: (i) are ignitable when in a mixture of 13 per cent or less by volume with air; or (ii) have a flammable range with air of at least 12 percentage points regardless of the lower flammable limit. [Text continues.] UN 2.2.2.1(a); ICAO 2-2.2

Class 2.1 - Flammable gases. Gases which at 20°C and a standard pressure of 101.3kPa: .1.1 are ignitable when in a mixture of 13% or less by volume with air; or .1.2 have a flammable range with air of at least 12 percentage points regardless of the lower flammable limit. [Text continues.] IMO Class 2, 1.6.1.1

Division 2.1 (Flammable gas). For the purpose of this subchapter, a flammable gas (Division 2.1) means any material which is a gas at 20°C (68°F) or less and 101.3 kPa (14.7 psi) of pres-sure (a material which has a boiling point of 20°C (68°F) or less at 101.3 kPa (14.7 psi)) which--(1) Is ignitable at 101.3 kPa (14.7 psi) when in a mixture of 13 percent or less by volume with air; or (2) Has a flammable range at 101.3 kPa (14.7 psi) with air of at least 12 percent regard-less of the lower limit. US 173.115(a)

Gases which at 20°C (68°F) and a standard pressure of 101.3 kPa (1.01 bar, 14.7 lb/in2): (a) are ignitable when in a mixture of 13% or less by volume with air; or (b) have a flammable range with air of at least 12 percentage points regardless of the lower flammable limit. [Text contin-ues.] IATA 3.2.2.1

Division 2.2

Non-flammable, non-toxic gases Gases which are transported at a pressure not less than 280 kPa at 20°C, or as refrigerated liquids, and which: (i) are asphyxiant - gases which dilute or replace the oxygen normally in the atmosphere; or (ii) are oxidizing - gases which may, gener-ally by producing oxygen, cause or contribute to the combustion of other material more than air does; or (iii) do not come under the other divisions. UN 2.2.2.1(b); ICAO 2-2.2

Class 2.2 - Non-flammable, non-toxic gases. Gases which are transported at a pressure not less than 280kPa at 20°C, or as refrigerated liquids, and which: .2.1 are asphyxiant - gases which dilute or replace the oxygen normally in the atmosphere; or .2.2 are oxidizing - gases which may, generally by providing oxygen, cause or contribute to the combustion of other material more than air does; or .2.3 do not come under the other classes. IMO Class 2, 1.6.1.2

*Division 2.2 (non-flammable, nonpoisonous compressed gas--*including compressed gas, liq-uefied gas, pressurized cryogenic gas, compressed gas in solution, asphyxiant gas and oxidizing gas). For the purpose of this subchapter, a non-flammable, nonpoisonous compressed gas (Division 2.2) means any material (or mixture) which--(1) Exerts in the packaging an absolute pressure of 280 kPa (40.6 psia) or greater at 20°C (68°F), and (2) Does not meet the definition of Division 2.1 or 2.3. US 173.115(b)

Gases which are transported at a pressure not less than 280 kPa at 20°C, or as refrigerated liquids, and which: (a) are asphyxiant - gases which dilute or replace the oxygen normally in the atmosphere; or (b) are oxidizing - gases which may, generally by providing oxygen, cause or contribute to the combustion of other material more than air does; or (c) do not come under the other divisions. IATA 3.2.2.2

Division 2.3

Toxic gases Gases which: (i) are known to be so toxic or corrosive to humans as to pose a hazard to health; or (ii) are presumed to be toxic or corrosive to humans because they have an LC_{50} value (as defined in 2.6.2.1) equal to or less than 5,000 ml/m^3 (ppm). UN 2.2.2.1(c)

Class 2.3 - Toxic gases. Gases which: .3.1 are known to be so toxic or corrosive to humans as to pose a hazard to health; or .3.2 are presumed to be toxic or corrosive to humans because they have an LC_{50} value equal to or less than 5,000ml/m^3 (ppm) when tested in accordance with 2.1.6.3 of the introduction to class 6.1. IMO Class 2, 1.6.1.3

Division 2.3 - Toxic gases. Gases which: a) are known to be so toxic or corrosive to humans as to pose a hazard to health; or b) are presumed to be toxic or corrosive to humans because they have an LC_{50} value equal to or less than 5000mL/m^3 (ppm) when tested in accordance with 6.2.1.2(c). ICAO 2-2.2

Division 2.3 (Gas poisonous by inhalation). For the purpose of this subchapter, a gas poisonous by inhalation (Division 2.3) means a material which is a gas at 20°C (68°F) or less and a pressure of 101.3 kPa (14.7 psi) (a material which has a boiling point of 20°C (68°F) or less at 101.3 kPa (14.7 psi)) and which--(1) Is known to be so toxic to humans as to pose a hazard to health during transportation, or (2) In the absence of adequate data on human toxicity, is presumed to be toxic to humans because when tested on laboratory animals it has an LC_{50} value of not more than 5000 ml/m^3 [text continues]. US 173.115(c)

Gases which: (a) are known to be so toxic or corrosive to humans as to pose a hazard to health; or (b) are presumed to be toxic or corrosive to humans because they have an LC_{50} value equal to or less than 5000 mL/m^3 (ppm) when tested in accordance with 3.6.1.5.3. IATA 3.2.2.3

Other Definitions

Asphyxiant gas means a gas which dilutes or replaces oxygen normally in the atmosphere. US 171.8

Atmospheric gases means air, nitrogen, oxygen, argon, krypton, neon and xenon. US 171.8

Calor Gas. Is a liquefied flammable hydrocarbon gas or a mixture of any of the liquefiable petroleum gases. IATA App. A

Compressed gas - a gas (other than in solution) which when packaged under pressure for transport is entirely gaseous at 20°C. UN 2.2.1.2(a); IMO Class 2, 1.2.1; ICAO 2-2.1.2(a)

Compressed gas - a gas (other than in solution) which, when packaged under pressure for transport, is entirely gaseous at 20°C (68°F). IATA 3.2.1.2(a)

Compressed gas in solution. A compressed gas in solution is a non-liquefied compressed gas which is dissolved in a solvent. US 173.115(f)

Cryogenic liquid. A cryogenic liquid means a refrigerated liquefied gas having a boiling point colder than -90°C (-130°F) at 101.3 kPa (14.7 psi) absolute. [Text continues.] US 173.115(g)

Cryogenic Liquids. Are low temperature liquefied gases, such as air, argon, helium, neon and nitrogen. IATA App. A

Gas in solution - compressed gas which when packaged for transport is dissolved in a solvent. UN 2.2.1.2(d); IMO Class 2, 1.2.4; ICAO 2-2.1.2(d); IATA 3.2.1.2(d)

Hazard Zone means one of four levels of hazard (Hazard Zones A through D) assigned to gases, as specified in 173.116(a) of this subchapter, and one of two levels of hazards (Hazard Zones A and B) assigned to liquids that are poisonous by inhalation, as specified in 173.133(a) of this subchapter. A hazard zone is based on the LC_{50} value for acute inhalation toxicity of gases and vapours, as specified in 173.133(a). US 171.8

Hydrocarbon Gas, Compressed. Hydrocarbon gas under high pressure, but not in the liquid condition. ICAO A2

Hydrocarbon Gas, Compressed. Consists of hydrocarbon gas under high pressure, but not in the liquid condition. IATA App. A

Hydrocarbon Gas, Liquefied. Hydrocarbon gas from natural gas or from distillation of petroleum which are liquefied by pressure. ICAO A2

Hydrocarbon Gas, Liquefied. Consists of hydrocarbon gas from natural gas or from distillations of petroleum which are liquefied by pressure. IATA App. A

Liquefied compressed gas. A liquefied compressed gas means a gas which in a packaging under the charged pressure, is partially liquid at a temperature of 20°C (68°F). US 173.115(e)

Liquefied gas - a gas which when packaged for transport is partially liquid at 20°C. UN 2.2.1.2(b); IMO Class 2, 1.2.2; ICAO 2-2.1.2(b)

Liquefied gas - a gas which, when packaged for transport, is partially liquid at 20°C (68°F). IATA 3.2.1.2(b)

Material poisonous by inhalation means: (1) A gas meeting the defining criteria in 173.115(c) of this subchapter and assigned to Hazard Zone A, B, C, or D in accordance with 173.116(a) of this subchapter; (2) A liquid (other than as a mist) meeting the defining criteria in 173.132(a)(1)(iii) of this subchapter and assigned to Hazard Zone A or B in accordance with 173.133(a) of this subchapter; or (3) Any material identified as an inhalation hazard by a special provision in Column 7 of the 172.101 Table. US 171.8

Non-liquefied compressed gas. A non-liquefied compressed gas means a gas, other than in solution, which in a packaging under the charged pressure is entirely gaseous at a temperature of 20°C (68°F). US 173.115(d)

Oxidizing gas means a gas which may, generally by providing oxygen, cause or contribute to the combustion of other material more than air does. US 171.8

Refrigerated liquefied gas - a gas which when packaged for transport is made partially liquid because of its low temperature. UN 2.2.1.2(c); IMO Class 2, 1.2.3; ICAO 2-2.1.2(c)

Refrigerated liquefied gas - a gas which, when packaged for transport, is partially liquid because of its low temperature. IATA 3.2.1.2(c)

Uncompressed Gas. Gas at a pressure not exceeding ambient atmospheric pressure at the time the containment system is closed. IATA App. A

REFERENCES
CTA, DGR, DOS, DOSATT, HCC, HOC, MEOS, MH14, REI, SSD

Halogenated Polyphenyls

P.c.b.s ■ PCBs ■ PCB's ■ Polychlorinated biphenyls, 9 ■ Polychlorinated biphenyls, liquid, 9 ■ Polychlorinated biphenyls, solid, 9 ■ Polyhalogenated biphenyls, liquid, 9 ■ Polyhalogenated biphenyls, solid, 9 ■ Polyhalogenated terphenyls, liquid, 9 ■ Polyhalogenated terphenyls, solid, 9

2315 ■ 3151 ■ 3152

Biphenyls and terphenyls are the lowest and only commercially significant members of the polyphenyl family of organic compounds in which two or more benzene rings (C_6H_6) are attached to each other in chains. Biphenyls contain two benzene rings for which there is only one configuration, while terphenyls contain three rings and three possible configurations: at the ortho-, meta-, and para- positions of the middle benzene. Polyphenyls are used as textile dye carriers and precursors to other products, the most notorious of which are the halogenated polyphenyls.

While any halogen may halogenate a polyphenyl, chlorine and bromine are by far the more usual. They result in a large number of possible *polychlorinated biphenyls* (*syn. p.c.b.s*, *PCBs*, *PCB's*), *polyhalogenated biphenyls*, and *polyhalogenated terphenyls*. No longer produced commercially, halogenated polyphenyls were widely used for their exceptional qualities as heat-transfer fluids, solvents, lubricants, dust suppressants, and fire retardants (some equipment that relies on these chemicals, like electrical transformers and capacitors, remains in use).

Due to their extensive use and resistance to decomposition, halogenated polyphenyls are almost universal environmental contaminants, present in waste sites, sediments, the human body, water supplies, and elsewhere. They are associated with many toxic and biochemical effects including wasting syndromes, bone marrow diseases, atrophy, chloracne, hyperplasin, liver damage, and cancer.

Most halogenated polyphenyls in transportation, then, derive not from product shipments but from decommissioned equipment, drained fluids, soils and debris from waste site and riverbed cleanups, demolition work, scrap metal disposal, contaminated textiles and absorbents, and other wastes.

RELATED TERMS

Halogens, see *Terminology, Elements*, p.235   Solid, see *Terminology, Solid*, p.247
Liquid, see *Terminology, Liquid*, p.241

REGULATORY DEFINITIONS

Polychlorinated Biphenyls; PCB's; Polyhalogenated Biphenyls, Liquid or Polyhalogenated Terphenyls, Liquid; Polyhalogenated Biphenyls, Solid or Polyhalogenated Terphenyls, Solid
This entry also covers appliances such as transformers and condensers containing polyhalo-

111

genated biphenyls or terphenyls and absorbent materials, such as rags, cotton waste, clothing, sawdust, etc., contaminated with polyhalogenated biphenyls or terphenyls. IMO 9035

REFERENCES
IMDG, KOE, THS

Hazardous Waste

Hazardous waste, liquid, n.o.s., 9 ■ Hazardous waste, solid, n.o.s., 9

3077 ■ 3082

Internationally, *wastes* are substances or objects which are disposed of or are intended or required to be disposed of by national law. In 1989, the *Basel Convention on the Control of Transboundary Movements of Hazardous Wastes and Their Disposal* (BCO) instituted environmentally sound management[11] as a prerequisite to any transboundary shipment of certain wastes. It was inspired in part by the perceived and real trend in which industrialized countries shipped wastes to less industrialized nations with fewer environmental controls or technologies. Many instances of adverse human health or environmental impact were reported from accidents or ill-managed wastes.

The Basel Convention requires importing countries to consent to the movement and disposal, as well as each state through which the waste passes. If the exporting country has reason to believe that the waste will not be handled soundly it must prohibit the shipment. Basel wastes are described in three Annexes:

- Annex I - Categories of wastes to be controlled: wastestreams such as waste paints, glues, wood preservatives, and drugs; wastes having certain constituents such as metal carbonyls, lead and its compounds, ethers, furans, and dioxins.
- Annex II - Categories of wastes requiring special consideration: wastes from households, residues from incineration of household wastes.
- Annex III - List of hazardous characteristics: explosives, flammables, oxidizers, toxics, corrosives, etc.

Most Basel Convention wastes are independently regulated under specific or generic entries in the lists of regulated materials in transportation. If a Basel waste is not otherwise listed, it becomes a Class 9 and is shipped under the entry *environmentally hazardous substance*.

UNITED STATES LAW

The United States Department of Transportation defines *hazardous wastes* as those requiring hazardous waste manifests (shipping document Form 8700-22) under the U.S. Environmental Protection Agency's requirements.[12]

---

[11] "Environmentally sound management of hazardous wastes or other wastes" means taking all practicable steps to ensure that hazardous wastes or other wastes are managed in a manner which will protect human health and the environment against the adverse effects which may result from such wastes. BCO, Article 2

[12] 40CFR262, Subpart B, under authority from Section 3001 of the *Resource Conservation and Recovery Act*, Public Law 98-616.

These state that generators that transport hazardous waste or offer it for transportation, offsite treatment, storage, or disposal must use a manifest so that the shipment is tracked from generation through to final disposition. The EPA's hazardous wastes are subsets of *solid wastes*, which include solids, liquids, semisolids, or gaseous materials that are discarded, abandoned, recycled, or inherently waste-like and are not otherwise excluded.

For solid wastes to warrant management as hazardous wastes, they must meet one of four characteristics (flammability, corrosivity, reactivity, and toxicity) or be listed on one of four lists, although some otherwise hazardous wastes by these definitions are excluded. Hazardous wastes carry one or more identifying waste codes: four digits comprised of a letter (D, F, K, P, or U) followed by three numbers. For example, a flammable hazardous waste, possibly waste naphtha, carries the waste code *D001*.

RELATED TERMS

Biomedical waste, see *Infectious Substances and Division 6.2*, p.115

Clinical waste, see *Infectious Substances and Division 6.2*, p.115

Environmentally hazardous substances, liquid or solid, n.o.s., see *Environmentally Hazardous Substances*, p.67

Liquid, see *Terminology, Liquid*, p.241

Medical waste, see *Infectious Substances and Division 6.2*, p.115

Regulated medical waste, see *Infectious Substances and Division 6.2*, p.115

Solid, see *Terminology, Solid*, p.247

REGULATORY DEFINITIONS

Hazardous waste, for the purposes of this chapter, means any material that is subject to the Hazardous Waste Manifest Requirements of the U.S. Environmental Protection Agency specified in 40 CFR part 262. US 171.8

Hazardous Waste. Any substance being disposed of which could adversely affect the environment. IATA 2.9.2

Wastes, for the purposes of this section, are substances, solutions, mixtures or articles containing or contaminated with one or more constituents which are subject to the provisions of this Code and for which no direct use is envisaged but which are carried for dumping, incineration or other methods of disposal. IMO 27.2.1

REFERENCES

40CFR261, BCO, LOSB

Infectious Substances and Division 6.2

Biological products ▪ Biomedical waste, n.o.s., 6.2 ▪ (Bio)medical waste, n.o.s., 6.2 ▪ Clinical waste, unspecified, n.o.s., 6.2 ▪ Cocculus ▪ Diagnostic specimens ▪ Etiologic agent ▪ Genetically modified micro-organisms, 9 ▪ Infectious substance, affecting animals only, 6.2 ▪ Infectious substance, affecting humans, 6.2 ▪ Medical waste, n.o.s., 6.2 ▪ Picrotoxin ▪ Regulated medical waste, n.o.s., 6.2 ▪ Toxins, liquid, extracted from living sources, n.o.s., 6.1 ▪ Toxins, solid, extracted from living sources, n.o.s., 6.1

2814 ▪ 2900 ▪ 3172 ▪ 3245 ▪ 3291

Biological materials may be transportation hazards because they contain or may reasonably be expected to contain *infectious substances syn. etiologic agents* or *pathogens*. These are microorganisms (e.g., bacteria, viruses, rickettsia, parasites, fungi) or hybrid or mutant recombinant microorganisms that cause infectious disease by direct infection or by production of a poisonous protein (*toxin*) that causes illness, although not all toxins cause harm to humans and animals. Toxins can be produced by plants (phytotoxins), certain animals (zootoxins), and by pathogenic bacteria (exotoxins, endotoxins, enterotoxins, neurotoxins, and toxic enzymes) which are particularly responsible for a wide range of diseases, including anthrax, botulism, cholera, the plague, diphtheria, and dysentery. *Cocculus* (124-87-8) *syn. picrotoxin* is a crystalline phytotoxin derived from fishberries (*Cocculus indicus*) used as a central nervous system stimulant and as an antidote for barbiturate poisoning.

SOURCES
The following materials may contain infectious substances:

- *Biological products*, products of animal or human origin including vaccines, serums, blood, animal bile derivatives, endocrine products, etc.
- *Diagnostic specimens*, materials of animal or human origin including excreta, secreta, blood and its components, tissue, and tissue fluids.
- *Genetically modified microorganisms*, modified microorganisms and organisms whose genetic material has been purposely altered through genetic engineering in a way that does not occur naturally.
- Infectious waste *syn. biomedical waste, clinical waste*, or *medical waste*.

Infectious waste is generated from hospitals, clinics, dentists, laboratories, veterinarians, and other research and healthcare facilities. It includes wastes associated with known infections or that are derived from the treatment or diagnosis of unknown or undiagnosed infections. Sources include

- Microbiological cultures and stocks of infectious agents.
- Human blood and blood products.

115

□ Isolation wastes, those generated by protecting others from communicable diseases.

□ Pathological wastes (human body parts, tissues, fluids, and organs).

□ Contaminated sharps (syringes, needles, scalpel blades, and glass). Uncontaminated sharps are sometimes included in this category.

□ Contaminated animal carcasses, bodies, and bedding.

□ Other contaminated wastes such as miscellaneous laboratory wastes, surgery and autopsy wastes, dialysis wastes, and equipment.

RISK GROUPS

Infectious substances are assigned to UN2814 or UN2900 according to which of four *Risk Groups* (1,2 3, and 4) they fall. These groups, developed by the World Health Organization (LBM) characterize a microorganism by its pathogenicity, mode and ease of transmission, risk to an individual and community, and availability of preventive measures and treatment. Risk Group 1 includes substances that are unlikely to cause human or animal disease (and are not classified as infectious substances in transportation). Risk Group 4 represents the highest risk.

RELATED TERMS

Liquid, see *Terminology, Liquid*, p.241   Solid, see *Terminology, Solid*, p.247

REGULATORY DEFINITIONS

Division 6.2 includes substances which are infectious to humans and/or animals and include genetically modified micro-organisms and organisms, biological products, diagnostic specimens and clinical and medical waste, as described in 3.6.2.1.1 to 3.6.2.1.5. IATA 3.6.2.1

Infectious Substances

Infectious substances These are substances known or reasonably expected to contain pathogens. Pathogens are defined as micro-organisms (including bacteria, viruses, rickettsia, parasites, fungi) or recombinant micro-organisms (hybrid or mutant), that are known or reasonably expected to cause infectious disease in humans or animals. UN 2.6.1(b)

Class 6.2 - Infectious substances. These are substances containing viable micro-organisms including bacterium, virus, rickettsia, parasite, fungus or a recombinant, hybrid or mutant that are known, or reasonably believed, to cause disease in animals or humans. IMO Class 6.2, 1.1.2

Infectious substances are those substances known or reasonably expected to contain pathogens. Pathogens are defined as micro-organisms (including bacteria, viruses, rickettsia, parasites, fungi) or recombinant micro-organisms (hybrid or mutant), that are known or reasonably expected to cause infectious disease in humans or animals. However, they are not subject to the provisions of this class if they are unlikely to cause human or animal disease. Infectious substances are subject to the provisions of this class if they are capable of spreading disease when exposure to them occurs. IMO Class 6.2, 1.1

Infectious substances are those substances known to contain, or reasonably expected to contain, pathogens. Pathogens are micro-organisms (including bacteria, viruses, rickettsia, parasites, fungi) or recombinant micro-organisms (hybrid or mutant) that are known or reasonably expected to cause infectious disease in humans or animals. Infectious substances are not subject to the provisions or these Instructions for Division 6.2 if they are unlikely to cause human or animal disease. Infectious substances are, however, subject to the provisions of these Instructions for Division 6.2 if they are capable of spreading disease when exposure to them occurs. ICAO 2-6.3.1

An infectious substance means a viable microorganism, or its toxin, that causes or may cause disease in humans or animals, and includes those agents listed in 42 CFR 72.3 of the regulations of the Department of Health and Human Services and any other agent that causes or may cause severe, disabling or fatal disease. The terms infectious substance and etiologic agent are synonymous. US 173.134(a)(1)

Infectious Substances are substances known to contain, or reasonably expected to contain, pathogens. Pathogens are micro-organisms (including bacteria, viruses, rickettsia, parasites, fungi) or recombinant micro-organisms (hybrid or mutant) that are known or reasonably expected to cause infectious disease in humans or animals. Infectious substances are not subject to the provisions of these Regulations for Division 6.2 if they are unlikely to cause human or animal disease. Infectious substances are, however, subject to the provisions of these Regulations for Division 6.2 if they are capable of spreading disease when exposure to them occurs. IATA 3.6.2.1.1

Biological Products

Biological products are those products derived from living organisms, that are manufactured and distributed in accordance with the requirements of national governmental authorities which may have special licensing requirements, and are used either for prevention, treatment, or diagnosis of disease in humans or animals, or for development, experimental or investigational purposes related thereto. They include, but are not limited to, finished or unfinished products such as vaccines and diagnostic products. UN 2.6.3.1.2; IMO Class 6.2, 1.4; ICAO 2-6.5.1

A biological product means a material that is prepared and manufactured in accordance with the provisions of 9 CFR part 102 (Licenses for biological products), 9 CFR part 103 (Experimental products, distribution, and evaluation of biological products prior to licensing), 9 CFR part 104 (Permits for biological products), 21 CFR part 312 (Investigational new drug application), or 21 CFR parts 600 to 680 (Biologics). US 173.134(a)(3)

Biological Products These are those products derived from living organisms, that are manufactured and distributed in accordance with the requirements of national governmental authorities which may have special licensing requirements, and are used either for prevention, treatment, or diagnosis of disease in humans or animals, or for development, experimental or investigational purposes related thereto. They include, but are not limited to, finished or unfinished products such as vaccines and diagnostic products. IATA 3.6.2.1.3

Diagnostic Specimens

Diagnostic specimens are any human or animal material including, but not limited to, excreta, secreta, blood and its components, tissue and tissue fluids being transported for diagnostic or investigational purposes, but excluding live infected animals. UN 2.6.3.1.3; IMO Class 6.2, 1.5; ICAO 2-6.5.3; IATA 3.6.2.1.4

A diagnostic specimen means any human or animal material including, but not limited to, excreta, secreta, blood, blood components, tissue, and tissue fluids, being shipped for purposes of diagnosis. US 173.134(a)(2)

Genetically Modified Microorganism

Genetically modified micro-organisms and organisms are micro-organisms in which genetic material has been purposely altered through genetic engineering in a way that does not occur naturally. [Text continues.] UN 2.6.3.1.4; IMO Class 6.2, 1.3; ICAO 2-6.4

Genetically Modified Micro-Organisms and Organisms These are micro-organisms and organisms in which genetic material has been purposely altered through genetic engineering in a way that does not occur naturally. [Text continues.] IATA 3.6.2.1.2

Waste

Wastes (transported under UN 3291) are wastes derived from the medical treatment of animals or humans or from bio-research where there is a relatively low probability that infectious substances are present. [Text continues.] UN 2.6.3.1.5; IMO Class 6.2, 1.6

A regulated medical waste means a waste or reusable material, other than a culture or stock of an infectious substance, that contains an infectious substance and is generated in-- (i) The diagnosis, treatment or immunization of human beings or animals; (ii) Research pertaining to the diagnosis, treatment or immunization of human beings or animals; or (iii) The production or testing of biological products. US 173.134(a)(4)

Clinical Waste and Medical Waste Wastes transported under UN3291 are wastes derived from the medical treatment of humans or animals or from bio-research, where there is a relatively low probability that infectious substances are present. Waste infectious substances which can be specified must be assigned to UN2814 or UN2900. Decontaminated wastes which previously contained infectious substances may be considered as not subject to these Regulations unless the criteria of another Class or Division are met. IATA 3.6.2.1.5

Other Definitions

Cocculus. Is the dried fruit or berry of an Oriental plant having toxic qualities. IATA App. A

REFERENCES

40CFR439, 42CFR72, BMD, BMS, DIM, FTR, HCC, HOT, MEOS, MWD, MWM, PME

Initiating Explosives

Blasting cap assemblies ▪ Blasting caps, electric ▪ Blasting caps, non-electric ▪ Boosters with detonator, 1.1B, 1.2B ▪ Boosters without detonator, 1.1D, 1.2D ▪ Cannon primers ▪ Caps, blasting ▪ Caps, primer ▪ Charges, supplementary, explosive, 1.1D ▪ Components, explosive train, n.o.s., 1.1B, 1.2B, 1.4B, 1.4S ▪ Cord, detonating, flexible, 1.1D, 1.4D ▪ Cord, detonating, metal clad, 1.1D, 1.2D ▪ Cord, detonating, mild effect, metal clad, 1.4D; ▪ Cordeau detonant fuse ▪ Cord, igniter, 1.4G ▪ Delay electric igniter ▪ Detonating relays ▪ Detonator assemblies, non-electric, for blasting, 1.1B, 1.4B, 1.4S ▪ Detonators, electric for blasting, 1.1B, 1.4B, 1.4S ▪ Detonators for ammunition, 1.1B, 1.2B, 1.4B, 1.4S ▪ Detonators, non-electric for blasting, 1.1B, 1.4B, 1.4S ▪ Electric squibs ▪ Fuse, detonating, metal clad, 1.1D, 1.2D ▪ Fuse, detonating, mild effect, metal clad, 1.4D ▪ Fuse, igniter, tubular, metal clad, 1.4G ▪ Fuse, non-detonating, 1.3G ▪ Fuse, safety, 1.4S ▪ Fuzes, combination, percussion or time ▪ Fuzes, detonating, 1.1B, 1.2B, 1.4B, 1.4S ▪ Fuzes, detonating, with protective features, 1.1D, 1.2D, 1.4D ▪ Fuzes, igniting, 1.3G, 1.4G, 1.4S ▪ Igniter fuse, metal clad ▪ Igniters, 1.1G, 1.2G, 1.3G, 1.4G, 1.4S ▪ Jet thrust igniters, for rocket motors or Jato ▪ Lighters, fuse, 1.4S ▪ Percussion caps ▪ Primers, cap type, 1.1B, 1.4B, 1.4S ▪ Primers, small arms ▪ Primers, tubular, 1.3G, 1.4G, 1.4S ▪ Quickmatch ▪ Safety fuse ▪ Safety squibs ▪ Squibs

0029 ▪ 0030 ▪ 0042 ▪ 0044 ▪ 0060 ▪ 0065 ▪ 0066 ▪ 0073 ▪ 0101 ▪ 0102 ▪ 0103 ▪ 0104 ▪ 0105 ▪ 0106 ▪ 0107 ▪ 0121 ▪ 0131 ▪ 0225 ▪ 0255 ▪ 0257 ▪ 0267 ▪ 0268 ▪ 0283 ▪ 0289 ▪ 0290 ▪ 0314 ▪ 0315 ▪ 0316 ▪ 0317 ▪ 0319 ▪ 0320 ▪ 0325 ▪ 0360 ▪ 0361 ▪ 0364 ▪ 0365 ▪ 0366 ▪ 0367 ▪ 0368 ▪ 0376 ▪ 0377 ▪ 0378 ▪ 0382 ▪ 0383 ▪ 0384 ▪ 0408 ▪ 0409 ▪ 0410 ▪ 0454 ▪ 0455 ▪ 0456 ▪ 0461 ▪ 0500

The primary safety considerations in the formulation, manufacture, and use of explosives are that they must never explode prematurely but always explode when desired. Often this balance is achieved with an insensitive main charge being initiated by the least amount of a more sensitive explosive. This concept is expanded to that of an *explosive train* in which a series of decreasingly sensitive explosives is ultimately sufficient to initiate an insensitive main charge. The sensitivity of explosives to initiation provides a common means of classification:

▫ *Primary syn. initiator explosives*, those most sensitive to heat, shock, electric spark, friction, and impact. They are usually high explosives such as the fulminates and lead azide, although lead azide may be sensitized with a small amount of lead styphnate.

▫ *Secondary explosives*, those that do not easily go from deflagration to detonation, do not initiate electrostatically with ease, and require larger shocks to detonate. They include nitrocellulose, PETN, and desensitized nitroglycerine.

▫ *Tertiary explosives*, those most difficult to detonate. They include extremely powerful high explosives like RDX and HMX.

Other explosives may be added to explosive trains to sensitize certain components or add a time delay. There are as many combinations as necessary to marry the function, power, speed, reliability, timing, and safety of a system. Many of the terms used to describe initiating explosives and devices are used interchangeably and without precision.

DETONATORS

Nonelectric Detonators

In mining and demolition, high explosive charges are initiated by *detonators* in carefully timed sequences to minimize vibration and damage to surrounding structures and control rock breakage and throw. Detonators are devices that contain a few grams of explosive that are placed on, against, or into the main charge (a blasting agent). Introduced in the 1860s, the first detonators were called *blasting caps* or *caps* (names still used synonymously with *detonator*). They were capsules of mercury fulminate which initiated a charge of nitroglycerine.

Mercury fulminate detonators were widely replaced due to their extreme sensitivity. Contemporary simple nonelectric detonators contain a small quantity of primary explosive (often lead azide) which, once ignited by a fuse, ignites the more powerful base charge of a secondary explosive (an additional low explosive may be used between the fuse and the primary explosive).

Electric Detonators

The introduction of bridgewire allowed blasting caps to be initiated electrically rather than by a fuse which greatly increased timing precision. A straightforward electric blasting cap contains a *matchhead*: a high resistance bridgewire buried in a low explosive. An electric current from an external power source heats the bridgewire which ignites the explosive. Newer detonators use a faster-acting and lower energy-demanding semiconductor bridge. The matchhead charge initiates the primary charge which initiates the secondary charge in sequence.

Often another controlled volume of low explosive is placed between the matchhead and the primary charge. This burns for a specified period introducing a delay (from a few milliseconds to a few seconds) into the detonator. However, any low explosive delay device (*syn. delay electric igniter*) introduces imprecision because the chemicals are affected by moisture, temperature, and time. High precision delay electric detonators implant the matchhead directly into the primary explosive and control the delay with an integrated circuit within the detonator.

Even more precise electric detonators use a capacitor to discharge an extremely strong current through a thin bridgewire embedded directly into the secondary explosive. The bridgewire is rapidly heated, evaporates, and explodes to initiate the charge. These also have the advantage of being extremely insensitive to all but strong electrical discharges, thus improving safety. Precision blasting is used in seismography, and military uses to initiate explosive trains in nuclear devices.

Ammunition Detonators

Detonators for ammunition include primary explosives to detonate the main charge. They may or may not be connected to a fuze.

IGNITERS

Igniters describe devices used to ignite low explosives, including the first elements of an explosive train. Igniter terminology is used interchangeably:

- *Fuse lighters* are handheld stiff wires covered with a slow-burning low explosive mixture lit by matches. They are used to ignite fuses.
- *Jet thrust igniters* are assemblies containing a low explosive ignited by a bridgewire, spark, or a catalyst and used to ignite the propellants in solid, liquid, or hybrid rocket motors.
- *Squibs* are cotton yarns or other fibres impregnated with black powder and which burn with an external flame to ignite pyrotechnics, although the term may apply to small pyrotechnic or explosive devices. *Safety squibs* are those squibs which do not burn with an external flame. *Electric squibs* are essentially synonymous with the matchheads (bridgewire-low explosive combinations) used in electric detonators as well as a type of jet thrust igniter.
- In a transportation context, *igniter fuses* are metal tubes containing black powder, although the term has also been applied to *quick-matches* and squibs.
- *Igniter cord* is a low explosive fuse which burns with an external flame. It is connected to a series of safety fuses, each connected to a detonator and a main charge which are ignited in sequence.

CORDS AND FUSES

Fuse

Fuses are cord-like igniting devices that contain a core of low explosive sheathed in a conduit of waterproofed flexible fabric or fibre. They are crimped or connected onto detonators or other explosive charges so that they may be initiated safely from a distance. Black powder fuses, like the classic and low cost *safety fuse*, burn at specified rates around 130s/m. Safety fuses burn with an internal flame (although a jet of flame should be visible on ignition) and are so called to distinguish them from other fuses that burn with an external flame.

Detonating Cord

Detonating cords contain high explosives and are used to detonate high explosive charges from a safe distance. The first *detonating cord syn. detonating fuse* was the *Cordeau detonant cord*, a lead tube filled with a TNT core introduced around 1907. In the 1930s TNT was replaced by PETN and the lead tube by more flexible materials. Modern detonating cords are coated, flexible lines of braided textile or yarn, wax, or plastic that sheath a powdered high explosive core (often RDX or PETN). Once initiated with a detonator, the explosion travels along the cord at speeds of around 6,000m/s. Being nonelectric, they remove the risk of premature initiation through the stray currents common in surface mining and quarries where lightning and radio or radar signals are frequent. They also serve in underwater applications where electrical insulation is difficult. Detonating cord is wrapped around or imbedded into the main charge.

The pursuit of safety has prompted the development of other detonating cords, many replacing traditional cord. Of particular note is the NONEL (*nonel*ectric) system comprising a narrow plastic tubing coated internally with a fine dusting (less than 0.02g/m) of explosive or a highly reactive material. Once initiated, an airborne shock wave propagates along the length of the tube by the heating and expansion of the gaseous reaction products.

Metal Clad Fuses

Some cord combines a high explosive sheathed in soft or flexible metal tubing (*metal clad detonating cord* or *metal clad detonating fuse*). Short lead tubes are filled with a mixture that produces mostly solid reaction products and is not, therefore, prone to the violent disruptions caused by expanding gases. This system, which might be termed *mild effect*, allows for highly accurate burning times. Similarly, shock tube, flexible plastic tubing thinly coated on the inside with HMX or other high explosive and aluminium powder, detonates around 2000m/s with little noise and leaves an unbroken cord. Other metal clad cords are dusted internally with an explosive or are clad with a woven metal braid to protect the cord from abrasion and other physical abuse.

Assemblies

Once initiating devices are connected to a cord or fuse they become *assemblies*, hence *blasting cap assembly* and *detonator assembly*. *Detonating relays* are assemblies of detonators connected in such a way that multiple charges are fired in parallel or sequentially as one explosive initiates another.

PRIMERS

Primers are devices that contain high explosives used as the first element in an explosive train. In a transportation context, however, the term applies to articles which ignite the propelling or expelling charge in a round of ammunition. *Primer caps syn. percussion caps* are small charges contained in metal cups at the firing end of and integral to small-arms and other ammunition cartridges. Most caps contain an anvil which when struck by the weapon's firing pin crushes and detonates the charge. In rim fire caps, the firing pin crushes the cartridge rim against the chamber of the weapon.

Other primers are detonated by electricity, heat, or friction. In larger ammunition, such as for cannon, *tubular primers* may include a supplementary low explosive and be used to fire bag charges.

FUZES

Fuzes, a term often used synonymously with *fuse*, are devices which initiate high explosives (*detonating fuze*) or low explosives (*igniting fuze*) in projectile ammunition or pyrotechnics after launch, or which activate the main charge of a mine, bomb, or other nonprojectile ammunition. In addition to an explosive, fuzes incorporate some kind of mechanical or electronic initiation device:

- Impact fuzes *syn. percussion fuzes* (in British usage) operate when the projectile hits the ground or target. They usually involve some kind of pin or striker which detonates the main charge by shock.
- *Time fuzes* which explode sometime after leaving the weapon.
- Proximity fuzes which detonate at a set distance from the target, usually through the use of radio, radar, or photoelectric equipment located in the fuze.
- Delay fuzes which detonate shortly after impact giving the projectile time to penetrate the target. These are often black powder charges that may take a few tenths of a second to burn through.
- Base fuzes are used in those instances when the head of the projectile must be unobstructed on impact, such as armour-piercing shells. A heavy firing pin at the base of the round is restrained by a spring until its momentum on impact thrusts it into the detonating charge.
- *Combination fuzes* incorporate two or more types of fuzes. In certain grenades, for example, a percussion fuze is detonated when a spring action firing pin is removed which ignites a time fuze allowing the aggressor to arm the grenade, throw it, and permitting it to reach its target (usually around 5 seconds) before it explodes. Alternatively, combination fuzes have two independent systems, possibly a time and an impact fuze, whose association greatly increases the certainty of detonation.

Fuzes usually incorporate protective features that enable the projectile to reach a safe distance from the weapon before they become active. Safety devices may lock the firing fuze mechanism until acceleration moves the lock backwards in a manner restricted by an escapement. Multiple locks may operate sequentially until the projectile is well under way before the projectile is armed. Proximity fuzes can be locked electronically, although these projectiles may be protected with a system that controls the time at which the detonator is placed in alignment with the main charge.

BOOSTERS

Boosters are high energy explosives used to supplement the initiating power of explosives. In blasting applications, detonators are placed into booster charges which together initiate ANFO and the aqueous blasting agents, all relatively insensitive. In this context, the booster and detonator may be called a *primer*. In military applications, boosters *syn.* **supplementary explosive charges**, often consist of pellets of tetryl that are initiated by the fuze and detonate the main explosive payload.

RELATED TERMS

Ammunition, see *Ammunition*, p.8

Blasting, see *Explosives and Class 1*, p.74

Explosive, see *Explosives and Class 1*, p.74

Fuses, tracer, see *Pyrotechnics and*

Signals, p.194

Jet thrust unit, see *Ammunition*, p.8

Small arms, see *Ammunition*, p.8

REGULATORY DEFINITIONS

Boosters Articles consisting of a charge of detonating explosive with or without means of initiation. They are used to increase the initiating power of detonators or detonating cord. UN App. B, ICAO A2, IATA App. A

Boosters. Articles consisting of a charge of detonating explosive without means of initiation. They are used to increase the initiating power of detonators or detonating cord. US 173.59

Charges, Supplementary, Explosive Articles consisting of a small removable booster used in the cavity of a projectile between the fuze and the bursting charge. UN App. B, ICAO A2, US 173.59, IATA App. A

Components, Explosive Train, n.o.s. Articles containing an explosive designed to transmit the detonation or deflagration within an explosive train. UN App. B, ICAO A2, US 173.59

Components, Explosive Train, Not Otherwise Specified. Devices containing an explosive, designed to transmit the detonation within an explosive train. IATA App. A

Cord Detonating, flexible Article consisting of a core of detonating explosive enclosed in spun fabric, with plastics or other covering unless the spun fabric is sift-proof. UN App. B, ICAO A2, IATA App. A

Cord, detonating, flexible. Articles consisting of a core of detonating explosive enclosed in spun fabric with plastics or other covering. US 173.59

Cord (Fuse), Detonating, metal clad Article consisting of a core of detonating explosive clad by a soft metal tube with or without protective covering. When the core contains a sufficiently small quantity of explosive, the words "Mild Effect" are added. UN App. B, ICAO A2, US 173.59, IATA App. A

Cord, Igniter Article consisting of textile yarns covered with black powder or another fast burning pyrotechnic composition and of a flexible protective covering; or it consists of a core of black powder surrounded by a flexible woven fabric. It burns progressively along its length with an external flame and is used to transmit ignition from a device to a charge or primer. UN App. B, ICAO A2, IATA App. A

Cord igniter. Articles consisting of textile yarns covered with black powder or another fast-burning pyrotechnic composition and a flexible protective covering, or consisting of a core of black powder surrounded by a flexible woven fabric. It burns progressively along its length with an external flame and is used to transmit ignition from a device to a charge or primer. US 173.59

Detonator Assemblies, Non-Electric for blasting Non-electric detonators assembled with and activated by such means as safety fuse, shock tube, flash tube or detonating cord. The[y] may be of instantaneous design or incorporate delay elements. Detonating relays incorporating detonating cord are included. Other detonating relays are included in "Detonators, non-electric". UN App. B, ICAO A2, US 173.59, IATA App. A

Detonators Articles consisting of a small metal or plastics tube containing explosives such as lead azide, PETN or combinations of explosives. They are designed to start a detonation train. They may be constructed to detonate instantaneously, or may contain a delay element. The term includes: Detonators for Ammunition and Detonators for blasting, both Electric and Non-Electric. Detonating relays without flexible detonating cord are included. UN App. B, ICAO A2

Detonators. Articles consisting of a small metal or plastic tube containing explosives such as lead azide, PETN, or combinations of explosives. They are designed to start a detonation train. They may be constructed to detonate instantaneously, or may contain a delay element. They may contain no more than 10g of total explosives weight, excluding ignition and delay charges, per unit. The term includes: detonators for ammunition; detonators for blasting, both electric and non-electric; and detonating relays without flexible detonating cord. US 173.59

Detonators Articles consisting of a small metal or plastics tube containing explosives such as lead azide, PETN or combinations of explosives. They may be constructed to detonate instantaneously, or may contain a delay element. The term includes: Detonators for Ammunition; Detonators for blasting, both electric and non-Electric; Detonating relays without flexible detonating cord. IATA App. A

Explosive, Initiating. Explosive substances which, even in very small quantities, detonate on contact with a flame, on mild or low impact or as a result of friction; they are able to transmit detonation to other explosives close to them. The main initiating explosives are mercury fulminate and lead azide. For transport purposes some explosives, such as lead styphnate, are considered as initiating explosives because of their great sensitivity to the contact of a flame, to impact or to friction. (Both these types of sensitive explosives are referred to as primary explosives.) IATA App. A

Fuse/Fuze (English text only) Although these two words have a common origin (French fusee, fusil) and are sometimes considered to be different spellings, it is useful to maintain the convention that fuse refers to a cord-like igniting device whereas fuze refers to a device used in ammunition which incorporates mechanical, electrical, chemical or hydrostatic components to initiate a train by deflagration or detonation. UN App. B

Fuse/Fuze. Although these two words have a common origin (French fusee, fusil) and are sometimes considered to be different spellings of the same word, it is useful to maintain the convention that FUSE refers to a cord-like igniting device whereas FUZE refers to a device used in ammunition which incorporates mechanical, electrical, chemical or hydrostatic components to initiate a train by deflagration or detonation. ICAO A2, IATA App. A

Fuse/Fuze. Although these two words have a common origin (French fusee, fusil) and are sometimes considered to be different spellings, it is useful to maintain the convention that fuse refers to a cord-like igniting device, whereas fuze refers to a device used in ammunition which incorporates mechanical, electrical, chemical, or hydrostatic components to initiate a train by deflagration or detonation. US 173.59

Fuse, igniter. Articles consisting of a metal tube with a core of deflagrating explosives. US 173.59

Fuse, Igniter, tubular, metal clad Article consisting of a metal tube with a core of deflagrating explosive. UN App. B, ICAO A2, IATA App. A

Fuse, Instantaneous, Non-Detonating (Quickmatch) Article consisting of cotton yarns impregnated with fine black powder (Quickmatch). It burns with an external flame and is used in ignition trains for fireworks, etc. UN App. B, ICAO A2

Fuse, instantaneous, non-detonating (Quickmatch). Article consisting of cotton yarns impregnated with fine black powder. It burns with an external flame and is used in ignition trains for fireworks, etc. US 173.59

Fuse, Instantaneous, Non-Detonating. Article consisting of cotton yarns impregnated with meal powder (quickmatch). It burns with an external flame and is used in ignition trains for fireworks, etc. IATA App. A

Fuse, Safety Article consisting of a core of fine grained black powder surrounded by a flexible woven fabric with one or more protective outer coverings. When ignited, it burns at a predetermined rate without any external explosive effect. UN App. B, ICAO A2, IATA App. A

Fuse, safety. Article consisting of a core of fine-grained black powder surrounded by a flexible woven fabric with one or more protective outer coverings. When ignited, it burns at a predetermined rate without any explosive effect. US 173.59

Fuzes Articles designed to start a detonation or a deflagration in ammunition. They incorporate mechanical, electrical, chemical or hydrostatic components and generally protective features. The term includes: Fuzes, Detonating; Fuzes, Detonating with protective features; Fuzes, Igniting. UN App. B, ICAO A2, US 173.59, IATA App. A

Igniters Articles containing one or more explosive substances used to start deflagration in an explosive train. They may be actuated chemically, electrically or mechanically. This term excludes the following articles which are listed separately: Cord, Igniter; Fuse, Igniter; Fuse, Instantaneous, Non-Detonating; Fuzes, Igniting; Lighters, Fuse; Primers, Cap Type; Primers, Tubular. UN App. B, ICAO A2

Igniters. Articles containing one or more explosive substance used to start deflagration of an explosive train. They may be actuated chemically, electrically, or mechanically. The term excludes: cord, igniter; fuse, igniter; fuse, instantaneous, non-detonating; fuze, igniting; lighters, fuse, instantaneous, non-detonating; fuzes, igniting; lighters, fuse; primers, cap type; and primers, tubular. US 173.59, IATA App. A

Ignition, means of A general term used in connection with the method employed to ignite a deflagrating train of explosive or pyrotechnic substances (for example: a primer for a propelling charge; an igniter for a rocket motor; an igniting fuze). UN App. B, ICAO A2, US 173.59, IATA App. A

Initiation, means of (1) A device intended to cause the detonation of an explosive (for example: detonator; detonator for ammunition: detonating fuze). (2) The term "with its own means of initiation" means that the contrivance has its normal initiating device assembled to it and this device is considered to present a significant risk during transport but not one great enough to be unacceptable. The term does not apply, however, to a contrivance packed together with its means of initiation provided the device is packaged so as to eliminate the risk of causing detonation of the contrivance in the event of accidental functioning of the initiating device. The means of initiating can even be assembled to the contrivance provided there are protective features such that the device is very unlikely to cause detonation of the contrivance in conditions which are associated with transport. (3) For the purposes of classification any means of initiation without two effective protective features shall be regarded as Compatibility Group B; an article with its own means of initiation, without two effective protective features, would be Compatibility Group F. On the other hand a means of initiation which itself possesses two effective protective features would be Compatibility Group D; and an article with a means of initiation which possesses two effective protective features would be Compatibility Group D or E. Means of initiation adjudged as having two effective protective features shall have been approved by the competent national authority. A common and effective way of achieving the necessary degree of protection is to use a means of initiation which incorporates two or more independent safety features. UN App. B

Initiation, Means of. (1) A device intended to cause the detonation of an explosive (e.g. detonator; detonator for ammunition: detonating fuze). (2) The term "with its own means of initiation" means that the contrivance has its normal initiating device assembled to it and this device is considered to present a significant risk during transport but not one great enough to be unacceptable. The term does not apply, however, to a contrivance packed together with its means of initiation provided the device is packaged so as to eliminate the risk of causing detonation of the contrivance in the event of accidental functioning of the initiating device. The means of initiating can even be assembled to the contrivance provided there are protective features such that the device is unlikely to cause detonation of the contrivance in conditions which are associated with transport. (3) For the purposes of classification any means of initiation without two effective protective features should be regarded as Compatibility Group B; an article with its own means of initiation, without two effective protective features, would be Compatibility Group F. However, a means of initiation which itself possesses two effective protective features would be Compatibility Group D; and an article with a means of initiation which possesses two effective protective features would be Compatibility Group D or E. Means of initiation adjudged as having two effective protective features should have been approved by the appropriate national authority. A common and effective way of achieving the necessary degree of protection is to use a means of initiation which incorporates two or more independent safety features. ICAO A2, IATA App. A

Initiation, means of. (1) A device intended to cause the detonation of an explosive (for example: detonator, detonator for ammunition, or detonating fuze). (2) The term with its own means of initiation means that the contrivance has its normal initiating device assembled to it and this device is considered to present a significant risk during transport but not one great enough to be unacceptable. The term does not apply, however, to a contrivance packed together with its means of initiation, provided the device is packaged so as to eliminate the risk of causing detonation of the contrivance in the event of functioning of the initiating device. The initiating device can even be assembled in the contrivance provided there are protective features ensuring that the device is very unlikely to cause detonation of the contrivance under conditions which are associated with transport. (3) For the purposes of classification, any means of initiation without two effective protective features should be regarded as Compatibility Group B; an article with its own means of initiation, without two effective protective features, is Compatibility Group F. A means of initiation which itself possesses two effective protective features is Compatibility Group D, and an article with its own means of initiation which possesses two effective features is Compatibility Group D or E. A means of initiation, adjudged as having two effective protective features, must be approved by the Associate Administrator for Hazardous Materials Safety. A common and effective way of achieving the necessary degree of protection is to use a means of initiation which incorporates two or more independent safety features. US 173.59

Lighters, Fuse Articles of various design actuated by friction, percussion or electricity and used to ignite safety fuse. UN App. B, ICAO A2, US 173.59, IATA App. A

Primers, Cap Type Articles consisting of a metal or plastics cap containing a small amount of primary explosive mixture that is readily ignited by impact. They serve as igniting elements in small arms cartridges, and in percussion primers for propelling charges. UN App. B, ICAO A2, US 173.59, IATA App. A

Primer, Tubular Articles consisting of a primer for ignition and an auxiliary charge of deflagrating explosive such as black powder used to ignite the propelling charge in a cartridge case for cannon, etc. UN App. B, ICAO A2, US 173.59, IATA App. A

REFERENCES

BH, D, DC, DOMS, DOMT, EOE, GAE, GOCE, HOE, KOC, MEOS, MH14, PND, RBA, REI, SSF, TBO, TE, TEA, TIE

Inorganic Compounds

Antimony compound, inorganic, liquid, n.o.s., 6.1 ▪ Antimony compound, inorganic, solid, n.o.s., 6.1 ▪ Arsenates, liquid, n.o.s., inorganic ▪ Arsenates, n.o.s. ▪ Arsenates, solids, n.o.s., inorganic ▪ Arsenic compound, liquid, n.o.s. ▪ Arsenic compound, liquid, n.o.s., inorganic, 6.1 ▪ Arsenic compound, solid, n.o.s., inorganic, 6.1 ▪ Arsenic sulphides, liquid, n.o.s., inorganic ▪ Arsenic sulphides, n.o.s. ▪ Arsenic sulphides, solid, n.o.s., inorganic ▪ Arsenites, liquid, n.o.s., inorganic ▪ Arsenites, n.o.s. ▪ Arsenites, solid, n.o.s., inorganic ▪ Bifluorides, n.o.s. ▪ Bifluorides, solid, n.o.s. ▪ Bifluorides, solution, n.o.s. ▪ Bisulphates, aqueous solution, 8 ▪ Bisulphites, aqueous solution, n.o.s., 8 ▪ Bromates, inorganic, aqueous solution, n.o.s., 5.1 ▪ Bromates, inorganic, n.o.s., 5.1 ▪ Chlorates, inorganic, aqueous solution, n.o.s., 5.1 ▪ Chlorates, inorganic, n.o.s., 5.1 ▪ Chlorites, inorganic, n.o.s., 5.1 ▪ Chlorosilanes, n.o.s., 3, 3.2, 4.3, 8 ▪ Cyanide mixture, inorganic, solid, n.o.s. ▪ Cyanide or cyanide mixture, dry ▪ Cyanides, inorganic, solid, n.o.s., 6.1 ▪ Cyanide solution, n.o.s., 6.1 ▪ Fluorosilicates, n.o.s., 6.1 ▪ Hydrogendifluorides, n.o.s., 8 ▪ Hydrogendifluorides, solid, n.o.s., 8 ▪ Hydrogendifluorides, solution, n.o.s., 8 ▪ Hypochlorites, inorganic, n.o.s., 5.1 ▪ Nitrates, inorganic, aqueous solution, n.o.s., 5.1 ▪ Nitrates, inorganic, n.o.s., 5.1 ▪ Nitrites, inorganic, aqueous solution, n.o.s., 5.1 ▪ Nitrites, inorganic, n.o.s., 5.1 ▪ Perchlorates, inorganic, aqueous solution, n.o.s., 5.1 ▪ Perchlorates, inorganic, n.o.s., 5.1 ▪ Permanganates, inorganic, aqueous solution, n.o.s., 5.1 ▪ Permanganates, inorganic, n.o.s., 5.1 ▪ Persulphates, inorganic, aqueous solution, n.o.s., 5.1 ▪ Persulphates, inorganic, n.o.s., 5.1 ▪ Selenium compounds, n.o.s., 6.1 ▪ Selenates, 6.1 ▪ Selenites, 6.1 ▪ Silicofluorides, n.o.s., 6.1 ▪ Tellurium compounds, n.o.s., 6.1

1450 ▪ 1461 ▪ 1462 ▪ 1477 ▪ 1481 ▪ 1482 ▪ 1549 ▪ 1556 ▪ 1557 ▪ 1588 ▪ 1598 ▪ 1740 ▪ 1791 ▪ 1908 ▪ 1935 ▪ 2627 ▪ 2630 ▪ 2693 ▪ 2837 ▪ 2856 ▪ 2985 ▪ 2986 ▪ 2987 ▪ 2988 ▪ 3141 ▪ 3210 ▪ 3211 ▪ 3212 ▪ 3213 ▪ 3214 ▪ 3215 ▪ 3216 ▪ 3218 ▪ 3219 ▪ 3283 ▪ 3284

Historically, *inorganic* described those compounds derived from minerals as distinguished from *organic* compounds, hydrocarbons and their derivatives with animal or plant origins. This distinction still exists although with the development of synthetic chemistry, *inorganic* has come to describe non-carbon compounds other than simple examples like the carbon oxides, carbon disulphide, carbonyls, carbonates, cyanides, and carbides. The transportation regulations cover many generic entries related to inorganic compounds.

METALLOID COMPOUNDS

Antimony Compounds

Antimony compounds are based on antimony, an element that exhibits both metal and nonmetal properties. Many of its compounds are toxic and corrosive, particularly the soluble salts. They include antimony iodide and antimony perchloride. Some antimony compounds decompose in water to produce toxic gases; e.g., antimony sulphate decomposes to sulphur dioxide while antimony bromide produces bromine gas.

Arsenic Compounds

Arsenic compounds are based on the metalloid arsenic. They are extremely toxic and include the *arsenates*, those that contain the AsO_4^{3-} group

128

(arsenic(V)), such as potassium arsenate; the *arsenites*, those that contain the AsO_2^- group (arsenic(III)), such as sodium arsenite; and the *arsenic sulphides*, such as arsenic disulphide, arsenic pentasulphide, and arsenic trisulphide.

Chlorosilanes

Silicon can form inorganic compounds called *silanes* in a manner analogous to carbon-based organic compounds. The simplest example is silane (SiH_4) analogous to methane (CH_4), the next is Si_2H_6 analogous to ethane (C_2H_6). *Chlorosilanes* are chlorinated silanes in which one or more of the hydrogen atoms has been displaced by a chlorine atom. Trichlorosilane is the most valuable being used in the production of high purity silicon and in coatings, adhesives, and resin technologies. Chlorosilanes react with water, steam, or moisture to evolve hydrogen chloride gas sometimes with sufficient heat to risk self-ignition that can lead to toxic and corrosive combustion products.

Tellurium Compounds

Tellurium compounds are toxic. They include tellurium sulphide and tellurium dioxide.

SALTS

Dissociation of many inorganic acids yields a variety of salts.

Bromates

Inorganic *bromates* are salts including the species BrO_3^- where bromine exists in an oxidation state of V (Br(V)) coupled with oxygen. The term can also include hypobromates (BrO^-) and perbromates (BrO_4^-) in which bromine exists as Br(I) and Br(VII), respectively. Common examples of these strong oxidizers include potassium bromate and magnesium bromate. Bromates can form explosive mixtures with organic materials, ammonium compounds, and metal powders. They react vigorously with acids.

Chlorates

Chlorine forms a number of ions with oxygen, all of which are strong oxidizing agents. They react similarly to bromates, evolving chlorine gas in contact with acids. They include

- *Hypochlorites* (ClO^-) in which the chlorine exists in the oxidation state of Cl(I) such as lithium hypochlorite.
- *Chlorites* (ClO_2^-), Cl(III), such as sodium chlorite.
- *Chlorates* (ClO_3^-), Cl(V), such as magnesium chlorate.
- *Perchlorates* (ClO_4^-), Cl(VII), such as barium perchlorate.

Cyanides

Cyanides are based on the species CN^- and are invariably extremely toxic. They include silver cyanide. In contact with acids they evolve the toxic gas hydrogen cyanide. Organic cyanides are called *nitriles*.

Bifluorides

The *bifluorides syn. hydrogendifluorides*, are crystalline solids containing the HF_2^- species. Examples include sodium bifluoride and potassium bifluoride. They are crystalline solids soluble in water or water-alcohol solutions and extremely corrosive to metals, glass, and other siliceous materials. They are used as etchants.

Bisulphates

Bisulphates are compounds with the HSO_4^- species in which sulphur exists as sulphur(VI) such as sodium bisulphate. *Bisulphites* are those with the HSO_3^- species (sulphur(IV)), such as sodium bisulphite. They are strong irritants to tissue.

Fluorosilicates

Fluorosilicates syn. silicofluorides are those salts containing either the SiF_6^{2-} or the SiF_5^- species involving silicon and fluorine including sodium fluorosilicate. Fluorosilicates are often toxic and react with acids to give hydrogen fluoride and silicon tetrafluoride.

Nitrates

Nitrates are the salts involving the nitrate species (NO_3^-), such as potassium nitrate. They are strong oxidizers and common components of explosives. Nitrate solutions may contain sufficient nitric acid to be corrosive. *Nitrites* involve the nitrite species (NO_2^-) and include sodium nitrite, a characteristically strong oxidizer. Nitrites react with acids to evolve hydrogen cyanide and, as liquids, evolve toxic vapours.

Permanganates

Permanganates are based on the MnO_4^- species and are strong oxidizers. They include potassium permanganate.

Persulphates

Persulphates are those compounds based on the species in which the sulphur exists in its highest state of oxidation, sulphur(VII), including sodium persulphate. They are strong oxidizers used as bleaches.

Selenates

Selenates are salts based on the species SeO_4^{2-} with selenium(VI). Selenites, involve SeO_3^{2-}, selenium(IV). Like all *selenium compounds*, these salts are toxic and irritating chemicals; sodium selenate, for example, is used as an insecticide.

RELATED TERMS

Alcohol, see *Alcohols*, p.5
Alkali metal alcoholates, self-heating, corrosive, n.o.s., see *Organometallics and Related Compounds*, p.167
Alkaline earth metal alcoholates, n.o.s., see

Organometallics and Related Compounds, p.167
Aqueous, see *Terminology, Aqueous*, p.231
Dry, see *Terminology, Dry*, p.235
Inorganic, see *Terminology, Inorganic*, p.241

Mixture, see *Terminology, Mixture*, p.243

Self-heating, see *Spontaneously Combustible Materials and Division 4.2*, p.226

Solid, see *Terminology, Solid*, p.247

Solution, see *Terminology, Solutions*, p.247

Water-reactive, see *Dangerous When Wet Materials and Division 4.3*, p.58

REGULATORY DEFINITIONS
None

REFERENCES
GOCT, HCC, IMDG, TPD

Leather

Dressing, leather ■ Leather bleach or dressing

Leather is the tough, wear-resistant material created from animal hides and skins by a number of processing steps:

1. Hair, fur, dirt, fat, and other unwanted materials are removed from the hide.
2. The hide is permanently altered as collagen, the principal protein present, is made into leather by being cross-linked with a *tanning* agent. Traditionally, vegetable tanning agents based on polyphenolics were (and still are) used. Commercially, chrome tanning agents, based on chromium sulphate, are used.
3. The tanned leather is stabilized and finished.

Other steps, including temporary preservation before tanning by curing or pickling, may be involved. Tanning makes hides resistant to bacterial decay, shrinkage, and drying out but leaves leather dark, reddish, or blue in colour. *Leather bleaches* are used to lighten or blanch the leather, removing part of its surface tan. Among the bleaches used are preparations of oxalic acid, sodium thiosulphate, sulphur dioxide, and hydrogen peroxide. Few are flammable, rather they are likely to be strong oxidizers that may be dissolved in flammable solvents, including oils, alcohols, and ethers.

The cleaning, conditioning, softening, bleaching, staining, dyeing, polishing, and waterproofing steps necessary to render roughly tanned leather into material suitable for commercial goods are collectively, and roughly, called *leather dressing*. Leather dressings include any number of chemicals used for this purpose:

- Wool grease or *degras*, a brown, waxy fat obtained as a by-product of scouring wool. It contains the leather softeners lanoceric acid, lanopalmic acid, and lanosterol.
- Beeswax which contains myricyl palmitate, paraffins, cerotic acid, and esters.
- Cottonseed oil, mostly glycerides or palmitic, oleic, and linoleic acids.

Many leather dressing materials are themselves combustible or otherwise hazardous and many are associated with flammable solvents as carriers to aid application.

RELATED TERMS
Bleach, see *Bleach*, p.34   *Liquids and Class 3*, p.96
Flammable liquid, n.o.s., see *Flammable*

Dressing, Leather.   A preparation which usually contains a solvent or other liquid with a low flash point. ICAO A2

Dressing, Leather. May contain liquids or solvents of low flash point, and hence be classified as flammable liquids. IATA App. A

REFERENCES
DOTU, HCC, KOE, LDD, MH14

Lifesaving Equipment

Air bag inflators, 9 ▪ Air bag inflators, compressed gas, 2.2 ▪ Air bag inflators, pyrotechnic, 9 ▪ Air bag modules, 9 ▪ Air bag modules, compressed gas, 2.2 ▪ Air bag modules, pyrotechnic, 9 ▪ Aircraft evacuation slides ▪ Aircraft survival kits ▪ Gas generator assemblies for aircraft escape slides containing a non-flammable, non-toxic gas and a propellant cartridge, 2.2 ▪ Life rafts ▪ Life saving appliance, not self-inflating, 9 ▪ Life-saving appliances, not self-inflating containing dangerous goods as equipment, 9 ▪ Life-saving appliances, self-inflating, 9 ▪ Mine rescue equipment containing carbon dioxide ▪ Rockets, line-throwing, 1.2G, 1.3G, 1.4G ▪ Seat-belt modules, 9 ▪ Seat-belt pretensioners, 9 ▪ Seat-belt pretensioners, compressed gas, 2.2 ▪ Seat-belt pretensioners, pyrotechnic, 9 ▪ Self-inflating passenger restraint systems (air bags) for motor vehicles

0238 ▪ 0240 ▪ 0453 ▪ 1013 ▪ 2990 ▪ 3072 ▪ 3268 ▪ 3353 ▪ 8013

Lifesaving appliances include a range of articles that prevent the loss of life, particularly through drowning, or increase the chances of survival. Many lifesaving appliances contain compressed gases:

□ Self-inflating appliances use compressed gases (usually nitrogen, carbon dioxide, or some other nontoxic, nonflammable gas or gas generant) to fill life jackets, *life rafts*, and marine and *aircraft evacuation slides*. *Gas generator assemblies* for larger equipment may also be transported independently as spare parts.

□ Standard *mine rescue equipment* includes motor vehicles, breathing apparatus, recharging pumps, oxygen cylinders, resuscitating equipment, gas masks, gas detection monitors, lamps, battery charging equipment, hoses, medical supplies, parts and supplies. Mine rescue equipment containing carbon dioxide gas (124-38-9) may include that which is used to purge sealed-off areas to control fires and create inert atmospheres following a mine explosion or fire.[13]

Other lifesaving equipment contains small and varied quantities of other hazardous materials:

□ Lifebuoys with self-igniting lights and signal flares.

□ *Line-throwing rockets* consisting of a weapon (gun, pistol, launcher) and cartridge attached to a light service or shot line which can be shot to shore or another ship.

□ Matches, usually strike-anywhere matches.

□ Pyrotechnic signal devices: smoke and illumination signal flares containing explosives.

□ *Aircraft survival kits* may contain any of the above equipment plus first aid supplies, life raft repair kits, batteries, etc.

---

[13] The author was unable to find any direct reference to the use of carbon dioxide in *mine rescue equipment*.

134

AUTOMOTIVE SAFETY

Air bags are vehicle-installed *passenger restraint systems* that absorb the force of a moving human body when the vehicle is involved in a side or frontal impact. They are most commonly mounted in the steering wheel hub, instrument panel, or door panels. Many designs exist, but they generally comprise an *air bag module* that includes an inflator assembly (*air bag inflator*), a bag, a reaction can (if necessary), and a protective cover. The inflator is connected to crash sensors and is the source of gas in compressed form or generated by chemical reaction. Normally these chemicals contain an oxidizer and fuel elements in powder or granular form compacted into a pellet (explosive and pyrotechnic systems). Nitrogen gas, for example, evolves from the ignition of sodium azide. The gas expands into the bag pushing it through pre-engineered stress seams in the cover.

Seatbelt assemblies or *seatbelt modules* include the straps, webbing, buckles, adjusting mechanisms, fasteners, and related hardware designed to secure people in a vehicle to minimize risk of harm in a collision. Most modules are mechanical, but some contain a pyrotechnic or compressed gas *seatbelt pretensioner* which, in an emergency, forces the retractor to take up any slack in the seatbelt.

RELATED TERMS

Cartridge, see *Ammunition*, p.8 or *Explosive Articles*, p.69

Compressed gas, see *Gases and Class 2*, p.104

Dangerous goods, see *Dangerous Goods and Hazardous Materials*, p.54

Gas, see *Terminology, Gas*, p.239

Motor vehicles, see *Self-Propelled Vehicles*, p.219

Non-flammable gas, see *Gases and Class 2*, p.104

Non-toxic gas, see *Gases and Class 2*, p.104

Propellant, see *Ammunition*, p.8

Pyrotechnic, see *Pyrotechnics and Signals*, p.194

REGULATORY DEFINITIONS

An air bag inflator (consisting of a casing containing an igniter, a booster material and a gas generant) is a gas generator used to inflate an air bag in a supplemental restraint system in a motor vehicle. An air bag module is the air bag inflator plus an inflator bag assembly. A seat belt pretensioner contains similar hazardous materials and is used in the operation of a seat-belt restraining system in a motor vehicle. A seat-belt module is the seat belt pretensioner plus seatbelt hardware. US 173.166(a)

Gas Generator Assemblies for Aircraft Escape Slides. A steel cylinder containing a charge of liquefied chlorodifluoromethane (R22) under pressure and a slow-burning solid propellant cartridge (safety type) in a specially designed breech block. The assembly is installed in certain types of aircraft to provide a means of generating a supply of high-pressure, low-temperature gas to the aspirators that inflate emergency escape slides. ICAO A2

Gas Generator Assemblies (for Aircraft Escape Slides). Consist of a steel cylinder containing a charge of non-flammable, non-toxic gas (usually Chlorodifluoromethane) under pressure and a slow-burning solid propellant cartridge (safety type) in a specially designed breech block. The assembly is installed in certain types of aircraft to provide a means of generating a supply of high pressure, low temperature gas to the aspirators that inflate emergency escape slides. Complete assemblies need to be transported for positioning as spare parts and thus comprise a non-flammable compressed gas and an explosive power device. IATA App. A

Life-Saving Appliances, Not Self-Inflating A life-saving appliance, other than a self-inflating one, which includes one or more of the following dangerous goods as equipment: signal devices (class 1); gases (class 2.2); small quantities of flammable substances (classes 3, 4.1 and 5.2); electric storage batteries (classes 8 and 9). IMO 9032

REFERENCES
46CFR160, ADM, EOTM, GOA, IMDG, MSA, SAS, SG, SVO, TSM

Lighters

Candles, gas ▪ Cer Mischmetall ▪ Cigar and cigarette lighter fluid ▪ Cigar and cigarette lighters, charged with fuel ▪ Ferrocerium, 4.1 ▪ Flammable gas in lighters ▪ Ignition element for lighter, containing pyrophoric liquid ▪ Lighter flints ▪ Lighter fluid ▪ Lighter refills, 2.1 ▪ Lighter refills (cigarettes) containing flammable gas, 2.1 ▪ Lighter refills containing flammable gas, 2.1 ▪ Lighter replacement cartridges containing lique-fied petroleum gases ▪ Lighters, 2.1 ▪ Lighters (cigarettes), containing flammable gas, 2.1 ▪ Lighters (cigarettes), containing pyrophoric liquid ▪ Lighters containing flamma-ble gas, 2.1 ▪ Lighters for cigars, cigarettes, etc., with lighter fluids, 3 ▪ Lighters, with lighter fluids (cigarettes) ▪ Mischmetall

1057 ▪ 1226 ▪ 1323

Lighters syn. gas candles are small mechanical or electrical flame-producing devices for igniting fuels, fires, cigars (rolls or tobacco wrapped in tobacco leaf), cigarettes (rolls of tobacco wrapped in a non-tobacco product such as paper), and pipes (devices used to burn and inhale tobacco). Electrical de-vices operate by generating a spark between two conductors as they are in-troduced into a fuel source; unless battery-powered, they have no internal fuel source of their own and do not represent a hazard in transportation. Other lighters, like cigarette lighters, contain a reservoir of *lighter fluid*, a liquefied flammable gas such as propane or butane, a valve, and an ignition device. Lighters may also contain pyrophoric liquids that do not require an ignition system but ignite when exposed to air. Some lighters are disposable, while others may be refilled or have their igniter fluid cartridge replaced.

Ignition systems create a spark in the gas stream or fuel vapours as they are discharged through the valve. The spark is either electrical or produced by the friction consequent to striking a *lighter flint* (a pyrophoric metal or alloy) with steel. This generates small hot particles that ignite the fuel. The term *flint* originates from the historical use of flintstone, a quartz that produces sparks when struck. Commercially, flints are likely to be made of *ferroce-rium*, a pyrophoric alloy of iron (35%) and *misch metal syn. Cer Misch-metall* or *mischmetall*. Misch metal is the primary commercial form of the extremely flammable mixed rare-earth metals, including cerium. Zinc-lead or zirconium-tin-lead alloys are also used.

RELATED TERMS

Flammable gas, see *Gases and Class 2*, p.104

Flammable liquid, n.o.s., see *Flammable Liquids and Class 3*, p.96

Fluid, see *Terminology, Fluid*, p.238

Gas, see *Terminology, Gas*, p.239

Lighters, fuse, see *Initiating Explosives*, p.119

Liquefied petroleum gases, see *Petroleum*, p.183

Pyrophoric liquid, see *Spontaneously Com-bustible Materials and Division 4.2*, p.226

Lighters

REGULATORY DEFINITIONS
Ferrocerium Alloy derived from cerium or mischmetal, with the addition of 10% to 65% iron. IMO 4144

REFERENCES
16CFR1210, 27CFR270, HCC, MH14

Magnetized Materials

Magnetized material, 9

Magnetic fields are generated in space around electric currents. And, like negative and positive electric charges, magnetic fields have two poles: north- and south-seeking. At the atomic level, electric current is generated by electrons and protons (electrically charged particles) as they spin. As each particle spins, a small magnetic field is generated whose effect is cancelled out on the super-atomic scale because in most materials these particles spin in random orientations. In *magnetized materials*, however, electron spins are organized and aligned to reinforce each other resulting in an observable magnetic field. Some materials, noticeably iron and its alloys, are more prone to magnetic effects than others.

All materials become magnetic in the presence of magnetic fields, but not all materials hold their magnetism; they may lose their strength immediately or if the atoms are realigned by being knocked or heated. Those that do hold their magnetism are called *permanent magnets*. *Coercivity* is the degree to which a magnet resists demagnetization; *remanence* is the degree to which magnetic field strength is retained. Magnetic fields also exist around an electric current as it passes through wire (this phenomenon and its inverse, that moving wire through magnetic fields induces an electrical current, is the basis of power generators and electric motors).

Magnets are used in magnetic latches, solenoids, in electric motors in thousands of consumer and industrial products, generators, and in many instruments. Navigational compasses rely on the tendency of free-moving magnets to align themselves to the magnetic field generated by the earth (itself a spinning electric charge), the intensity of which is greater at the poles than at the equator.

HAZARDS
Magnets or magnetized materials placed in the vicinity of a compass will affect the instrument. Other materials, such as metal construction material, pipes, and automobiles may contain sufficient ferro-magnetic material to affect aircraft instruments.

RELATED TERMS
None

REGULATORY DEFINITIONS
Magnetized Material: Any material which, when packed for air transport, has a magnetic field strength of 0.159 A/m or more at a distance of 2.1 m from any point on the surface of the assembled package [text continues]. ICAO 2-9.1

Magnetic material. For carriage by aircraft, any package which has a magnetic field of more than 0.00525 gauss measured at 4.5 m (15 feet) from any surface of the package. US 173.21(d)

Magnetized Materials. Cover material with relatively high magnetic field strength such as magnetrons and non-shielded permanent magnets without keeper bars installed. Masses of ferro-magnetic metals such as automobiles, automobile parts, metal fencing, piping and metal construction materials, even if not meeting the definition of magnetized materials, may be subject to operator's special stowage regulations since they may affect aircraft instruments, particularly the compasses. IATA App. A

Any material which, when packed for air transport, has a magnetic field strength of 0.159 A/m (0.002 gauss) or more at a distance of 2.1 m (7ft) from any point on the surface of the assembled package [text continues]. IATA 3.9.1.2

REFERENCES
HCC, MEOP, MH14

Marine Pollutants

Marine pollutants, liquid or solid, n.o.s.

The *International Convention for the Prevention of Pollution from Ships* (MARPOL) is a global agreement to control accidental and operational discharges of pollution from ships. It contains five annexes:

□ Annex I, transport of oil.
□ Annex II, transport of chemicals in bulk.
□ Annex III, transport of harmful substances (Regulations for the Prevention of Pollution by Harmful Substances Carried by Sea in Packaged Forms or in Freight Containers, Portable Tanks or Road and Rail Tank Wagons).
□ Annex IV, ship-generated sewage.
□ Annex V, ship-generated garbage.

Annex III recognizes that certain harmful substances (***marine pollutants***) kill or retard the growth of marine life and bioaccumulate in marine organisms causing problems in the food chain that present risk to humans, birds, and other wildlife. These marine pollutants must be packaged (to survive sea immersion for a time reasonably sufficient to effect recovery), marked, documented, limited in quantity, and stowed to minimize risk to the marine environment. Port authorities must be notified of marine pollutant shipments.

In 1985 it was decided that Annex III would be implemented internationally through the International Maritime Dangerous Goods Code (IMDG). Individual states have since incorporated the requirements into their national standards (e.g., the United States Department of Transportation has promulgated marine pollutant regulations in 49CFR171.4). In doing so, over 600 chemicals, mixtures, and solutions, including a large number of pesticides, have been categorized[14] as marine pollutants or severe marine pollutants, those that present an extreme pollution potential (i.e., they strongly bioaccumulate or are particularly toxic). Marine pollutants include barium compounds, copper metal powder, dieldrin, mercuric oxide, and PCBs.

Many marine pollutants are already listed by chemical or generic name in the transportation regulations. Those that are not are assigned to Class 9 and shipped under the entry *environmentally hazardous substances*.

RELATED TERMS

Environmentally hazardous substance, liquid or solid, n.o.s. see *Environmentally Hazardous Substances*, p.67

Liquid, see *Terminology, Liquid*, p.241
Solid, see *Terminology, Solid*, p.247

---

[14] 49CFR172.101 App. B, *List of Marine Pollutants.*

141

REGULATORY DEFINITIONS
Marine pollutant, means a material which is listed in appendix B to Sec. 172.101 of this sub-chapter and, when in a solution or mixture of one or more marine pollutants, is packaged in a concentration which equals or exceeds: (1) Ten percent by weight of the solution or mixture for materials listed in the appendix; or (2) One percent by weight of the solution or mixture for materials that are identified as severe marine pollutants in the appendix. US 171.8

REFERENCES
49CFR, IMDG, LOSA

Matches

Fusee, matches ■ Matches, block ■ Matches, fusee, 4.1 ■ Matches, safety (book, card or strike on box), 4.1 ■ Matches, "strike anywhere," 4.1 ■ Matches, wax "vesta," 4.1

1331 ■ 1944 ■ 1945 ■ 2254

A *match* consists of a shank or stick of short, thin combustible material such as wood or resin-stiffened cardboard, tipped with flammable, solid materials, often phosphorus-based, along with binders, friction enhancers, and inert materials. When heated through friction or shock as the match is struck, the tips easily ignite with sufficient heat to cause the shank to combust. Strips of cardboard matches are stapled into blocks, books, or cards, while wooden matches are often distributed in boxes:

- *Safety matches* require a specially prepared surface on which to be struck. The surface compositions are the same as those used on the match tip.
- *Strike-anywhere matches* syn. *block matches* do not require a prepared surface and can be ignited on almost any dry surface. They often contain phosphorus sesquisulphide, which ignites easily by friction, or potassium chlorate, a strong oxidizer, which encourages the shank to burn. *Fusee matches* are strike-anywhere matches with bulbous heads that burn intensely with little flame and are difficult to blow out (e.g., matchsticks impregnated with potassium nitrate and tipped with sulphur).
- The shanks of *wax "vesta" matches* are wicks of twisted cotton coated with combustible paraffin wax, stearic acid, or gum to aid burning. They are ignited by friction and may require a prepared surface (the term originates from *Vesta* the Roman goddess of the hearth).

RELATED TERMS

Matches, trick, see *Pyrotechnics and Signals*, p.194

Fireworks, see *Pyrotechnics and*

Signals, p.194

Quickmatch, see *Initiating Explosives*, p.119

REGULATORY DEFINITIONS

Fusee matches are matches the heads of which are prepared with a friction-sensitive igniter composition and a pyrotechnic composition which burns with little or no flame, but with intense heat. US 173.186(b)(1)

Matches, Safety. Matches, contained in a book, card or box, which are only ignited when struck on a prepared surface. ICAO A2

Matches, Safety (Book, Card, or Strike-On-Box). Are matches intended to be struck on a prepared surface. IATA App. A

Matches, "Strike anywhere" or Fusee. They usually contain phosphorus sesquisulphide, potassium chlorate and other ingredients. The "strike-anywhere" matches are readily ignited by friction on almost any dry surface. ICAO A2

Matches, Strike-Anywhere. Usually contain phosphorus sesquisulphide, potassium chlorate and other ingredients. The strike-anywhere matches are readily ignited by friction on almost any dry surface. When a closed package of strike-anywhere matches is ignited by impact or friction the head composition burns off the matches and the fire then usually goes out unless the package is broken. If the package is broken, allowing access of air, the fire will continue. Packages of these matches that have been wetted for any reason and subsequently dried should be handled with extreme caution. IATA App. A

Safety matches are matches combined with or attached to the box, book or card that can be ignited by friction only on a prepared surface. US 173.186(b)(2)

Strike anywhere matches are matches that can be ignited by friction on a solid surface. US 173.186(b)(3)

Wax "Vesta" matches are matches that can be ignited by friction either on a prepared surface or on a solid surface. US 173.186(b)(4)

REFERENCES
E2, IMDG, KOC, MH14, MST, WNN

Mercury

Cinnabar ■ Electron tubes containing mercury ■ Mercuric salt ■ Mercurous compound ■ Mercury compound, liquid, n.o.s, 6.1 ■ Mercury compound, solid, n.o.s., 6.1 ■ Mercury contained in manufactured articles, 8 ■ Mercury vapour tubes ■ Phenylmercuric compound, n.o.s., 6.1 ■ Quicksilver

2024 ■ 2025 ■ 2026 ■ 2809

Mercury is a silvery-white liquid metal with high electrical conductivity. It is the only common metal liquid at room temperature. It is unique in that it has the highest surface tension of any liquid (480 dynes/cm^2) causing it to pool in beads when allowed. Its colour and properties led to its being called *quicksilver*. It is recovered from *cinnabar* (natural mercury sulphide occurring in veins in volcanically active areas), other ores, and as a by-product of gold-refining. It is also extensively reclaimed in part because of its toxicity from scrap and waste sources. Mercury evolves vapour in ambient conditions and is absorbed through the skin, by inhalation, and ingestion.

Mercury is widely used in the chemical industry as a cathode to manufacture chlorine and sodium hydroxide; as a laboratory reagent; in *phenylmercuric compounds* (e.g., phenylmercuric acetate and phenylmercuric oleate); and other *mercury compounds* (e.g., merbromin, ammoniated mercury, thimerosal) which act as fungicides, biocides, paint preservatives, seed disinfectants, wood preservatives, pharmaceuticals, cosmetics, pigments (mercuric sulphide as cinnabar, a red pigment, which has been used for thousands of years); as a plastics catalysts; and in explosives (e.g., mercury fulminate and mercuric 5-nitrotetrazole). It is also used in many manufactured articles and items, particularly measuring instruments which take advantage of its conductivity and uniform expansion over its entire liquid range:

barometers (pressure measurement)
batteries
dental amalgams
fluorescent tubes
high intensity discharge lamps
magnetometers (magnetic measurement)
manometers (flow measurement)
mercury electron tubes (measuring devices show the deflection of electric charge)
mercury vapour tubes (incandescent lamps rich in UV light)
motor meters
neutron capture shielding
pumps
rectifiers
switchgear
thermometers (temperature measurement)
tilt switches
wick liquids in heat transfer pipes
wiring devices

Mercury compounds retain the toxicity of their parent metal. Many are unstable, including *mercurous compounds* in which mercury exists as Hg$^+$ or Hg(I); or as a mercury salt like *mercuric salts* in which mercury exists as Hg^{2+} or Hg(II). Organometallic phenylmercurics are commercially important, but alkyl mercury compounds are rarely manufactured due to their toxicity. Mercury readily reacts with most metals to form amalgams, thus mercury is corrosive to all but a few resistant metals, most noticeably iron.

RELATED TERMS

Amalgams, see *Metals, Alloy*, p.151

Compound, see *Terminology, Compound*, p.234

Fulminate; Fulminating, see *Terminology, Fulminating*, p.238

Liquid, see *Terminology, Liquid*, p.241

Mercury-based pesticide, see *Pesticides*, p.179

Solid, see *Terminology, Solid*, p.247

REGULATORY DEFINITIONS

None

REFERENCES

ACM, AND, HCC, ISD, KOC, KOE, MS1, POG

Metallurgical By-Products

Aluminium dross ■ Aluminium dross, hot ■ Aluminium dross, wet ■ Aluminium dross, wet or hot ■ Aluminium remelting by-products, 4.3 ■ Aluminium residues ■ Aluminium skimmings ■ Aluminium smelting by-products, 4.3 ■ Arsenical dust, 6.1 ■ Arsenical flue dust ■ Cerium, turnings or gritty powder, 4.3 ■ Ferrous metal borings in a form liable to self-heating, 4.2 ■ Ferrous metal cuttings in a form liable to self-heating, 4.2 ■ Ferrous metal shavings in a form liable to self-heating, 4.2 ■ Ferrous metal turnings in a form liable to self-heating, 4.2 ■ Flue dusts, arsenical ■ Flue dusts, poisonous ■ Flue dusts, toxic ■ Iron swarf ■ Lead dross ■ Magnesium dross, hot ■ Magnesium dross, wet ■ Magnesium scrap ■ Steel swarf ■ Zinc ashes, 4.3 ■ Zinc dross, 4.3 ■ Zinc dust, 4.2, 4.3 ■ Zinc dust, pyrophoric ■ Zinc residue, 4.3 ■ Zinc skimmings, 4.3 ■ Zirconium scrap, 4.2

1435 ■ 1436 ■ 1562 ■ 1932 ■ 2793 ■ 3078 ■ 3170

Metallurgy encompasses a wide range of processes including extractive metallurgy (techniques that extract metal from commercially viable sources: ore, scrap, intermediates, and metallurgical residues) and process metallurgy. Each generates many different types of by-products for which the descriptive terminology is often used interchangeably.

EXTRACTIVE METALLURGY BY-PRODUCTS

Extractive metallurgy is the chemical and physical removal of metal from raw materials, the primary source of which are the metal deposits present in the earth's crust where metals exist elementally or as compounds (often oxides and sulphides), or in seawater and brines where they exist as metal salts in solution. *Ore* is any deposit which may be mined profitably. Gold ores may contain as little as 0.001% by weight of gold while iron ore may range up to 60% by weight iron. The remaining materials may be valueless minerals, impurities, or commodities. Generally, extractive metallurgy involves a series of physical steps (e.g., mining, crushing, sorting, flotation) that serve to sort and concentrate the ore prior to one of three metallurgic processes that chemically remove impurities:

- Hydrometallurgy in which wet processes recover metal ions from solution.
- Electrometallurgy in which electrochemistry is used to refine metals.
- Pyrometallurgical processes, such as roasting, smelting, and remelting in which heat is used to induce chemical change.

The processing of bauxite aluminium ore (hydrated alumina oxides) exemplifies these processes. In the Bayer process hot concentrated sodium hydroxide dissolves alumina which is subsequently separated and crystallized as aluminium hydroxide (hydrometallurgy). This is calcined to anhydrous alumina (pyrometallurgy) before being reduced in the Hall process (electrometallurgy) wherein alumina serves as the electrode from which elemental aluminium is deposited on the cathode.

147

Pyrometallurgical smelting occurs in blast, reverbatory, or electric furnaces in which the metal is chemically reduced and separated in elemental or fused form. Remelting is a similar process in which metal is reclaimed from its scrap. Chemicals, including other metals, are added to achieve these separations. As the melt forms, unwanted by-products (i.e., *dross*, slag, and *skimmings*) rise to the surface from which they are drawn, skimmed, or spilled. These by-products solidify on cooling and contain metal oxides, inorganic compounds, flux, and a proportion of the desired metal with which slag is unavoidably associated. Other by-products vaporize and collect in the duct work, bag houses, other air pollution control devices, or are emitted to the environment.

Extractive metallurgical by-products include:

□ *Aluminium smelting by-products* and *aluminium remelting by-products* include dross, slag, skimmings, and spillage which also originate from electrometallurgical melts.

□ *Aluminium residue* includes electrochemical waste such as sludges, potliners, and spent cathodes (carbon and steel), each of which contain some fraction of elemental aluminium.

□ *Salt slags* are generated from aluminum and other metal recycling wherein scrap metal is melted in ovens where molten salts, like sodium and potassium chloride, float on top of the metal to prevent its oxidation. The dross (salt slag) includes metal oxide, salt, and elemental metals.

□ *Lead dross* from heating lead ore (principally lead sulphide) in an oxidizing environment is high in lead sulphate. In fact, this process is one of the methods used commercially to manufacture this acid.

□ *Magnesium dross* from remelting and the pyrometallurgical Pidgeon reduction process can contain residual magnesium along with inert compounds and mixtures of calcium, silicon, aluminium, and magnesium oxides. Magnesium drosses may also be generated when magnesium is added to remove impurities (e.g., bismuth) in other metal melts, such as lead refining.

□ Zinc melts at a relatively low temperature ($907^{\circ}C$) and is removed from its zinc oxide ores pyrometallurgically by being distilled and condensed, or by being roasted, leached, then electrodeposited. *Zinc residues* and *zinc drosses* can be generated from these, as well as remelting processes.

□ Any pyrometallurgical processing of zinc will likely volatilize zinc. In steel-making, for example, where the temperatures are around the melting point of iron ($1536^{\circ}C$), zinc present in the raw materials is condensed as *zinc dust* in the ductwork and bag houses above the furnace.

▫ During copper and lead smelting, arsenic (mostly as arsenic trioxide) vaporizes and condenses in the ductwork to form *flue dust*. Arsenical flue dust is a commercial source of arsenic.

PROCESS METALLURGICAL BY-PRODUCTS

Process metallurgy involves working or shaping metals and alloys through casting, hot and cold working, machining, electroforming, and powder metallurgy. Each of these may produce any number of scrap parts as well as by-products. Metal *scrap* includes chips, turnings, and other waste pieces of material of no further use in a particular process:

▫ *Metal turnings* are the chips or curls of metal machined or cut from a circular piece as it revolves on a lathe.

▫ Boring is the process of cutting or trimming out a hole in a casting. Metal *borings* are the chips resulting from the action of the boring tool when opening a cavity or holes on a metal part.

▫ *Cuttings* are generated when a revolving disk impregnated with an abrasive such as Carborundum,[15] alumina, or diamond is fed edgewise into a metal piece.

▫ Metal *shavings* are generated when a tool used to finish rough-machined work is presented at a tangent to the piece and travels over it removing a thin layer or metal.

▫ *Swarf* is a general term applied to the fine metal particles removed from any metalworking techniques. It can contain metal, abrasive particles from grinding tools, and coolant.

Metal scrap also includes enormous quantities of post-use items such as industrial machinery, vehicles, metal furniture, washing machines, and other consumer equipment, as well as metal residues from electroplating baths, wastewater treatment sludges, and the like.

Dross, slags, residues, sludges, and other by-products may be generated from a wide variety of electroplating, metal coating, alloying, finishing, remelting, and other operations. For example, galvanizing is the process in which zinc is electrodeposited on steel and other ferrous metals as they pass through a bath of molten zinc. Zinc oxide forms on the surface which is periodically skimmed off. This is variously called *zinc skimmings, zinc dross, zinc residue*, and *zinc ash*, terms that might also apply to bottom dross removed from the bath. These by-products may also contain impurities, fluxes, and high concentrations of elemental zinc (20 to 50% zinc in skimmings is common). *Zinc ash* also refers to the less dense zinc oxide fraction separated from elemental zinc in skimmings by being milled.

---

[15] Trademark, Foote Mineral Company.

HAZARDS

Metallurgical by-products present a number of hazards:

- Following processing, reactive metals may exist in fine states of subdivision as ashes, dross, swarf, etc. As such they may oxidize violently, particularly in the presence of water, heat, oils, and cutting fluids. They represent a combustion hazard and can evolve toxic and flammable gases as they oxidize.

- Similarly, finely divided mixed metals undergo electrochemical reactions in contact with one another, sometimes with sufficient heat to ignite surrounding combustible materials or with a dangerous depletion of surrounding oxygen in confined spaces.

- Lead dross (mostly lead sulphate) is acidic.

- Arsenical flue dust and other metal by-products are toxic.

- Residues may be mixed with other compounds and metals such as lead, cadmium, mercury, and uranium.

RELATED TERMS

Ferrous metals, see *Terminology, Metals*, p.242

Magnesium alloys, see *Metals, Alloys*, p.151

Metal, see *Terminology, Metals*, p.242

Powder, see *Terminology, Powder*, p.246 and *Metals, Elemental Products*, p.153

Self-heating, see *Spontaneously Combustible Materials and Division 4.2*, p.226

REGULATORY DEFINITIONS

Aluminium Processing By-Products. The material, consisting of skimmings of virgin aluminium, rising to the surface of impure molten aluminium metal. ICAO A2

Aluminium Smelting/Remelting By-Products. Are the materials, consisting of skimming of virgin aluminium, rising to the surface of impure molten aluminium metal. IATA App. A

Arsenical Dust. Smelter dust which contains large proportions of arsenic. These dusts are hazardous due to their toxic characteristics. ICAO A2

Arsenical Dust (Arsenical Flue Dust). Consists of smelter dust containing large proportions of arsenic. These dusts are hazardous due their toxic characteristics. IATA App. A

Magnesium Scrap. Borings, clippings, scalpings, shavings, sheets or turnings from machining operations or cuttings from thin magnesium metal sheets. The scrap can be ignited by external flame and burns intensely and persistently. It does not heat spontaneously. The scrap may have a bright metal lustre or may be dull and sometimes have a painted surface. ICAO A2

Magnesium Scrap. Is borings, clippings, shavings, sheets, turnings, or scalpings from machining operations or cuttings from thin magnesium metal sheets. The scrap can be ignited by external flame and burns intensely and persistently. It does not heat spontaneously. The scrap may have a bright metal lustre or may be dull and sometimes have a painted surface. IATA App. A

REFERENCES

ADM, ADO, AEO, DOM, DOMA, DOMST, GM, HCC, IMDG, KOE, MEOS, MFP, RW

Metals, Alloys

Alkali metal alloys, liquid, n.o.s., 4.3 ▪ Alkali metal amalgam, liquid, 4.3 ▪ Alkali metal amalgams, 4.3 ▪ Alkali metal amalgam, solid, 4.3 ▪ Alkaline earth metal alloy, n.o.s., 4.3 ▪ Alkaline earth metal amalgam, 4.3 ▪ Barium alloys ▪ Barium alloys, non-pyrophoric ▪ Barium alloys, pyrophoric, 4.2 ▪ Barium amalgams ▪ Caesium alloy (liquid) ▪ Caesium amalgams ▪ Calcium alloy, non-pyrophoric ▪ Calcium alloys ▪ Calcium alloys, pyrophoric, 4.2 ▪ Calcium amalgams ▪ Lithium alloy (liquid) ▪ Lithium amalgams ▪ Magnesium alloys, 4.1 ▪ Magnesium alloys, powder, 4.2, 4.3 ▪ Magnesium amalgams ▪ Potassium alloys, metal ▪ Potassium amalgams ▪ Potassium metal alloys, 4.3 ▪ Potassium metal, liquid alloy ▪ Potassium sodium alloys, 4.3 ▪ Pyrophoric alloy, n.o.s., 4.2 ▪ Rubidium alloy (liquid) ▪ Rubidium amalgam ▪ Sodium alloys (liquid) ▪ Sodium amalgam ▪ Sodium metal, liquid alloy ▪ Strontium alloy, non-pyrophoric ▪ Strontium alloy, pyrophoric ▪ Strontium amalgam

1383 ▪ 1389 ▪ 1392 ▪ 1393 ▪ 1418 ▪ 1420 ▪ 1421 ▪ 1422 ▪ 1854 ▪ 1855 ▪ 1869

Alloys are metallic substances containing two or more elements which are miscible when molten and do not separate when solidified. They may be liquid or solid. This mixture of elements, usually but not necessarily metals, allows careful manipulation of strength, melting point, corrosion resistance, magnetic, thermal, electrical, and other properties; steel, for example, is an alloy of iron and carbon often present with nickel, chromium, copper, aluminium, boron, tungsten, manganese, cobalt, silicon, and other elements.

Alkali metal alloys and *alkaline earth metal alloys* have a wide range of applications. The degree to which the alloys retain the pyrophoric or water-reactive properties of their parent metals depends on their concentration in the alloy, the modifying nature of the alloyed components, and the state of subdivision. Many commercial alloys of this type, present no hazard. Others, such as the potassium-sodium alloys used in heat exchangers, present significant concern. Other alkali and alkaline earth metal alloys include

- Rubidium alloys used in research to mimic the properties of caesium, a rarer metal.
- Strontium-aluminium alloys used in aluminium-silicon casting.
- Magnesium-aluminium alloys (97%:3%) used as photoengraving plates.
- Calcium alloyed to strengthen lead.
- Sodium alloyed to lead to make it more brittle and, therefore, more able to be crushed and powdered. Sodium-lead alloys are used extensively to supply the lead in the manufacture of anti-knock compounds.
- Lithium and other metals alloyed to aluminium, zinc, or manganese to produce lightweight metals for many uses.
- Potassium-sodium alloys used in organic and inorganic synthesis.

AMALGAMS

Amalgams are alloys of mercury with other metals used extensively as chemical reagents and catalysts. The proportion of mercury dictates whether the amalgam is solid or liquid. They include sodium amalgam (Na_xHg_y), used to manufacture sodium hydroxide, and dental amalgams (alloys of mercury with some combination of silver, copper, tin, gold, or silver) used to fill dental cavities.

RELATED TERMS

Alkali metal, see *Terminology, Metals*, p.242

Alkaline earth metal, see *Terminology, Metals*, p.242

Liquid, see *Terminology, Liquid*, p.241

Metal, see *Terminology, Metals*, p.242

Powder, see *Terminology, Powder*, p.246

Pyrophoric, see *Spontaneously Combustible Materials and Division 4.2*, p.226

Solid, see *Terminology, Solid*, p.247

REGULATORY DEFINITIONS

Potassium Sodium Alloys. Mixtures of metallic sodium and potassium that are solid at ordinary temperatures. All mixtures, regardless of physical state, will react vigorously with water and may be self-igniting. The mixtures are all combustible. ICAO A2

Potassium Sodium Alloys. Are mixtures of metallic sodium and potassium that are solid at ordinary temperatures. All mixtures, regardless of physical state, will react vigorously with water and may be self-igniting. The mixtures are all combustible. IATA App. A

REFERENCES

40CFR63, HCC, KOE, MEOS, MH

Metals, Elemental Products

Alkali metal dispersions, 4.3 ▪ Alkaline earth metal dispersion, 4.3 ▪ Aluminium, powder ▪ Aluminium powder, coated, 4.1 ▪ Aluminium powder, pyrophoric ▪ Aluminium powder, uncoated, 4.3 ▪ Aluminium powder, uncoated, non-pyrophoric, 4.3 ▪ Antimony powder, 6.1 ▪ Barium dispersions ▪ Barium, powder ▪ Barium powder, pyrophoric, 4.2 ▪ Beryllium powder, 4.1, 6.1 ▪ Calcium dispersions ▪ Caesium dispersions ▪ Caesium, powder ▪ Caesium powder, pyrophoric ▪ Cerium powder, pyrophoric ▪ Cerium slabs, ingots, or rods, 4.1 ▪ Copper metal powder ▪ Hafnium powder, dry, 4.2 ▪ Hafnium powder, wetted, 4.1 ▪ Iron powder ▪ Iron powder, pyrophoric ▪ Lithium dispersions ▪ Lithium in cartouches ▪ Lithium in cartridges ▪ Magnesium dispersions ▪ Magnesium granules, coated, 4.3 ▪ Magnesium in pellets, turnings, or ribbons, 4.1 ▪ Magnesium powder, 4.2, 4.3 ▪ Metal powder, n.o.s., 4.1, 4.2 ▪ Potassium dispersions ▪ Rubidium dispersion ▪ Sodium dispersion ▪ Strontium, powder ▪ Strontium dispersion ▪ Titanium powder, dry, 4.2 ▪ Titanium powder, wetted, 4.1 ▪ Titanium sponge granules, 4.1 ▪ Titanium sponge powders, 4.1 ▪ Zinc powder, 4.2, 4.3 ▪ Zinc powder, non-pyrophoric, 4.3 ▪ Zinc powder, pyrophoric ▪ Zirconium, dry, coiled wire, finished metal sheets, strip, 4.1 ▪ Zirconium, dry, finished sheets, strip or coiled wire, 4.2 ▪ Zirconium powder ▪ Zirconium powder, dry, 4.2 ▪ Zirconium powder, wetted, 4.1 ▪ Zirconium suspended in a flammable liquid, 3, 3.1, 3.2, 3.3

1308 ▪ 1309 ▪ 1326 ▪ 1333 ▪ 1352 ▪ 1358 ▪ 1383 ▪ 1391 ▪ 1396 ▪ 1418 ▪ 1567 ▪ 1869 ▪ 2008 ▪ 2009 ▪ 2545 ▪ 2546 ▪ 2858 ▪ 2871 ▪ 2878 ▪ 2950 ▪ 3089 ▪ 3189

METAL PARTICLES

Small metal particles are produced by mechanical comminution, chemical reaction, electrolytic deposition, and liquid-metal atomization:

□ A *powder* is a finely divided solid smaller than about 1mm in maximum dimension.

□ Powders can be compressed into *pellets*.

□ *Granules* are those particles having approximately equidimensional, nonspherical shapes. The term can also apply to small, coarse metal particles often formed by pouring molten metal through a screen into water and agitating violently as the metal solidifies.

□ *Dusts* are synonymous with powders that are sufficiently fine to become airborne.

Metal powders are used in powder metallurgy chiefly to manufacture precision metal parts that are close to the final dimensions of the finished product and require little machining. Powder metallurgy subjects metals and alloys mixed with a lubricant to pressure and heat wherein bonds between the particles are made. Powdered metals have many other uses including functioning as catalysts, fuels in pyrotechnics and explosives (e.g., aluminium and magnesium), ceramics, emission control technologies, and chemical reagents.

PROTECTIVE MEASURES

Alkali metals and alkaline earth metals are often prepared in commerce as dispersions in which the metal is suspended in some liquid, often flammable, to inhibit reaction with atmospheric oxygen or moisture. For example, sodium is transported in toluene, xylene, naphtha, and kerosene. Metals may also be hermetically insulated during handling. *Lithium cartouches syn. lithium cartridges* are the hermetically sealed cartridges made of copper in which lithium is supplied under vacuum or in an inert, dry atmosphere.

METAL PRODUCTS

Powder metallurgy and other types of process metallurgy (i.e., casting, working, machining, and electroforming) result in a vast number of finished or intermediate products:

- *Ingots*, masses of metal cast in a mould into shapes convenient for subsequent handling, storage, or transportation. They differ from castings in that they must be remelted, rolled, or forged to be made useful. They are generally of square section for rolling into bars or billets or for use in small forgings; or rectangular section to be rolled into slabs for sheets or strips. Larger ingots may be octagonal or duo-hexagonal. They are typified by having two opposing surfaces of different cross-sections with slightly tapering sides.
- *Slabs*, semi-finished blocks cut from a rolled ingot with a width generally at least twice its thickness (around 40mm minimum). Slabs are intended for further processing into plate, sheet, foil (sheet metal around 0.15 to 6.35mm thick), strip, or other flat products.
- *Sheets*, rolled products flattened or stretched to appropriate standard thicknesses and cross-sectional areas. *Strips* are slightly smaller than sheets. For example, a piece of hot rolled carbon steel 5 to 6mm thick and between 15 and 30cm wide is a strip, but greater than 30 to 120cm becomes a sheet. (At less than 15cm it is a *bar* and at greater than 120cm it is *plate steel*.)
- *Rods*, metal cast, rolled, or extruded into lengths between wire and shaftings; they may be any shape but are usually round.
- *Wires* are long, slender, flexible threads of metal drawn or extruded from their source.
- *Ribbons*, long, thin strips of metal, often irregular and curved.

TITANIUM SPONGE

Titanium is produced commercially by converting titanium oxide ore to titanium tetrachloride, then reducing it to produce a crude titanium product, *titanium sponge*. Titanium sponge can react with carbon dioxide to generate oxygen.

HAZARDS

Many metals oxidize in contact with the air or moisture. For reactive metals, this reaction may be particularly violent and evolve high heat. When in contact with water, hydrogen gas is liberated which will ignite if temperatures are sufficiently high. Alternatively other combustibles may be similarly ignited. The finer the state of subdivision of a reactive metal, the greater the surface area exposed upon which reactions can take place. Thus powders and dust present significant hazards. In the presence of oxidizers, they may form explosive mixtures.

RELATED TERMS

Alkali metal, see *Terminology, Metals*, p.242
Alkaline earth metal, see *Terminology, Metals*, p.242
Dispersion, see *Terminology, Dispersion*, p.235
Flammable, see *Flammable Solids and Division 4.1*, p.99
Metal, see *Terminology, Metals*, p.242

Microns, see *Units of Measure*, p.258
Powder, see *Terminology, Powder*, p.246
Pyrophoric, see *Spontaneously Combustible Materials and Division 4.2*, p.226
Self-heating, see *Spontaneously Combustible Materials and Division 4.2*, p.226
Sponge, see *Terminology, Sponge*, p.248
Wetted, see *Terminology, Wetted*, p.253

REGULATORY DEFINITIONS

Alkali Metal Dispersion; Alkaline Earth Metal Dispersion Finely divided alkali or alkaline earth metal, e.g. metallic sodium suspended in a flammable liquid such as toluene, xylene, naphtha, kerosene, etc. Reacts violently with moisture, water or acids, evolving hydrogen, which may be ignited by the heat of reaction. IMO 4324

Aluminium Powder. The uncoated powder may evolve hydrogen in contact with water, and finely divided dust may be ignited by naked lights or sparks. Coated aluminium powders which have been treated with oils or wax for printing or paint purposes are generally not dangerous. ICAO A2

Aluminium Powder. The unpolished powder may evolve hydrogen in contact with water, and finely divided dust may be ignited by naked lights or sparks. Polished aluminium powders which have been treated with oils or wax for printing or paint purposes are not generally dangerous. IATA App. A

Zirconium Suspended in a Flammable Liquid. Very finely divided metallic zirconium which is usually suspended in some highly volatile and flammable liquid. If spilled, the material is liable to self-ignition. ICAO A2

Zirconium Suspended in a Flammable Liquid. Consists of very finely divided metallic zirconium which is usually suspended in some highly volatile and flammable liquid. If spilled, the material is liable to self-ignition and therefore can be shipped only in very limited quantities when specially packed. IATA App. A

REFERENCES

15CFR700, ADM, AEO, AMR, DOM, DOMST, HCC, HOTC, IMDG, KOE, MH, MH14

Metals, Inorganic Compounds

Alkali metal amides, 4.3 ■ Barium compound, n.o.s., 6.1 ■ Beryllium compound, n.o.s., 6.1 ■ Cadmium compound, 6.1 ■ Copper compounds ■ Hydrides, metal, water-reactive, n.o.s., 4.3 ■ Lead compound, soluble, n.o.s., 6.1 ■ Metal carbonyls, n.o.s., 6.1 ■ Metal hydrides, n.o.s., 4.1, 4.3 ■ Metallic substance, n.o.s., 4.2, 4.3 ■ Metal sulphides concentrates ■ Thallium compound, n.o.s., 6.1 ■ Vanadium compound, n.o.s., 6.1

1390 ■ 1409 ■ 1564 ■ 1566 ■ 1707 ■ 2291 ■ 2570 ■ 3208 ■ 3209 ■ 3281 ■ 3285

Metals form an enormous number of compounds (inorganic and organometallic) with a wide variety of hazards of which many retain the toxic nature of their parent metal:

- Alkali metals react with ammonia to yield *alkali metal amides* such as sodium amide, a reactive substance used as a reagent. They decompose in water or acid to evolve ammonia and produce very caustic alkali solutions.
- *Metal carbonyls* are compounds of carbon monoxide with a metal. Metal carbonyls such as nickel carbonyl easily lose the carbon monoxide, a flammable and poisonous gas, particularly on heating.
- *Metal hydrides* include lithium aluminium hydride used extensively in organic synthesis. They are highly reactive.
- *Metal sulphides* are naturally occurring forms of many metals (e.g., lead sulphide and nickel sulphide). Prior to direct chemical treatment of these ores to yield the metal they are concentrated by crushing, screening, and other physical methods.
- Many other metal compounds exist including the following often toxic examples: *barium compounds*, eg., barium bromide; *beryllium compounds*, eg., beryllium hydroxide; *cadmium compounds*, e.g., cadmium iodide; *copper compounds* like copper hydroxide, used as a pigment, in paper manufacturing, and as a pesticide; *lead compounds*, like the soluble lead fluorosilicate; *thallium compounds*, e.g., thallium sulphide; and *vanadium compounds*, e.g., vanadium dichloride.
- *Metallic substances* include any metal compound, be it inorganic or organometallic.

RELATED TERMS

Alkali metal, see *Terminology, Metals*, p.242

Aluminum liquid, see *Paints and Coatings*, p.175

Compounds, see *Terminology, Compound*, p.234

Concentrates, see *Terminology, Concentrates*, p.234

Inorganic, see *Terminology, Inorganic*, p.241

Iron oxide, spent (obtained from coal gas purification), see *Coal*, p.44

Iron sponge, spent (obtained from coal gas purification), see *Coal*, p.44

Metal, see *Terminology, Metals*, p.242

Metal sulphides, see also *Solid Bulk Materials*, p.221

Powder, see *Terminology, Powder*, p.246

Soluble, see *Terminology, Solutions*, p.247

Water-reactive, see *Dangerous When Wet Materials and Division 4.3*, p.58

Miscellaneous Dangerous Goods (including Class 9)

Aviation regulated liquid, n.o.s., 9 ▪ Aviation regulated solid, n.o.s., 9 ▪ Consumer commodity, ORM-D, 9 ▪ Dangerous goods in apparatus ▪ Dangerous goods in machinery ▪ Other regulated substance, liquid, n.o.s., 9 ▪ Other regulated substance, solid, n.o.s., 9 ▪ Security type attaché cases incorporating dangerous goods, for example lithium batteries or pyrotechnic material

3077 ▪ 3082 ▪ 3334 ▪ 3335 ▪ 8000 ▪ 8001 ▪ 8027

A large number of dangerous goods do not fit neatly into hazard classes 1 through 8, they contain a variety of different dangerous materials, or, because of their size or use, are transported with minimal or no regulation.

CONSUMER COMMODITY

Consumer commodities include dangerous goods packaged in limited small quantities that are intended or suitable for use by the public, as opposed to the larger quantities of dangerous goods in use by industry, government, hospitals, etc. Because of their small size, packages of these materials, which may include small cans of paint, tubes of glue, containers of cosmetics, and household cleaning supplies, are subject to less stringent transportation regulations. The acronym *ORM* stands for *other regulated material* and is a term used by the United States Department of Transportation. Originally, the U.S. DOT used five types of ORM (A through E)[16] although the only classification remaining in use is *ORM-D* for consumer commodities.

DANGEROUS GOODS IN MACHINERY AND APPARATUS

Certain pieces of *apparatus* and *machinery* contain dangerous goods that are integral to their function; e.g., thallium amalgams used as mercury substitutes in electrical switches for low temperature situations; alcohol and mercury in thermometers; and batteries in pacemakers. If the quantities of dangerous goods are below specified limits and if they are compatible with other contents, the items may be shipped under a reduced body of regulation.

Security-type attaché cases or briefcases are equipped with devices that deter theft, including batteries that give powerful electric shocks to would-be thieves or provide power to run remote control activation systems, alarms, and sirens; tear gas devices; and small pyrotechnics designed to burst and mark either the thief or the contents with traceable or visible dyes and paints.

---

[16] ORM-As included anaesthetic, irritating, and noxious materials that cause annoyance or discomfort; ORM-B covered materials that could cause significant damage to a vehicle from leakage; ORM-C included items such as life rafts, escape slides, bleaching powder; ORM-E included hazardous wastes, hazardous substances, and PCBs.

OTHER REGULATED SUBSTANCES

Other regulated substances include those chemicals which cause annoyance or discomfort to the passengers or crew of a transport vehicle, particularly in the confined spaces of aircraft (*aviation regulated materials*). While these chemicals may not be toxic according to the transportation definitions they may have pungent odours, irritate eyes, or exhibit other noxious effects. Often the chemical is emitted as a vapour or gas from a leaking package or by a chemical reaction that can occur during normal transportation. Were some of these chemicals to be transported in greater concentrations or under different physical conditions they might be regulated under hazard classes 1 through 8. These substances include

- Zinc dithionite, a widely-used bleach component which, on contact with moisture, generates sulphur dioxide, a sharp-smelling and irritating gas.
- Acetaldehyde ammonia which can decompose and emit ammonia on contact with moisture.
- Dibromodifluoromethane, a highly volatile liquid with irritating vapours.
- Allergens in castor beans and their products.
- Benzaldehyde which in low concentrations is not toxic by definition but it is extremely pungent, smelling of almonds, and is often used in perfumes.

RELATED TERMS

Dangerous goods, see *Dangerous Goods and Hazardous Materials*, p.54
Liquid, see *Terminology, Liquid*, p.241
Lithium batteries, see *Batteries*, p.27
Other regulated substance, aromatic extract or flavouring, see *Extracts*, p.86
Pyrotechnic material, see *Spontaneously Combustible Materials and Division 4.2*, p.226
Solid, see *Terminology, Solid*, p.247

REGULATORY DEFINITIONS

Class 9 substances and articles (miscellaneous dangerous substances and articles) are substances and articles which during transport present a danger not covered by other classes. This class includes, inter alia, substances that are transported or offered for transport at temperatures equal to or exceeding 100°C in a liquid state or at temperatures equal to or exceeding 240°C in a solid state. UN 2.9.1

In this Code, *class 9* comprises: .1 substances and articles not covered by other classes which experience has shown, or may show, to be of such a dangerous character that the provisions of part A of chapter VII of the International Convention for the Safety of Life at Sea, 1974, as amended, should apply; these include substances that are transported or offered for transport at temperatures equal to or exceeding 100°C and in a liquid state, and solids that are transported or offered for transport at temperatures equal to or exceeding 240°C; and .2 harmful substances not subject to the provisions of part A of chapter VII of the aforementioned Convention, but to which the provisions of Annex III of the International Convention for the Prevention of Pollution from Ships, 1973, as modified by the Protocol of 1978 relating thereto (MARPOL 73/78), apply. IMO Class 9, 1.1

Articles and substances which during air transport present a danger not covered by other classes. [Text continues.] ICAO 2-9.1

Some examples of articles in Class 9 are: engines, internal combustion; life-saving appliances, self-inflating; battery-powered equipment or vehicle. Some examples of substances in Class 9 are: blue, brown or white asbestos; carbon dioxide (Dry ice); Environmentally hazardous substance, liquid/solid, n.o.s.; Zinc dithionite. ICAO 2-9.1

For the purposes of this subchapter, *miscellaneous hazardous material (Class 9)* means a material which presents a hazard during transportation but which does not meet the definition of any other hazard class. This class includes: (a) Any material which has an anesthetic, noxious or other similar property which could cause extreme annoyance or discomfort to a flight crew member so as to prevent the correct performance of assigned duties; or (b) Any material that meets the definition in Sec. 171.8 of this subchapter for an elevated temperature material, a hazardous substance, a hazardous waste, or a marine pollutant. US 173.140

Articles and substances which during air transport present a danger not covered by other classes. Included in this class are: Aviation regulated solids or liquids, Magnetized material and miscellaneous articles and substances. IATA 3.9.1

Examples included in this class: Asbestos; Carbon dioxide, solid (dry ice); Consumer commodity; Chemical and First aid kits; Environmentally hazardous substance; Life-saving appliances; Engines, internal combustion; Vehicles (flammable gas powered), Vehicles (flammable liquid powered); Polymeric beads; Battery-powered equipment or vehicles; Zinc dithionite; Genetically modified organisms and micro-organisms which are not infectious substances but which are capable of altering animals, plants, or microbiological substances in a way which is not normally the result of natural reproduction. IATA 3.9.14

Aviation-Regulated

Aviation regulated solid or liquid: Any material which has narcotic, noxious or other properties such that, in the event of spillage or leakage on an aircraft, extreme annoyance or discomfort could be caused to crew members so as to prevent the correct performance of assigned duties. ICAO 2-9.1

Aviation Regulated Solid or Liquid Any material which has narcotic, noxious, irritating or other properties such that, in the event of spillage or leakage on an aircraft, could cause extreme annoyance or discomfort to crew members so as to prevent the correct performance of assigned duties. The materials included under this proper shipping name must not meet any of the definitions for Classes 1 through 8. IATA 3.9.1.1

Consumer Commodity

Consumer Commodity. A material which is packed and distributed in a form intended or suitable for retail sales for the purposes of personal care or household use. ICAO A2

Consumer commodity means a material that is packaged and distributed in a form intended or suitable for sale through retail sales agencies or instrumentalities for consumption by individuals for purposes of personal care or household use. This term also includes drugs and medicines. US 171.8

ORM means other regulated material. See Sec. 173.144 of this subchapter. US 171.8

Consumer commodities are materials that are packaged and distributed in a form intended or suitable for retail sale for purposes of personal care of household use. These include items administered or sold to patients by doctors or medical administrations. IATA Pack. Instr. 910

References
49CFR, EAC, HCC, HMW

Nitrocellulose Products

Box toe gum ■ Celluloid, in blocks, rods, rolls, sheets, tubes, etc. (except scrap), 4.1 ■ Celluloid scrap, 4.2 ■ Collodion ■ Collodion cottons ■ Fabrics impregnated with weakly nitrated nitrocellulose, n.o.s., 4.1 ■ Fabrics impregnated with weakly nitrated nitrocellulose, n.o.s. (including toe puffs, nitrocellulose base), 4.1 ■ Fibres impregnated with weakly nitrated nitrocellulose, n.o.s., 4.1 ■ Fibres impregnated with weakly nitrated nitrocellulose, n.o.s. (including toe puffs, nitrocellulose base), 4.1 ■ Film scrap ■ Films, nitrocellulose base from which gelatin has been removed ■ Films, nitrocellulose base, gelatin coated, except scrap, 4.1 ■ Lacquer base or lacquer chips, nitrocellulose, dry ■ Lacquer base or lacquer chips, plastic, wet with alcohol or solvent ■ Lacquer, liquid ■ Nitrated paper (unstable) ■ Nitrocellulose membrane filters, 4.1 ■ Plastics, nitrocellulose-based, self-heating, n.o.s., 4.2 ■ Pyroxylin plastic ■ Pyroxylin solution ■ Pyroxylin solution or solvent ■ Pyroxylin solvent, n.o.s. ■ Toe puffs, nitrocellulose base

1324 ■ 1353 ■ 2000 ■ 2002 ■ 2006 ■ 3270

Cellulose (9004-34-6) is a polymer consisting of the basic unit $C_6H_{10}O_5$ repeated a thousand times or more. It is the principal constituent of vegetable cell walls accounting for 30% of all vegetable matter. Nitrating cellulose with mixed acid (nitric acid catalyzed with sulphuric acid) results in a series of products collectively called *nitrocellulose* (9004-70-0) *syn. nitrocotton* (in reference to a major commercial source of cellulose: cotton) in which from one to three nitro groups ($-NO_2$) replace one or more of the three hydroxyl groups ($-OH$) present on each polymeric unit of cellulose. Nitrocelluloses range from around 7.65% to a theoretical maximum of 14.14% nitrogen by weight which corresponds to cellulose dodecanitrate. Low nitrogen forms (*pyroxylin*) are used extensively as the basis of coatings, plastics, and fibres while the higher nitrogen forms are used as explosives. Nitrocellulose is insoluble in water but easily dissolved in many organic solvents.

PLASTICS
Nitrocellulose plastics are quite resistant to common acid and alkali attack at ambient temperatures, are noncompressible, transparent in thin laminae, difficult to twist, and extremely resistant to tearing. The most well-known nitrocellulose plastic is *celluloid*,[17] the first plastic, made by plasticizing nitrocellulose with camphor. Celluloid softens at 80°C allowing easy shaping and moulding. Rigidity returns on cooling. As a consequence of these excellent properties, the manufacture of celluloid blocks, rolls, rods, sheets, and tubes for subsequent transformation into hundreds of consumer items was common particularly around the end of the 19th century. Nitrocellulose may be transported mixed with pigment for manufacture in plastics. Plasticizers may be added to aid in processing and handling.

---

[17] Trademark, Hoechst Celanese Corporation.

Nitrocellulose films for photographic purposes include a nitrocellulose plastic (originally celluloid) coated with photosensitive chemicals suspended in emulsions or gelatins. Celluloid scraps (including *film scrap*) are the trimmings, clippings, and other waste generated during manufacture or processing.

OTHER USES

Nitrocellulose has many other applications including

- Explosives.
- Fibres and fabrics which are impregnated with weakly nitrated nitrocellulose to hold pyrotechnics, bag charges, and explosives.
- *Nitrated papers* which are based on highly purified wood pulp and other fibres used to manufacture the highest grades of nitrocellulose for use as celluloid, films, and lacquers.
- *Nitrocellulose membrane filters*, nitrated papers used as membranes in protein engineering and the electrophoresis (electrical separation of particles in a liquid) of DNA and polymeric samples.
- Adhesives. *Box toe gum*, for example, is a nitrocellulose adhesive used to stiffen the toe area of shoe uppers (*toe puffs*).
- As a lacquer base for automotive and other coatings.
- *Collodion syn. collodion cotton*, a solution of nitrocellulose in ether or alcohol used as a solvent and to coat wounds.
- As a precursor for resins in nail polish.

HAZARDS

Nitrocellulose is extremely flammable as pyroxylin and explosive in its higher nitrogen forms. In the early days of manufacture, many celluloid factories and warehouses were destroyed by fire. There were reports of the finished products (billiard balls, combs, buttons, film, etc.) spontaneously igniting. In reality, nitrocellulose does not spontaneously ignite at ambient temperatures; it readily ignites in contact with an ignition source when combustion proceeds extremely rapidly to the point of explosion. In addition, it decomposes at temperatures achievable in normal manufacturing, storage, and use, the heat of which is sufficient to ignite surrounding materials, particularly packaging materials such as paper, string, etc. In turn, these materials supply an ignition source for nitrocellulose.

As always for reactive materials, a high relative surface area exposed to the air, moisture, etc., increases the chance of reaction. Decomposition of nitrocellulose proceeds with age and is encouraged by exposure to ultraviolet (UV) radiation from sunlight and other sources. Nitrocellulose wetted or in mixtures or solutions with other materials often presents a decreased risk. Dry nitrocellulose or nitrocellulose products with exposed nitrocellulose surfaces present the greatest risk of reaction.

RELATED TERMS

Fabrics; Fibres, see *Fibres and Fibrous Products*, p.90

Flammable liquid, n.o.s., see *Flammable Liquids and Class 3*, p.96

Lacquer base or lacquer chips, plastic, wet with alcohol or solvent, see also *Paints and Coatings*, p.175

Lacquer liquid, see also *Paints and Coatings*, p.175

Paint, see *Paints and Coatings*, p.175

Pyroxylin cement, see *Adhesives*, p.1

Self-heating, see *Spontaneously Combustible Materials and Division 4.2*, p.226

Solution, see *Terminology, Solutions*, p.247

Solvent, see *Solvents*, p.224

Wetted, see *Terminology, Wetted*, p.253

REGULATORY DEFINITIONS

Fibres or Fabrics Impregnated with Weakly Nitrated Nitrocellulose, n.o.s. Toe board used in the manufacture of boots and shoes. When involved in a fire, evolves toxic fumes; in closed compartments, these fumes may form an explosive mixture with air. IMO 4144-1

Films, Nitrocellulose Base. A type of film which consists mainly of nitrocellulose. As such the material has a low ignition temperature and burns rapidly when ignited, evolving gases which are toxic. When new and in good condition the film is reasonably stable and free from liability to spontaneous heating and combustion. Film that has deteriorated badly becomes very unstable and may be liable to spontaneous heating unless kept under water. ICAO A2

Films, Nitrocellulose Base. This type of film consists essentially of nitrocellulose. As such the material has a comparatively low ignition temperature and burns with great rapidity when ignited. Also, when burning, the material evolves gases that are toxic. When new and in good condition, the film is reasonably stable and free from liability to spontaneous heating and combustion. Film that has deteriorated badly becomes quite unstable and may be liable to spontaneous heating unless kept under water. IATA App. A

Lacquer Base or Lacquer Chips, Nitrocellulose, Dry. It may consist of a colloided solid mixture of nitrocellulose, pigment, gums, and plasticizer. ICAO A2

Lacquer Base or Lacquer Chips, Dry. May consist of a colloided solid mixture of nitrocellulose, pigment, gums, and plasticizer. Those containing nitrocellulose are highly flammable. IATA App. A

Pyroxylin Solution. Pyroxylin (nitrocellulose) or soluble cotton dissolved in amyl acetate or other organic solvents. Pyroxylin solution is used as a basis for the manufacture of lacquer, leather coating compounds, leather substitutes, cements, etc. It is generally more viscous than ordinary lacquers. ICAO A2

Pyroxylin Solution. Consists of pyroxylin (nitrocellulose) or soluble cotton dissolved in amyl acetate or other organic solvents. Pyroxylin solution is used as a basis for the manufacture of lacquer, leather coating compounds, leather substitutes, cements, etc. It is generally more viscous than ordinary lacquers. IATA App. A

REFERENCES

29CFR520, CIM, GOCT, HCC, KOC, KOE, MH14, TDO, TOE, VNS

Organic Compounds

Aldehydes, n.o.s., 3, 3.1, 3.2, 3.3, 6.1 ■ Alkylphenols, liquid, n.o.s. (including C_2-C_{12} homologues), 8 ■ Alkylphenols, solid, n.o.s. (including C_2-C_{12} homologues), 8 ■ Amines, liquid, n.o.s., 3, 8 ■ Amines, n.o.s., 3, 3.1, 3.2, .3.3, 8 ■ Amines, solid, n.o.s., 8 ■ Butylphenols, liquid, n.o.s., 8 ■ Butylphenols, solid, n.o.s., 8 ■ Chlorinated paraffins (C_{10} - C_{17}) ■ Chlorocarbonates, n.o.s., 3, 6.1, 8 ■ Chloroformates, n.o.s., 3, 6.1, 8 ■ Chloropicrin mixture, n.o.s., 6.1 ■ Cyanides, organic, flammable, toxic, n.o.s. ■ Cyanides, organic, toxic, flammable, n.o.s. ■ Cyanides, organic, toxic, n.o.s. ■ Esters, n.o.s., 3, 3.2, 3.3 ■ Ethers, n.o.s., 3, 3.1, 3.2, 3.3 ■ Isocyanates, n.o.s., 3, 3.1, 3.2, 3.3, 6.1 ■ Isocyanate solution, n.o.s., 3, 3.1, 3.2, 3.3, 6.1 ■ Ketones, liquid, n.o.s., 3, 3.2, 3.3 ■ Mercaptan mixture, liquid, n.o.s., 3, 3.1, 3.2, 3.3, 6.1 ■ Mercaptans, liquid, n.o.s., 3, 3.1, 3.2, 3.3, 6.1 ■ Nitriles, liquid, n.o.s., 6.1 ■ Nitriles, n.o.s., 3, 3.1, 3.2, 6.1 ■ Nitriles, solid, n.o.s., 6.1 ■ Polyamines, n.o.s., 3.1, 3.2, 3.3, 8

1988 ■ 1224 ■ 1228 ■ 1583 ■ 1989 ■ 2206 ■ 2430 ■ 2478 ■ 2733 ■ 2734 ■ 2735 ■ 2742 ■ 3071 ■ 3080 ■ 3145 ■ 3259 ■ 3271 ■ 3272 ■ 3273 ■ 3275 ■ 3276 ■ 3277

Historically, *organic* chemicals were those hydrocarbons and hydrocarbon derivatives from animal or plant origins, in contrast to *inorganic* compounds with a mineral origin. The distinction between organic and inorganic classes of chemical substances remains even though many organic and inorganic compounds are manufactured synthetically.

ALDEHYDES

Aldehydes are those chemicals with the RCHO group, where *R* is any organic group. They are generally reactive, present a combustion hazard, and can be toxic. They include benzaldehyde and acetaldehyde.

ALKYLPHENOLS

Alkylphenols are compounds in which noncyclic carbon chains are attached to a phenol (C_6H_5OH) group. The simplest members are the isomers of methyl phenol *syn.* cresol (CH_3-C_6H_4OH). The *butylphenols* are the group of isomeric alkylphenols in which the butyl group (C_4H_9-) is attached to the phenol group, such as o-butylphenol.

AMINES

Amines contain a functional group based on ammonia (NH_3) in which any of the three hydrogens are replaced by a bond to an organic species. *Polyamines* are those with more than one substituted ammonia group. Amines and polyamines, which include diethylamine and diethylenetriamine, are characteristically basic. They are corrosive to most metals, especially copper.

CHLORINATED PARAFFINS

Paraffins are the alkane hydrocarbons (straight or branch chained, noncyclic, aliphatic hydrocarbons) the simplest of which is methane, CH_4, a gas. With more carbons, paraffins become liquid and eventually waxy solids. *Chlo-*

rinated paraffins are those in which one or more hydrogens are replaced by a chlorine. C_{10}-C_{17} homologues are not flammable and are used as flame retardants, plasticizers, and detergents.

CHLOROFORMATES

The *chloroformate* group (-O-COCl) *syn.* the *chlorocarbonate* group, results in compounds such as ethylchloroformate and chloromethylchloroformate which are often flammable, highly irritating and corrosive to the skin, and toxic. They react and decompose with water or heat to form hydrogen chloride gas.

CHLOROPICRIN

Chloropicrin (76-06-2) is an extremely toxic chemical with a very strong, irritating odour. In addition to its use in organic synthesis, pesticides, and tear gas it is mixed in small quantities with other toxic but nonodoriferous gases to provide a warning of their presence if they are released.

ESTERS

Esters are organic compounds derived from organic or inorganic acids in which the hydrogen has been replaced by an organic group, as exemplified by the catalyzed reaction of acetic acid with ethanol to produce ethylacetate. Esters based on lower molecular weight precursors are flammable liquids.

ETHERS

Ethers include the group ROR' where *R* and *R'* are any organic group. The simplest ethers are dimethyl ether and ethyl methyl ether. The lighter molecules are often gaseous, while all are highly flammable.

ISOCYANATES

Isocyanates include the group -NCO in which both the oxygen and the nitrogen are bound to the carbon with a double bond. They include ethyl isocyanate and toluene diisocyanate which are both characteristically toxic and flammable. Often toxic, isocyanates react with water to form carbon dioxide.

KETONES

Ketones are liquid compounds in which the carbon in a carbonyl group (-C=O) is attached to two alkyl groups. The simplest member of these is acetone. Ketones are usually flammable and present a degree of toxicity.

MERCAPTANS

Mercaptans, or more properly *thiols*, are those organic compounds with a sulphur-hydrogen group (-SH) attached to a carbon chain. The simplest in the group is methyl mercaptan. Mercaptans are often flammable, toxic, and extremely pungent.

NITRILES

Nitriles syn. organic cyanides are organic compounds with the -CN group containing a triple bond between the carbon and nitrogen. They are toxic and often flammable. The simplest member of the group is acetonitrile. Inorganic nitriles, called *cyanides*, are common. On contact with acids, hydrogen cyanide, a highly toxic gas, is evolved.

RELATED TERMS

Corrosive, see *Corrosives and Class 8*, p.47

Cyanides, see *Inorganic Compounds*, p.128

Flammable, see *Flammable Liquids and Class 3*, p.96

Homologue, see *Terminology, Homologues*, p.239

Liquid, see *Terminology, Liquid*, p.241

Mixture, see *Terminology, Mixture*, p.243

Organic, see *Terminology, Organic*, p.244

Paraffin, see *Terminology, Hydrocarbons*, p.240

Solid, see *Terminology, Solid*, p.247

Toxic, see *Toxic Substances and Division 6.1*, p.255

Water-reactive, see *Dangerous When Wet Materials and Division 4.3*, p.58

REGULATORY DEFINITIONS

Isocyanates, n.o.s. or Isocyanate Solution, n.o.s. These include a number of chemical products used in the manufacture of plastic foams, synthetic rubber, etc. Some are sufficiently toxic or lachrymatory to need classification as toxic substances, particularly isocyanates in pure form. Others may need to be classified as flammable liquids, depending on their characteristics, and a number may not be subject to these Instructions. ICAO A2

Isocyanates include a number of chemical products used in the manufacture of plastic foams, synthetic rubber, etc. Some are sufficiently toxic or lachrymatory to need classification as toxic substances, particularly isocyanates in pure form. Others may need to be classified as flammable liquids, depending on their characteristics, and a number may not be subject to these regulations. IATA App. A

REFERENCES

HCC, KOE

Organometallics and Related Compounds

Alcoholates solution, n.o.s. in alcohol, 3, 3.1, 3.2, 8 ▪ Alkali metal alcoholates, n.o.s., 4.2, 8 ▪ Alkaline earth metal alcoholates, n.o.s., 4.2 ▪ Alkyl aluminium halides ▪ Aluminium alkyl halides, 4.2, 4.3 ▪ Aluminium alkyl hydrides, 4.2, 4.3 ▪ Aluminium alkyls, 4.2, 4.3 ▪ Grignard solution ▪ Lithium alkyls, 4.2, 4.3 ▪ Magnesium alkyls, 4.2, 4.3 ▪ Metal alkyl halides, n.o.s., 4.2, 4.3 ▪ Metal alkyl hydrides, n.o.s., 4.2, 4.3 ▪ Metal alkyls, n.o.s., 4.2, 4.3 ▪ Metal alkyl, solution, n.o.s., 3 ▪ Metal aryl halides, n.o.s., 4.2, 4.3 ▪ Metal aryl hydrides, n.o.s., 4.2, 4.3 ▪ Metal aryls, n.o.s., 4.2, 4.3 ▪ Metal salts of organic compounds, n.o.s., 4.1 ▪ Organic compounds of arsenic, n.o.s., 6.1 ▪ Organoarsenic compound, n.o.s., 6.1 ▪ Organometallic compound dispersion, n.o.s., 3, 4.3 ▪ Organometallic compound, n.o.s., 3, 4.3, 6.1 ▪ Organometallic compound solution, n.o.s., 3, 4.3 ▪ Organophosphorus compound, n.o.s., 3, 6.1 ▪ Organotin compound, liquid, n.o.s., 6.1 ▪ Organotin compound, solid, n.o.s., 6.1 ▪ Pyrophoric organometallic compound (liquid), n.o.s., 4.2, 4.3 ▪ Pyrophoric organometallic compound, n.o.s., 4.2, 4.3 ▪ Pyrophoric organometallic compound (solid), n.o.s., 4.2, 4.3 ▪ Triaryl phosphates, n.o.s.

2003 ▪ 2445 ▪ 2788 ▪ 3049 ▪ 3050 ▪ 3051 ▪ 3052 ▪ 3053 ▪ 3076 ▪ 3146 ▪ 3181 ▪ 3182 ▪ 3203 ▪ 3205 ▪ 3206 ▪ 3207 ▪ 3274 ▪ 3278 ▪ 3279 ▪ 3280 ▪ 3282

A large number of compounds exist between the branches of organic and inorganic chemistry that involve metals and metalloids bonding with organic compounds, including organometallics, soaps, and alkoxides.

ORGANOMETALLICS

Organometallic compounds are compounds in which a metal is attached directly to a carbon atom in an organic molecule. The metal can be thought of as a substitute for a hydrogen atom. Almost all metals form organometallic compounds, thus their number renders it impossible to generalize about their properties and hazardous characteristics. They are used extensively as catalysts and as chemical reagents. Examples include:

- *Grignard* reagents are a class of organomagnesium compounds used in organic syntheses. They take the form RMgX where *R* is an organic group and *X* a halogen. They react vigorously with water and are, for that reason, often handled in ether or other solvents.
- *Metal alkyls* involve alkyl groups attached to metal such as the *aluminium alkyl* triethylaluminium which is a pyrophoric liquid used as a catalyst; the *magnesium alkyl* diethylmagnesium; and the *lithium alkyl* methyllithium. These alkyls ignite in air or carbon dioxide and react violently with water and acids, halogens, alcohols, and amines to evolve flammable gas.
- *Metal aryls* such as phenyllithium, an extremely flammable material, involve aryl groups attached to metals. They react similarly to metal alkyls.
- Metal halides include the *metal alkyl halide* ethylaluminium sesquichloride and the *metal aryl halide* phenylmagnesium chloride both

of which are dangerous combustion risks. They react similarly to metal alkyls. Grignard reagents are metal halides.

- Metal hydrides, which normally involve strongly electropositive metals such as sodium, lithium, boron, and aluminium, form hydrides like lithium hydride. *Metal alkyl hydrides* or *metal aryl hydrides* are elaborate molecules called *complexes* including $HIr(CO)(CH_3)$-$(P(C_6H_5)_3)_2Cl$.

- *Organotin compounds*, a series of toxic alkyl tin compounds, such as butyl tin chloride and dibutyltin oxide used as plastic stabilizers and catalysts.

RELATED COMPOUNDS

Certain metalloids, like phosphorus and arsenic, when compounded with organic groups, are classed as organometallics. *Organophosphorus compounds* include four organic groups:

- Phospholipids which include proteins and nucleic acids.
- Esters of phosphinic and phosphonic acids.
- Pyrophosphates, such as tetraethyl pyrophosphate, of which all are extremely toxic.
- Other phosphoric esters such as *triaryl phosphates* which take the form $PO(OR)_3$ where R is an aryl group (a condensed ring of carbons). A simple triaryl phosphate is triphenyl phosphate, a toxic chemical used as a plasticizer and fire-retarding agent.

Organic compounds of arsenic syn. organoarsenic compounds are all toxic. They include phenyldichloroarsine, a tear gas and solvent.

SOAPS

Metal salts of organic compounds are formed by metals replacing the hydrogen of an organic acid. While these are combinations of metal and organic compounds, they are not included in the definition of organometallics. They are called *soaps* because they are able to lower the surface tension of water, allowing it to emulsify oil and grease. Fatty acid (long chain alkyl organic acid) soaps are used commercially in detergents. Some, like magnesium palmitate, are combustible.

ALKOXIDES

Alcoholates, more properly called *alkoxides*, are compounds in which a metal (or sometimes a metalloid such as boron or silicon) is attached to alkyl groups by an oxygen atom (the alkyl-oxygen linkage is derived from an alcohol). Examples also exists that describe an alcoholate as an alcohol acting as the *water of hydration* but not directly attached by the oxygen of the group. Many alcoholates such as calcium alkoxide and sodium methylate are soluble in alcohol. They are often shipped in this form to avoid contact with air, with which they can react violently. They are used widely as catalysts in a range of synthesis reactions.

Metal alkyl and aryl *oxyhydrides* are similar to alkoxides in that they involve a metal hydride attached to the oxygen from an alkyl or aryl alcohol. Examples include the **aluminium metal hydride** sodium bis(2-methoxyethoxy) aluminohydride and lithium aluminium-tri-tert-butoxyhydride. They are dangerous fire risks in contact with water and can decompose to evolve hydrogen gas.

RELATED TERMS

Alkali metal, see *Terminology, Metals*, p.242

Alkaline earth metal, see *Terminology, Metals*, p.242

Alkyl, see *Terminology, Hydrocarbons*, p.240

Aryl, see *Terminology, Hydrocarbons*, p.240

Compounds, see *Terminology, Compound*, p.234

Dispersion, see *Terminology, Dispersion*, p.235

Flammable liquid, see *Flammable Liquids and Class 3*, p.96

Liquid, see *Terminology, Liquid*, p.241

Metal, see *Terminology, Metals*, p.242

Organic, see *Terminology, Organic*, p.244

Pyrophoric, see *Spontaneously Combustible Materials and Division 4.2*, p.226

Salts, see *Terminology, Salts*, p.247

Self-heating, see *Spontaneously Combustible Materials and Division 4.2*, p.226

Solid, see *Terminology, Solid*, p.247

Solution, see *Terminology, Solutions*, p.247

Toxic, see *Toxic Substances and Division 6.1*, p.255

Water-reactive, see *Dangerous When Wet Materials and Division 4.3*, p.58

REGULATORY DEFINITIONS

None

REFERENCES

EOI, HCC, IMDG, KOE, TPD

Oxidizers and Class 5

Organic peroxide types B, C, D, E, F, liquid, 5.2 ■ Organic peroxide types B, C, D, E, F, liquid, temperature controlled, 5.2 ■ Organic peroxide types B, C, D, E, F, solid, 5.2 ■ Organic peroxide types B, C, D, E, F, solid, temperature controlled, 5.2 ■ Oxidizing liquid, n.o.s., 5.1, 6.1, 8 ■ Oxidizing solid, n.o.s., 4.1, 4.2, 4.3, 5.1, 6.1, 8 ■ Peroxides, inorganic, n.o.s., 5.1 ■ Peroxide, organic

1479 ■ 1483 ■ 3085 ■ 3087 ■ 3098 ■ 3099 ■ 3100 ■ 3101 ■ 3102 ■ 3103 ■ 3104 ■ 3105 ■ 3106 ■ 3107 ■ 3108 ■ 3109 ■ 3110 ■ 3111 ■ 3112 ■ 3113 ■ 3114 ■ 3115 ■ 3116 ■ 3117 ■ 3118 ■ 3119 ■ 3120 ■ 3121 ■ 3137 ■ 3139

OXIDIZERS

Oxidizing solids and *oxidizing liquids* are chemicals that lose electrons in a reaction. The most common oxidizers are those that make oxygen, an extremely reactive atom, available for reaction although many other strong oxidizers like fluorine are possible. In contact with a fuel (organic materials or other combustibles) many oxidizers are sufficiently powerful to cause highly exothermic reactions like combustion and explosion. For example, the reaction of oxygen gas present in air is sufficient to cause phosphorus to spontaneously ignite. Other combustion reactions progress only on heating, on introduction of an ignition source, or by catalyzation. Many oxidation reactions take place without violent evolution of energy.

Common oxidizers include the *peroxides* and superoxides (those compounds containing the peroxy group, -O-O-, such as peroxysulphuric acid), bisulphates, bromates, chlorates, nitrates, permanganates, and persulphates. Organic peroxides are common.

ORGANIC PEROXIDES

Organic peroxides are organic oxidizers that contain the peroxy group. They are thermally unstable and are often desensitized or refrigerated for transportation. They are extensively used as bleaches. They are assigned to one of seven *Types* (A through G) based on their tendency to detonate or explode, deflagrate, undergo thermal expansion, and show violent effect when heated or confined.

The assignment to one of these categories indicates the degree to which they are regulated in transportation. *Type A*, for example, are usually forbidden while *Type G* are sufficiently benign to be exempt from the organic peroxide classification. Assignment changes with concentration; the presence of desensitizers, water, and inert materials; and the transportation temperature. For example, 90% or higher concentrations of 1,1-di-(tert-butylperoxy)-3,3,5-trimethylcyclohexane are extremely hazardous and assigned to Type B; concentrations less than 32%, suitably desensitized, are assigned to Type E.

170

RELATED TERMS

Corrosive, see *Corrosives and Class 8*, p.47

Flammable see *Flammable Solids and Division 4.1*, p.99

Inorganic, see *Terminology, Organic*, p.244

Liquid, see *Terminology, Liquid*, p.241

Organic, see *Terminology, Organic*, p.244

Self-heating, see *Spontaneously Combustible*

Materials and Division 4.2, p.226

Solid, see *Terminology, Solid*, p.247

Toxic, see *Toxic Substances and Division 6.1*, p.255

Water-reactive, see *Dangerous When Wet Materials and Division 4.3*, p.58

REGULATORY DEFINITIONS

Class 5 is divided into two divisions: Division 5.1 Oxidizing substances. Division 5.2 Organic peroxides. IATA 3.5.0

Division 5.1

Oxidizing substances Substances which, while in themselves not necessarily combustible, may, generally by yielding oxygen, cause, or contribute to, the combustion of other material. Such substances may be contained in an article. UN 2.5.1(a)

Class 5.1 - Oxidizing substances (agents). These are substances which, although in themselves not necessarily combustible, may, either by yielding oxygen or by similar processes, increase the risk and intensity of fire in other materials with which they come into contact. IMO Gen. Intro. 5.1.5

Division 5.1 - Substances which, in themselves not necessarily combustible, may generally, by yielding oxygen, cause or contribute to the combustion of other material. ICAO 5.1

For the purpose of this subchapter, *oxidizer* (Division 5.1) means a material that may, generally by yielding oxygen, cause or enhance the combustion of other materials. [Text continues.] US 173.127(a)

Oxidizing substances are substances which, in themselves are not necessarily combustible, but may generally cause or contribute to the combustion of other material by yielding oxygen. IATA 3.5.1.1

Division 5.2

Organic peroxides Organic substances which contain the bivalent -O-O- structure and may be considered derivatives of hydrogen peroxide, where one or both of the hydrogen atoms have been replaced by organic radicals. Organic peroxides are thermally unstable substances, which may undergo exothermic self-accelerating decomposition. In addition, they may have one or more of the following properties: (i) be liable to explosive decomposition; (ii) burn rapidly; (iii) be sensitive to impact or friction; (iv) react dangerously with other substances; (v) cause damage to the eyes. UN 2.5.1(b)

Class 5.2 - Organic peroxides. Organic substances which contain the bivalent -O-O- structure and may be considered derivatives of hydrogen peroxide, where one or both of the hydrogen atoms have been replaced by organic radicals. Organic peroxides are thermally unstable substances, which may undergo exothermic self-accelerating decomposition. In addition, they may have one or more of the following properties: liable to explosive decomposition; burn rapidly; sensitive to impact or friction; react dangerously with other substances; cause damage to the eyes. IMO Gen. Intro. 5.1.5

Organic peroxides are liable to exothermic decomposition at normal or elevated temperatures. The decomposition can be initiated by heat, contact with impurities (e.g. acids, heavy-metal compounds, amines), friction or impact. The rate of decomposition increases with temperature and varies with the organic peroxide formulation. Decomposition may result in the evolution of harmful, or flammable, gases or vapours. For certain organic peroxides the temperature shall be controlled during transport. Some organic peroxides may decompose explosively, particularly if confined. This characteristic may be modified by the addition of diluents or by the use of appropriate packagings. Many organic peroxides burn vigorously. UN 2.5.3.1.1

Contact of organic peroxides with the eyes should be avoided. Some organic peroxides will cause serious injury to the cornea, even after brief contact, or will be corrosive to the skin. UN 2.5.3.1.2, ICAO 2-5.3.2.2

Organic peroxides are liable to exothermic decomposition which can be started by heat, contact with impurities (e.g. acids, heavy metal compounds, amines), friction or impact. The rate of decomposition increases with temperature and varies with the peroxide formulation. Decomposition may result in the evolution of harmful or flammable gases or vapours. Some organic peroxides decompose explosively, particularly if confined. Many organic peroxides burn vigorously. ICAO 2-5.3.2.1

Division 5.2 - Organic substances which contain the bivalent -O-O- structure and may be considered derivatives of hydrogen peroxide, where one or both of the hydrogen atoms have been replaced by organic radicals. Organic peroxides are thermally unstable substances, which may undergo exothermic, self-accelerating decomposition. In addition, they may have one or more of the following properties: be liable to explosive decomposition; burn rapidly; be sensitive to impact or friction; react dangerously with other substances; cause damage to the eyes. ICAO 2-5.1

For the purposes of this subchapter, *organic peroxide (Division 5.2)* means any organic compounds containing oxygen (O) in the bivalent -O-O- structure and which may be considered a derivative of hydrogen peroxide, where one or more of the hydrogen atoms have been replaced by organic radicals [text continues]. US 173.128(a)

This division is made up of organic substances which contain the bivalent structure -O-O- and may be considered derivatives of hydrogen peroxide in which one or both of the hydrogen atoms have been replaced by organic radicals. *Note: Hydrogen peroxide is made up of two hydrogen atoms and two oxygen atoms connected in a chain thusly: H-O-O-H.* Organic peroxides are thermally unstable substances which may undergo exothermic, self-accelerating decomposition. In addition, they may have one or more of the following properties: be liable to explosive decomposition; burn rapidly; be sensitive to impact or friction; react dangerously with other substances; cause damage to the eyes. IATA 3.5.2.1

REFERENCES
HCC

Oxygen Generators

Oxygen generator, chemical, 5.1

3356

Oxygen generators syn. oxygenators produce oxygen for purposes of respiratory support in submarines, aircraft, spacecraft, bomb shelters, and breathing apparatus (e.g., emergency response, personal use medical devices, and anaesthesiology). They are based on chemical reactions that yield oxygen. For example, almost pure oxygen (97 to 98%) is evolved by heating a chlorate candle (a solid mixture of a fuel with a lithium, sodium, or potassium chlorate or perchlorate) some of which propagates the reaction by oxidizing more fuel while the rest provides respiratory support. Fuels are usually powdered iron or other metal powders.

Many other oxygen generators are in use or being researched:

- Peroxide and superoxide systems (peroxide systems are not particularly efficient and are often supplemented with compressed oxygen gas).
- Liquid oxygen systems.
- Carbon dioxide systems which are of particular interest in space applications because they regenerate themselves, converting the carbon dioxide respired by humans back into oxygen.

The majority of commercial oxygenators must be sealed from the environment to prevent oxidation of organic materials and other adverse reactions.

RELATED TERMS
Oxidizing substances, see *Oxidizers and Class 5*, p.170

REGULATORY DEFINITIONS
Oxygen Generators, Chemical Oxygen generators, chemical, are devices containing chemicals which upon activation releases oxygen as a product of chemical reaction. Chemical oxygen generators are used for the generation of oxygen for respiratory support, e.g. in aircraft, submarines, spacecraft, bomb shelters and breathing apparatus. Oxidizing salts such as chlorates and perchlorates of lithium, sodium and potassium, which are used in chemical oxygen generators, evolve oxygen when heated. These salts are mixed (compounded) with a fuel, usually iron powder, to form a chlorate candle, which produces oxygen by continuous reaction. The fuel is used to generate heat by oxidation. Once the reaction begins, oxygen is released from the hot salt by thermal decomposition (a t[h]ermal shield is used around the generaot[to]r). A portion of the oxygen reacts with the fuel to produce more heat which produces more oxygen, and so on. Initiation of the reaction can be achieved by a percussion device, friction device or electric wire. UN App. B; IMO 5165-1

Oxygen Generator, Chemical. A device containing chemicals which upon activation releases oxygen as a product of chemical reaction. Chemical oxygen generators are used for the generation of oxygen for respiratory support, e.g. in aircraft, submarines, spacecraft, bomb shelters and breathing apparatus. Oxidizing salts such as chlorates and perchlorates of lithium, sodium and potassium, which are used in chemical oxygen generators, evolve oxygen when heated.

These salts are mixed (compounded) with a fuel, usually iron powder, to form a chlorate candle, which produces oxygen by continuous reaction. The fuel is used to generate heat by oxidation. Once the reaction begins, oxygen is released from the hot salt by thermal decomposition (a thermal shield is used around the generator). A portion of the oxygen reacts with the fuel to produce more heat which produces more oxygen, and so on. Initiation of the reaction can be achieved by a percussion device, friction device or electric wire. ICAO A2

Oxygen generator (chemical) means a device containing chemicals that upon activation release oxygen as a product of chemical reaction. US 171.8

Oxygen Generator, Chemical. Devices containing chemicals which upon activation releases oxygen as a product of chemical reaction. Chemical oxygen generators are used for generators of oxygen for respiratory support e.g. in aircraft, submarines, spacecraft, bomb shelters and breathing apparatus. Oxidizing salts such as chlorates and perchlorates of lithium, sodium and potassium, which are used in chemical oxygen generators, evolve oxygen when heated. These salts are mixed (compounded) with a fuel, usually iron powder to form a chlorate candle, which produces oxygen by continuous reaction. The fuel is used to generate heat by oxidation. Once the reaction begins, oxygen is released from the hot salt by thermal decomposition (a thermal shield is used around the generator). A portion of oxygen reacts with fuel to produce more heat which produces more oxygen, and so on. Initiation of the reaction can be achieved by a percussion device, friction device or electric wire. IATA App. A

REFERENCES
21CFR868, DOMA, KOC, KOE

Paints and Coatings

Aluminium liquid ■ Aluminium paint ■ Coating solution, 3, 3.1, 3.2, 3.3 ■ Compound, enamel ■ Driers, paint or varnish, liquid, n.o.s. ■ Driers, paint or varnish, solid, n.o.s. ■ Enamel ■ Filler, liquid ■ Ink, printer's, flammable ■ Lacquer ■ Lacquer base, liquid ■ Lacquer base or lacquer chips, plastic, wet with alcohol or solvent ■ Lacquer base, solution, 3.1, 3.2, 3.3 ■ Lacquer, liquid ■ Paint driers ■ Paint (including paint, lacquer, enamel, stain, shellac, varnish, polish, liquid filler and liquid lacquer base), 3, 3.1, 3.2, 3.3, 8 ■ Paint related material (including paint thinning or reducing compounds), 3, 3.1, 3.2, 3.3, 8 ■ Polish ■ Printing ink, 3, 3.1, 3.2, 3.3 ■ Shellac ■ Shellac solution ■ Stain ■ Varnish ■ Varnish drier, liquid ■ Varnish drier, solid

1139 ■ 1210 ■ 1263 ■ 3066

Coatings *syn. coating solutions* are liquid, liquefiable, or mastic compositions which harden to a film to decorate or protect a variety of materials against corrosion or oxidation. Coatings, including *paints* (pigmented coatings), varnishes, lacquers, enamels, stains, polishes, fillers, and shellacs have one or more of the following components:

▫ Binders that hold the ingredients of the coating together and form tough, solid, elastic films on drying. Binders include natural or synthetic resins (e.g., rosin and balsam) which harden on evaporation of a solvent; drying oils (e.g., linseed, tung, soybean, fish, castor, perilla), organic liquids that polymerize to hard films through air oxidation; or other film-forming agents. Hardening is accelerated by the addition of *paint driers* or *varnish driers*, metals, and other compounds that speed oxidation, evaporation, or polymerization. Heat or other radiation sources may increase the rate of hardening.

▫ Pigments and dyes, including titanium dioxide, lead chromate, and aluminium flakes (which are used in *aluminium paint syn. aluminium liquid* to give a highly reflective metallic finish).

▫ Dispersants, preservatives, flame retardants, fillers, and other additives.

▫ Solvents or *paint thinners* which carry the binder, pigments, and additives and facilitate application. These evaporate on application and are also known as *reducing compounds* (compounds used to reduce viscosity).

COATING TYPES

Traditionally, the taxonomy of paints was straightforward:

▫ Paints were based on drying oils, water or organic solvents, and pigments.

▫ Varnishes were resin-based without pigment.

▫ Enamels were pigmented varnishes known for their hard and glossy characteristics.

175

▫ Lacquers were based on cellulose without pigment which dried through solvent evaporation.

Today, coatings technology has advanced to the point where these distinctions are no longer meaningful: paints can contain resins, varnishes may be pigmented, enamels can contain oils, and lacquers can contain resins:

▫ Paints may contain all typical coatings components, although the use of organic solvents is being reduced to rely more heavily on water or high-solids paints.

▫ *Varnishes* contain gums, resins (spirit varnishes), or drying oils (oil varnishes) and solvents, but generally no pigments. *Shellac* derives from a resin secreted from the insect *Kerria lacca* collected from trees native to certain Asian countries. Dissolved in alcohol or other organic solvent, shellac becomes a spirit varnish, although it has many uses dry. *Stains* are often transparent varnishes with very low pigment or dye concentrations that penetrate the surface leaving little film.

▫ *Enamels* are characterized by their smooth and glossy appearance and hard nature. They are primarily pigmented varnishes or varnish-paint mixtures.

▫ *Lacquers* are clear, synthetic coatings, most often cellulose-based (such as nitrocellulose often modified with a resin, cellulose acetate, methyl cellulose, ethyl cellulose, and benzyl cellulose) that dry through evaporation of a solvent. Lacquers can be made from a *liquid lacquer base*, a starting point for speciality lacquers (e.g., automotive) which most often contain the film-forming material in concentrated solution with or without other components. The base may be up to 25% nitrocellulose.

▫ *Printing inks* describe the carbon black or lamp black pigments carried in a drying oil (e.g., linseed) with resins, solvents, adhesives, and driers. The resins may add gloss and hardness and increase the speed of drying.

▫ *Polishes* are used to decorate, produce a gloss, or provide surface protection. They can be abrasive, deposit a film, or remove tarnish. Depending on the application (e.g., furniture, metal, shoes), polishes may contain resins, oils, waxes, and solvents as well as abrasives, surfactants, water, pigments, fillers, and tarnish-removing chemicals. Similar in composition to polishes, *liquid fillers* disperse pigment and inert materials (silicates, clays, calcium carbonate, carbon black) to fill pores and imperfections in surfaces prior to further coating.

HAZARDS

Many binders (nitrocellulose derivatives, drying oils, and resins) are hazardous and many pigments and additives contribute hazards, but it is the host of flammable solvents and thinners used, like turpentine, naphtha, toluene, acetone, and xylene, that usually determines the overall hazard of a coating. Water-based paints may contain only water as the solvent or they may contain flammable solvents with sufficient water and surfactants to fully disperse the binder and render the mixture nonhazardous. Paint removing and stripping compounds have traditionally included chlorinated organic or alkali compounds, including dichloromethane and sodium hydroxide.

Paints and coatings are not corrosive in the usual sense. Any extreme pH will damage the substrate being painted. Paint removers, including some corrosives, exploit this property and are applied as dips and pastes to lift or destroy the painted surface. Acids and alkalis are also used as additives in some coatings to inhibit bacterial decomposition. The solvents in other coatings may cause structural corrosive damage to packagings, particularly some plastics. Similarly, some coatings constituents burn or otherwise affect human skin. The effects of methylene chloride, a common paint stripper, can be immediately felt on the skin.

RELATED TERMS

Alcohol, see *Alcohols*, p.5

Compound, see *Terminology, Compound*, p.234

Flammable liquids, n.o.s., see *Flammable Liquids and Class 3*, p.96

Flammable solid, inorganic, n.o.s., see *Flammable Solids and Division 4.1*, p.99

Flammable solid, organic, n.o.s., see *Flammable Solids and Division 4.1*, p.99

Lacquer base or lacquer chips, nitrocellulose, dry, see *Nitrocellulose Products*, p.161

Lacquer base or lacquer chips, plastic, wet with alcohol or solvent, see also *Nitrocellulose Products*, p.161

Lacquer, liquid, see also *Nitrocellulose Products*, p.161

Liquid, see *Terminology, Liquid*, p.241

Solid, see *Terminology, Solid*, p.247

Solution, see *Terminology, Solutions*, p.247

Solvent, see *Solvents*, p.224

REGULATORY DEFINITIONS

Coating Solution. Material such as automobile undercoating, drum or barrel lining material, etc., which cannot properly be described as cement, but presents similar hazards during transport. It usually contains flammable solvents. ICAO A2

Coating Solutions. Are materials such as automobile undercoatings, drum or barrel lining materials, etc., which cannot properly be described as cements, but present similar hazards in transportation. They usually contain flammable solvents. IATA App. A

Paint is the proper shipping name for paint, lacquer, enamel, stain, shellac, varnish, liquid aluminium, liquid bronze, liquid gold, liquid wood filler, and liquid lacquer base. US 173.173(a)

Paint related material is the proper shipping name for a paint thinning, drying, reducing or removing compound. US 173.173(a)

Paints and Coatings

REFERENCES
CP, GOCT, HCC, KOC, KOE, MEOS, MH14, MII, SSD, STRPV, TAOP, TEA, TNP, WNN

Pesticides

Alkaloids and alkaloid salts (pesticides) ■ Aluminium phosphide pesticide, 6.1 ■ Arsenical pesticide, liquid, 3, 3.2, 6.1 ■ Arsenical pesticide, solid, 6.1 ■ Arsenic compounds (pesticides) ■ Benzoic derivative pesticide, liquid, 3, 6.1 ■ Benzoic derivative pesticide, solid, 6.1 ■ Bipyridilium pesticide, liquid, 3, 3.2, 6.1 ■ Bipyridilium pesticide, solid, 6.1 ■ Carbamate pesticide, liquid, 3, 3.2, 6.1 ■ Carbamate pesticide, solid, 6.1 ■ Cargo transport unit under fumigation, 9 ■ Compounds, tree killing, liquid, 3, 6.1, 8 ■ Compounds, weed killing, liquid, 3, 6.1, 8 ■ Container under fumigation ■ Copper based pesticide, liquid, 3, 3.2, 6.1 ■ Copper based pesticide, solid, 6.1 ■ Coumarin derivative pesticide, liquid, 3, 3.2, 6.1 ■ Coumarin derivative pesticide, solid, 6.1 ■ Dithiocarbamate pesticide ■ Dithiocarbamate pesticide, liquid, 3, 6.1 ■ Dithiocarbamate pesticide, solid, 6.1 ■ Fluorine compounds (pesticides) ■ Fumigant ■ Fungicide ■ Insecticide ■ Insecticide gas, n.o.s., 2.1, 2.2, 2.3 ■ Insecticide solid or liquid ■ London Purple, 6.1 ■ Mercury based pesticide, liquid, 3, 3.2, 6.1 ■ Mercury based pesticide, solid, 6.1 ■ Organochlorine pesticide, liquid, 3, 3.2, 6.1 ■ Organochlorine pesticide, solid, 6.1 ■ Organophosphorus pesticide, liquid, 3, 3.2, 6.1 ■ Organophosphorus pesticide, solid, 6.1 ■ Organotin pesticide, liquid, 3, 3.2, 6.1 ■ Organotin pesticide, solid, 6.1 ■ Pesticide, liquid, n.o.s., 3, 3.2, 6.1 ■ Pesticide, solid, n.o.s., 6.1 ■ Pesticide, toxic, under compressed gas, n.o.s. ■ Phenoxyacetic acid pesticide, liquid, 3, 3.2, 6.1 ■ Phenoxyacetic acid pesticide, solid, 6.1 ■ Phenoxy pesticides, liquid, 3, 6.1 ■ Phenoxy pesticides, solid, 6.1 ■ Phenyl urea pesticide, liquid, 3, 6.1 ■ Phenyl urea pesticide, solid, 6.1 ■ Phthalimide derivative pesticide, liquid, 3, 6.1 ■ Phthalimide derivative pesticide, solid, 6.1 ■ Pyrethroid pesticide, liquid, 3, 3.2, 6.1 ■ Pyrethroid pesticide, solid, 6.1 ■ Substituted nitrophenol pesticide, liquid, 3, 3.2, 6.1 ■ Substituted nitrophenol pesticide, solid, 6.1 ■ Thiocarbamate pesticide, liquid, 3, 3.2, 6.1 ■ Thiocarbamate pesticide, solid, 6.1 ■ Triazine pesticide, liquid, 3, 3.2, 6.1 ■ Triazine pesticide, solid, 6.1 ■ Wood preservatives, liquid, 3, 3.2, 3.3

1306 ■ 1621 ■ 1760 ■ 1967 ■ 1968 ■ 1993 ■ 2588 ■ 2757 ■ 2758 ■ 2759 ■ 2760 ■ 2761 ■ 2762 ■ 2763 ■ 2764 ■ 2765 ■ 2766 ■ 2767 ■ 2768 ■ 2769 ■ 2770 ■ 2771 ■ 2772 ■ 2773 ■ 2774 ■ 2775 ■ 2776 ■ 2777 ■ 2778 ■ 2779 ■ 2780 ■ 2781 ■ 2782 ■ 2783 ■ 2784 ■ 2786 ■ 2787 ■ 2810 ■ 2902 ■ 2903 ■ 2991 ■ 2992 ■ 2994 ■ 2995 ■ 2996 ■ 2997 ■ 2998 ■ 2999 ■ 3000 ■ 3001 ■ 3002 ■ 3003 ■ 3004 ■ 3005 ■ 3006 ■ 3007 ■ 3008 ■ 3009 ■ 3010 ■ 3011 ■ 3012 ■ 3013 ■ 3014 ■ 3015 ■ 3016 ■ 3017 ■ 3018 ■ 3019 ■ 3020 ■ 3021 ■ 3024 ■ 3025 ■ 3026 ■ 3027 ■ 3048 ■ 3345 ■ 3346 ■ 3347 ■ 3348 ■ 3349 ■ 3350 ■ 3351 ■ 3352 ■ 3354 ■ 3355

Pests include any plant or animal (insect, rodent, nematode, fungus, terrestrial or aquatic algae or plant, virus, bacteria, or other microorganism) that is detrimental to man. Agricultural pests claim approximately 14% of worldwide food production. *Pesticides* are contact, digestive, or respiratory poisons that prevent, destroy, inhibit the growth of, repel, or mitigate any pest. Plant pesticides include algicides (algae), *fungicides* (mould, mildew, mushrooms, yeast, and other fungi), and herbicides (defoliants, *weed killers* and *tree killers*). Animal pesticides include

- *Insecticides*, those agents used to combat insects, taken to include larvicides (larval pesticides), acaricides or miticides (mites), and ixodicides (ticks).

179

☐ A series of species-specific pesticides: avicides (birds), bactericides (bacteria), molluscicides (molluscs), nematicides (nematodes), piscicides (fish), rodenticides (rodents), and viricides (viruses).

PESTICIDE TYPES

There are thousands of available pesticides: broad-based to combat a large number of pests or narrowly focused to combat individual species and leave others unharmed. They include a range of chemicals, almost all of which are toxic to both the pest and humans, including:

☐ *Alkaloid pesticides*, many naturally occurring alkaloids like nicotine in tobacco function as natural pesticides.

☐ *Aluminium phosphide pesticides*, those based on aluminium phosphide (20859-73-8), an insecticide.

☐ *Arsenical pesticides*, those based on arsenic, such as arsenic disulphide, a rodenticide.

☐ *Benzoic derivatives pesticides*, those based on the fungicide benzoic acid (65-85-0), such as tricamba, a herbicide.

☐ *Bipyridilium pesticides*, those including or based on bipyridinium such as diquat or paraquat, both herbicides.

☐ *Carbamate pesticides*, salts and esters of carbamic acid (463-77-4), including the insecticide carbaryl.

☐ *Copper-based pesticides*, those that include copper such as the fungicide copper acetate.

☐ *Coumarin derivative pesticides*, those based on coumarin (91-64-5), including warfarin, coumaphos, and coumafuril.

☐ *Dithiocarbamate pesticides syn. thiocarbamate pesticides*, salts and complexes including and based on dithiocarbamic acid, a seed disinfectant, like the deer poison zinc dimethyldithiocarbamatecyclohexylamine complex, metam-sodium, nabam, and maneb.

☐ *Fluorine pesticides*, those based on fluorine, including fluoroacetic acid, a rodenticide.

☐ *London Purple*, an insecticide compounded of arsenic, aniline, lime, and ferrous oxide.

☐ *Mercury-based pesticides*, those that contain mercury, including the bactericide mercuric chloride.

☐ *Organochlorine pesticides*, including organic compounds containing chlorine such as chlordane, used as a fumigant for termite control.

☐ *Organophosphorus pesticides*, including tetraethyl pyrophosphate, a miticide.

☐ *Organotin pesticides*, including tributyl tin oxide, a fungicide for above-ground applications.

☐ *Phenoxyacetic acid pesticides*, those based on the fungicide phenoxyacetic acid (122-59-8).

☐ *Phenyl urea pesticides*, those including phenyl urea (64-10-8).

□ *Phthalimide derivative pesticides*, those based on phthalimide (85-41-6), itself a fungicide.

□ *Pyrethroid pesticides*, a group of neurotoxic insecticides that intervene in the maturation of an insect.

□ *Substituted nitrophenol pesticides*, those based on nitrophenol (25154-55-6), a fungicide, including the acaricide parathion.

□ *Triazine pesticides*, herbicides derived from triazine (101-05-3), including atrazine.

APPLICATION

Pesticides may be solid, liquid, or gaseous. They may be applied to crops, soils, seeds and grain, fabrics, etc., as dust, smoke, dispersions, emulsions, or solutions. *Fumigants* are gases, volatile chemicals, or mixtures of pesticides in a volatile liquid, often petroleum or coal tar distillates. They are most effective in enclosed environments where penetration to all points is assured. Fumigants include hydrogen cyanide, methyl bromide, carbolic acid, hydrocyanic acid, and chloropicrin.

During transportation of human and animal foodstuffs, particularly during lengthy ship voyages, fumigants are used to poison or asphyxiate existing pests and ensure that no onboard pests spoil the cargo. In extreme cases, infestation of bulk cargoes like cereals can lead to severe and dangerous heat buildup. These *cargo transport units under fumigation* and *containers under fumigation* include containers that are reasonably gas-tight, and although the fumigant may be applied under pressure to force it into the cargo, the container is not generally under residual gas pressure. Ventilation is required to disperse both the flammable carrier and the toxic fumigant before accessing the cargo.

WOOD PRESERVATIVES

Historically, almost every chemical or compound known to be toxic has been tried and used to inhibit the action of wood-destroying pests. Among the many chemicals used as *wood preservatives* are compounds based on creosote, mercury, arsenic, thallium, borate, cyanide, chromium, copper, zinc, nickel, fluorides, and pentachlorophenol. The majority fall into 3 classes:

□ Coal tar derivatives, including creosote.
□ Oil-borne preservatives, including pentachlorophenol.
□ Waterborne preservatives, normally water soluble salts, which then undergo fixation on action.

Petroleum distillates play an extensive role in wood preservation as solvents and diluents for the active ingredient but appear to have little or no preservative value. Many light petroleum solvents and gases deliver the preservative and help it penetrate into the wood before evaporating.

Pesticides

RELATED TERMS

Aerosol, see *Aerosols*, p.3
Alkaloids, see *Alkaloids*, p.7
Compounds, see *Terminology, Compound*, p.234
Compressed gas, see *Gases and Class 2*, p.104
Derivative, see *Terminology, Derivative*, p.234
Disinfectant, liquid, corrosive, n.o.s., see *Drugs and Medicines*, p.61
Disinfectant, liquid, toxic, n.o.s., see *Drugs and Medicines*, p.61
Disinfectant, solid, toxic, n.o.s., see *Drugs and*

Medicines, p.61
Flammable, see *Flammable Liquids and Class 3*, p.96
Flash point, see *Terminology, Flash Point*, p.237
Liquid, see *Terminology, Liquid*, p.241
Salts, see *Terminology, Salts*, p.247
Solid, see *Terminology, Solid*, p.247
Substituted, see *Terminology, Substituted*, p.251
Toxic, see *Toxic Substances and Division 6.1*, p.255

REGULATORY DEFINITIONS

Cargo Transport Unit Under Fumigation; Container Under Fumigation A closed cargo transport unit loaded with cargoes under fumigation. The fumigant gases are either poisonous or asphyxiant. The gases are evolved from solid or liquid preparations within the closed cargo transport unit. IMO 9025-1

London Purple Mixture of arsenic trioxide, lime and ferric oxide, used as an insecticide. IMO 6171

REFERENCES

40CFR455, HSR, IMDG, ISD, KOC, MEOS, MH14, ROTS, SDO, WDA

Petroleum

Acid sludge ▪ Asphalt ▪ Asphalt, at or above its flashpoint, 3 ▪ Aviation gasoline ▪ Case oil ▪ Casinghead gasoline ▪ Crude naphtha ▪ Diesel fuel, 3, 3.3 ▪ Fuel, aviation, turbine engine, 3, 3.1, 3.2, 3.3 ▪ Fuel oil ▪ Fuel oil No. 1 ▪ Fuel oil (No. 1, 2, 4, 5, or 6), 3 ▪ Gas drips, hydrocarbon ▪ Gasohol, 3 ▪ Gas oil, 3, 3.3 ▪ Gasoline, 3, 3.1 ▪ Gasoline, casinghead ▪ Heating oil, light, 3, 3.3 ▪ Jet fuel ▪ Kerosene, 3, 3.3 ▪ Ligroin ▪ Liquefied natural gas ▪ Liquefied petroleum gas ▪ LNG ▪ LPG ▪ Lythene ▪ Motor spirit, 3, 3.1 ▪ Naphtha ▪ Naphtha, petroleum ▪ Naphtha, solvent ▪ Natural gas, compressed with high methane content, 2.1 ▪ Natural gases ▪ Natural gasoline ▪ Natural gas, refrigerated liquid with high methane content, 2.1 ▪ Oil gas, compressed, 2.3 ▪ Oil well sampling device, charged ▪ Paraffin ▪ Petrol, 3, 3.1 ▪ Petroleum coke, calcined or uncalcined ▪ Petroleum crude oil, 3, 3.1, 3.2, 3.3 ▪ Petroleum distillates, n.o.s., 3, 3.1, 3.2, 3.3 ▪ Petroleum ether ▪ Petroleum gases, liquefied, 2.1 ▪ Petroleum naphtha ▪ Petroleum oil, 3 ▪ Petroleum products ▪ Petroleum products, n.o.s., 3, 3.1, 3.2, 3.3 ▪ Petroleum raffinate ▪ Petroleum spirit ▪ Shale oil, 3, 3.2, 3.3 ▪ Sludge acid, 8 ▪ Turpentine substitute, 3, 3.2, 3.3 ▪ White spirit

1071 ▪ 1075 ▪ 1202 ▪ 1203 ▪ 1223 ▪ 1267 ▪ 1268 ▪ 1288 ▪ 1300 ▪ 1863 ▪ 1906 ▪ 1971 ▪ 1972 ▪ 1993 ▪ 1999

Petroleum is the mixture of gaseous, liquid, and solid hydrocarbons (although the term is often restricted to the non-gases) derived from chemical reactions on prehistoric animal and vegetable matter which occurs as reservoirs in sedimentary rocks from which it is extracted. Petroleum (8002-05-9) *syn. **crude oil** or **petroleum oil*** also contains inorganics such as compounds of sulphur, nitrogen, oxygen, metals, etc.

During production, steel pipes (casings) are used in oil wells[18] to seal off fluids from the borehole and to prevent the walls from sloughing off or caving in. Casingheads, to which control pipes and valves are attached, protrude from the top of the pipes. *Casinghead gasoline syn. natural gasoline* are the liquid hydrocarbons resulting from any gases and vapours that accumulate and condense at the top of the well. It requires processing to remove the unstable and flammable light fractions before use as a commercial gasoline.

Natural gas (8006-14-2) is the gaseous component of petroleum. It is mostly methane with some ethane and smaller quantities of propane, butane, carbon dioxide, hydrogen sulphide, nitrogen, helium, and other gases. It is distributed in commerce as compressed or liquefied natural gas (*LNG*) for fuel and other purposes.

Crude oil is processed in refineries by distillation, cracking, reforming, alkylation, polymerization, and other methods to generate gases, distillates, residuum, and sludges from which an enormous number of *petroleum products*

---

[18] The author was unable to find a reference for why an *oil well sampling device* might be charged with a compressed or liquefied gas.

are derived: fuels, petrochemicals, and other products. (The term *paraffin* (8002-74-2) applies to the family of alkyl hydrocarbons that range from petroleum gases to residuum waxes.)

PETROLEUM GASES

In addition to natural gas a number of gases are associated with petroleum:

- *Liquefied petroleum gases* (68476-85-7) *syn. LPG* are by-products of petroleum refining. They are predominantly mixtures of propane, butane, and pentane.
- *Oil gas syn. Pintsch gas* is mostly methane with ethane, propane, butane, ethylene, and propylene made by exposing petroleum oils to high temperature. Although superseded in importance by natural gas, it is used in cutting torches to give a low temperature flame.
- *Gas drips* are the extremely volatile and flammable liquids that condense as a result of the compression of oil gas and other gas fuels distributed through pipelines and gas mains.

PETROLEUM DISTILLATES

Initially, petroleum constituents are separated by distillation into a variety of fractions. The names of these *petroleum distillates*[19] are unsystematic and can vary depending on the source. (The temperatures given here are approximate.)

RANGE	PRODUCT
-1 to 250°C	Light distillates including *crude naphtha* and some oils.
-1 to 150°C	*Petroleum naphtha syn. petroleum spirits, petroleum ether*, mineral spirits, and *naphtha* (8030-30-6): a volatile colourless liquid used extensively as a solvent (*naphtha solvent*) and a fuel sometimes blended with casinghead gasoline to produce gasoline.
-1 to 180°C	*Gasoline* (8006-61-9) *syn. petrol*, a volatile mixture of liquid hydrocarbons blended with additives and used as a fuel in spark-ignition internal combustion engines in automobiles and aircraft (special blends of gasoline produce a high performance fuel). *Motor spirit syn. case oil*, a highly volatile fraction is blended with gasoline. *Gasohol*, a mixture of ethanol and gasoline, is an alternate fuel designed for its environmental properties: ethanol generates relatively favourable combustion products and is a renewable resource. At around 60 to 110°C is *ligroin* (8032-32-4), a volatile fraction used as a solvent in paints and coatings.
150 to 205°C	Heavy naphtha including *turpentine substitutes* (9005-90-7) *syn. white spirit*, products that compete with and contain many of the same hydrocarbons as turpentine derived from pine.

---

[19] The author was unable to find a reference for the petroleum distillate *lythene*.

RANGE	PRODUCT (CONT.)
205 to 260°C	***Kerosene*** (8008-20-6), used mainly as light household ***heating oil*** and as ***aviation turbine engine fuels*** *syn.* ***jet fuel*** which describes any liquid suitable for the generation of power by combustion in aircraft gas turbine engines. These liquids are specified by performance and property rather than chemical composition, and vary primarily in their volatility. They are based on well-defined, complex mixtures of four types of hydrocarbons: paraffins, cyclo-paraffins or naphthenes, aromatics, and olefins.
200 to 425°C	Intermediate petroleum distillates which include ***fuel oil*** (although fuel oil can mean any of a large number of petroleum liquids used to generate heat or power):

□ No.1 fuel oil is slightly heavier than kerosene and is used almost exclusively for domestic heating and light diesel engines (***diesel fuel,*** 68334-30-5).

□ No.2 fuel oil comes from two sources: first, it is the fraction distilled immediately after No.1 fuel oil; second, from catalytically cracked distillate. The former is often called ***gas oil*** (68476-30-2) because it was used to manufacture heating gas; it is still used widely as domestic heating fuel. The cracked distillate is used in industrial furnaces for heating, and as a heavier diesel fuel, particularly in engines where constant loads are common, like railroad engines.

up to 600°C	Heavy distillates including lubricating oils, paraffin waxes, and greases. Lubricating oils are solvent-refined from heavy distillates; the portion that is not soluble in the solvent is called ***petroleum raffinate***.

RESIDUUM

After the volatile hydrocarbons have been removed through distillation the remaining residuum is processed for use, primarily as heavy fuel oils and ***asphalts*** (8052-42-4). Among the properties of asphalt are its ability to flow when heated and to solidify on cooling. It is used extensively as a road coating, roofing, and waterproofing. Residuum is also the source of other fuel oils:

□ No.4 fuel oil is a blend of distillate and around 20 to 50% residuum.

□ No.5 fuel oil is also a blend, but with a higher proportion of residuum (55 to 80%).

□ No.6 fuel oil is residuum.

Higher grades of fuel oil are less expensive and contain more energy than low grades, but are also more viscous and require special handling. The lighter grades are used in small- and medium-use residential, commercial, and industrial boilers, while a No.6 fuel oil may be used for the constant, heavy loads necessary to power naval or commercial shipping.

REFINERY SLUDGES

Refinery sludges remain after distillation and residuum processing. These include *petroleum coke* (64741-79-3), oils burned in the refinery, and *acid sludges* containing sulphuric acid used in the refining processes. Acid sludges are used commercially in fertilizer manufacture.

SHALE

Shale, a sedimentary rock, can contain kerogen, a complex, fossilized, organic polymerized material. By distilling shale, the kerogen yields *shale oil* (68308-34-9) and other petroleum products (shale can also be impregnated with petroleum, but this is not shale oil).

RELATED TERMS

Asphalt, cut back, see *Bituminous Products*, p.32

Calcined, see *Terminology, Calcined*, p.233

Cartridges, oil well, see *Explosive Articles*, p.69

Coal tar naphtha, see *Coal*, p.44

Coal tar, crude and solvent, see *Coal*, p.44

Compressed gas, flammable, n.o.s, see *Gases and Class* 2, p.104

Flashpoint, see *Terminology, Flash Point*, p.237

Fracturing devices, explosive, for oil wells, without detonators, see *Explosive Articles*, p.69

Gas, see *Terminology, Gas*, p.239

Hydrocarbon, see *Terminology, Hydrocarbons*, p.240

Liquefied gas, see *Gases and Class 2*, p.104

Liquid, see *Terminology, Liquid*, p.241

M86 fuel, see *Batteries*, p.27

Oil, see *Terminology, Oil*, p.244

Petroleum coke, calcined or uncalcined, see also *Solid Bulk Materials*, p.221

Paraffin, see *Terminology, Hydrocarbons*, p.240

Refrigerated liquid, see *Gases and Class 2*, p.104

Solvent, see *Solvents*, p.224

Tars, liquid, including road asphalt and oils, bitumens and cut backs, see *Bituminous Products*, p.32

Uncalcined, see *Terminology, Uncalcined*, p.252

REGULATORY DEFINITIONS

Gas Drips, Hydrocarbon. The liquid that condenses on compression of Pintsch Gas or the condensate from gas mains. It consists principally of a mixture of benzene and unsaturated hydrocarbons. ICAO A2

Gas Drips, Hydrocarbon. Is the liquid that condenses on compression of Pintsch Gas or the condensate from gas mains. It consists principally of a mixture of benzene and unsaturated hydrocarbons. It is very combustible and has a low flash point. IATA App. A

Oil Gas, Compressed. A gas made by the reaction of steam at high temperatures on gas oil or similar fractions of petroleum, or by high-temperature cracking of gas oil. The gas is flammable, but it is classified as a toxic gas because it contains a high proportion of carbon monoxide. ICAO A2

Oil Gas Compressed. Is a gas made by the reaction of steam at high temperatures on gas oil or similar fractions of petroleum, or by high temperature cracking of gas oil. The gas is flammable but is classified as a toxic gas because it contains a high proportion of Carbon Monoxide. IATA App. A

Sludge Acid. The acid waste resulting from oil refining, or from nitrating processes. It generally has somewhat the same hazards as the original acid. ICAO A2

Sludge, Acid. Is the acid waste resulting from oil refining, or from nitrating processes. It generally has somewhat the same hazards as the original acid. IATA App. A

Turpentine Substitute. A petroleum distillate which might contain some aromatic components and which usually has a flash point of approximately 40°C. White spirit is a synonym for turpentine substitute. ICAO A2, IATA App. A

<u>REFERENCES</u>
10CFR436, 40CFR60, CAB, DOSATT, FOM, GOCT, HCC, HOA, HOO, KOC, MEOS, MH14, PSF, STRPP, TCAOP, TEA, TIP

Pine Products

Pine oil, 3, 3.3 ■ Resinates, liquid ■ Resinates, solid ■ Rosin oil, 3, 3.2, 3.3 ■ Turpentine, 3, 3.3

1272 ■ 1286 ■ 1299

Pine trees, chiefly *Pinus palustris* and *Pinus caribaea*, contain an oleoresin (a mixture of essential oils and resins) which can be tapped or removed by extraction or distillation to yield a number of commercially significant products:

- Distillation of the oleoresin yields turpentine oil *syn.* ***turpentine*** (8006-64-2) which is mostly a mixture of pinene and diterpene. Turpentine is used extensively as a thinner and solvent for paints and coatings although naphtha and turpentine substitutes are replacing its use. It remains important in the manufacture of resins and synthetic chemicals. The residue from this separation is *rosin.*
- Destructive distillation of rosin yields a fraction above 360°C of ***rosin oil*** (8002-16-2). It is used to blend with turpentine and as a plasticizer for rubber.
- Fractionation of pine oleoresin or turpentine yields ***pine oil*** (8002-09-3), mostly secondary and tertiary terpene alcohols, at 93°C to 107°C. Pine oil is used as a disinfectant, insecticide, and odorant.

RESINATES
Rosin is chiefly abietic and pimaric acids, organic acids of the formula $C_{10}H_{29}COOH$ with a phenanthrene group. Salts of rosin acids (***resinates***) are easily formed with metals such as aluminium resinate. They are used as paint driers and as catalysts.

HAZARDS
Pine oil, rosin oil, and turpentine are all flammable and combustible liquids. Resinates are also dangerous fire risks.

RELATED TERMS
Flammable solid, organic, n.o.s., see *Flammable Solids and Division 4.1*, p.99
Flammable; Flammable liquid, n.o.s., see *Flammable Liquids and Class 3*, p.96

Liquid, see *Terminology, Liquid*, p.241
Oil, see *Terminology, Oil*, p.244
Solid, see *Terminology, Solid*, p.247
Turpentine Substitute, see *Petroleum*, p.183

REGULATORY DEFINITIONS
None

REFERENCES
GOT, HCC, MH14, SSD

Polymers and Resins

Fibreglass repair kit ■ Gutta percha solution ■ Indiarubber ■ Plastics moulding compound in dough, sheet or extruded rope form evolving flammable vapour, 9 ■ Polyester resin kit, 3, 3.2, 3.3 ■ Polymeric beads, expandable evolving flammable vapour, 9 ■ Polystyrene beads, expandable ■ Polystyrene beads, expandable, evolving flammable vapour ■ Resin solution, flammable, 3, 3.1, 3.2, 3.3 ■ Rubber scrap, powdered or granulated, 4.1 ■ Rubber shoddy, powdered or granulated, 4.1 ■ Rubber solution, 3, 3.2, 3.3

1287 ■ 1345 ■ 1866 ■ 2211 ■ 3269 ■ 3314

Polymers are large molecules made up of multiple *monomers*, molecules that combine repeatedly with themselves or other monomers (*copolymers*). Dimers, trimers, and tetramers are made up of two, three, and four monomers, respectively. Any molecule of five or more monomers is a polymer. For example, the styrene monomer ($C_6H_5CHCH_2$) readily polymerizes to make polystyrene ($C_6H_5CHCH_2)_n$ where n denotes the number of repeating monomeric units, which in this case ranges between 2000 to 3000. Cellulose, $(C_6H_{10}O_5)_n$, a natural polymer, can consist of 3500 repeating units. *High polymers* are those with a molecular weight (the sum of the atomic weights of its constituents) of over 5000 or 6000. Polymers can be divided into those that occur naturally, those that are synthetic, and semisynthetics (manufacturing manipulations of natural cellulose and starches).

NATURAL POLYMERS
Natural polymers include polysaccharides (including cellulose, starch, and gums), polypeptides (proteins like casein, albumin, keratin, and DNA), and hydrocarbons (rubber and gutta percha). Rubber is polyisoprene ($C_5H_8)_n$, 9003-31-0), a high polymer of isoprene (78-79-5) derived from the latex (aqueous suspensions of hydrocarbon polymers) that exudes in commercial quantities from *Hevea braziliensis* trees, although rubber can be tapped from 200 other plant species. *Indiarubber* was the early name given to rubber owing to its ability to erase pencil marks by rubbing, and its West Indian origin. Before use, crude rubber is often vulcanized to give it useful strengthening properties by having its hydrocarbon chains cross-linked with sulphur on heating. It is used in over 50,000 different products. Crude or prevulcanized rubber is soluble in most organic solvents, many of which are flammable including acetone and alcohols. *Rubber solutions* are used as cements, adhesives, and coatings.

Rubber scrap, largely from tires and factory by-products (sheets, slabs, pellets, or powder), can be reclaimed by chopping or shredding the rubber into granules (non-spherical particles that span a broad range of particle dimension to around 12mm) or even powder form. Like most organic polymers, rubber is combustible, a property enhanced when a greater surface area is exposed to atmospheric oxygen. *Rubber shoddy* is reclaimed rubber, so named to suggest inferior quality. Scrap tires are depolymerized at high temperature and converted to oil, carbon black, steel (from the radial belting), and ash.

Gutta percha is an isomer (transpolyisoprene) of natural rubber made from the latex of various trees of the genus *Dichopsis*. It is not as elastic as rubber, but becomes highly plastic on gentle heating. It is used chiefly for insulation of submarine cable. It is similarly soluble in organic solvents and carbon disulphide.

SYNTHETIC POLYMERS

An enormous variety and volume of synthetic polymers are manufactured including synthetic versions of rubber and other natural polymers. Synthetic polymers are used in isolation or as constituents of *plastics* which, strictly speaking, are mixtures of high polymers (usually synthetic, but sometimes natural) and other ingredients, like fillers, colorants, plasticizers (organic compounds that blend with some of the polymer chains helping processing and flexibility), curatives, reinforcing agents, etc. (Note that the terms *plastic* and *polymer* are sometimes used interchangeably.)

Plastics are solid in their finished shape but can be classified into two types:

- Thermoplastics, those that be resoftened on heating, including nylon, polyvinyl chloride, polyethylene, polystyrene (9003-53-6), polypropylene, polyurethane, etc.
- Thermoset, those that cannot be resoftened, including polyethylene (crosslinked), phenolics, and alkyds like the polyesters (polymers based on maleic, terephthalic, or other carboxylic acids and glycol, propylene glycol or similar alcohols).

Polymeric raw materials for plastics can be supplied in *polymeric bead* form (spherical particles of polymer about 1mm in diameter) such as *polystyrene beads* that are expandable into foams and other cellular forms by thermal, chemical, or mechanical means: incorporation of blowing agents (gases or flammable liquids, containing pentane or other hydrocarbons, that decompose to liberate a gas); mechanical whipping of the polymer in a gas; or addition of an agent which is leached out after forming to leave voids. The volatile fractions of blowing agents evolve during storage, handling, and use. *Plastic moulding compounds* include putty-like mixtures of any thermosetting plastic mixture prior to its being thermoset by heat or other means. They are often available in the form of dough, sheets, or ropes. Moulding compounds may contain flammable liquids to aid processing.

RESINS

Strictly speaking, *resins* are naturally occurring, nonpolymeric mixtures of carboxylic acids, essential oils, and terpenes that exude from various trees and shrubs and some insects that polymerize on exposure to atmospheric oxygen or other agents. They include rosin, linseed and other drying oils, and shellac. Natural resins and some synthetic resins are soluble in organic solvents. **Resin solutions** are often available in commerce as coatings, adhesives, laminates, etc.

However, the term *resin* has become synonymous with any of the synthetic polymers made from reacting two or more substances including acrylic resins (polymers based on acrylic acid), vinyl resin (polymers based on vinyl), melamine resin (polymers based on phenols and aldehydes), and others. For example, **polyester resin kits** incorporate two substances that contain the components that form a polymer when mixed. Copolymers, including styrene copolymers, are possible. **Fibreglass repair kits** include such a two-part resin which is applied to reinforcing fibreglass sheets that, on hardening of the resin, provide a strong, tough, flexible coating that is resistant to corrosion, chemicals, water, and solvents.

<u>RELATED TERMS</u>

Compound, see *Terminology, Compound*, p.234

Explosive, blasting, type D, see *Explosives and Class 1*, p.74

Flammable; Flammable vapour; Flammable liquid, n.o.s., see *Flammable Liquids and Class 3*, p.96

Lacquer base or lacquer chips, plastic, wet with alcohol or solvent, see *Nitrocellulose*

Products, p.161 or *Paints and Coatings*, p.175

Plastic explosives, see *Explosives and Class 1*, p.74

Plastics, nitrocellulose-based, self-heating, n.o.s., see *Nitrocellulose Products*, p.161

Pyroxylin plastic, see *Nitrocellulose Products*, p.161

Solution, see *Terminology, Solutions*, p.247

<u>REGULATORY DEFINITIONS</u>

Polyester Resin Kit. The proper shipping name "Polyester resin kit" covers different kits such as filler, bonding and sealing compounds, chemical anchors and fibreglass repair kits. A polyester resin kit commonly consists of an unsaturated polyester resin mixed with styrene and a separate hardener (usually a phlegmatized organic peroxide) as a minor component. The main component (viscous liquid or paste) is inherently flammable due to the styrene content (flash point 29°C to 32°C). ICAO A2, IATA App. A

Polyester resin kits consist of two components: a base material (Class 3, Packing Group II or III) and an activator (organic peroxide), each separately packed in an inner packaging. US 172.102(c)(1),40

Polymeric Beads, Expandable. Semi-processed products used to manufacture polymeric articles, and which have been impregnated with a flammable gas or liquid as a blowing agent. They may evolve small quantities of flammable gas during transport. ICAO A2

Polymeric Beads, Expandable; Polystyrene Beads, Expandable; Plastics Moulding Compound A moulding material in bead or granular form consisting predominantly of polystyrene, poly(methyl methacrylate) or other polymeric material and containing 5% to 8% of a volatile hydrocarbon which is predominantly pentane. During storage a small proportion of this pentane is released to the atmosphere; this proportion increases at elevated temperatures. IMO 9036

Polymeric Beads, Expandable. Are semi-processed products used to manufacture polymeric articles. When impregnated with a flammable gas or liquid as a blowing agent, they may evolve small quantities of flammable gas during transportation. IATA App. A

191

Polymers and Resins

Polymerizable Material. Any liquid, solid, or gaseous material which, under conditions incident to transportation, may polymerize (combine or react with itself) so as to cause dangerous evolution of gas or heat. IATA App. A

REFERENCES
7CFR319, CEO, GOCT, GOT, HCC, HOHM, IMDG, MH14, STRP, STRPV, STRR, WDO

Pressurized Articles

Accumulators, pressurized, hydraulic, 2.2 ■ Accumulators, pressurized, pneumatic, 2.2 ■ Articles, pressurized, hydraulic, 2.2 ■ Articles, pressurized, pneumatic, 2.2

1956 ■ 3164

Pressurized articles syn. pressurized accumulators include devices that store fluids under pressure to facilitate mechanical work or absorb energy. Of three possible pressure sources, weights, springs, or compressed gas (usually nonflammable, nontoxic, and nonliquefied gas, such as nitrogen), it is the last that constitutes a potential hazard during transportation. Contemporary definitions of both *hydraulic* and *pneumatic* describe the mechanics of fluids (i.e., gases and liquids) that yield and move freely when subjected to pressure. In this context, the term *hydraulic* applies primarily to the storage of liquids, usually oil or water under pressure, while *pneumatic* applies primarily to the storage of gases under pressure.

Pressurized accumulators contain compressed gas on one side of a piston, flexible diaphragm, or bladder. On the other side is the fluid which is discharged as necessary to an external system, normally to provide power. If a piston rod is added to a piston-type accumulator and the system is closed, it becomes a shock absorber using the fluid under pressure to absorb repeated shocks or sudden impulses.

Accumulators can be found in motor, construction, machinery, and other equipment as shock absorbers, motion dampeners, and fluid sources for actuators (devices that convert hydraulic or pneumatic power into useful mechanical work).

RELATED TERMS
Non-flammable gas, see *Gases and Class 2*, p.104

REGULATORY DEFINITIONS
None

REFERENCES
DOSATT, GAE, HPT, MEOS, VNS, WNN

Pyrotechnics and Signals

Aeroplane flares ■ Ammunition, illuminating with or without burster, expelling charge or propelling charge, 1.2G, 1.3G, 1.4G ■ Ammunition, smoke (water-activated contrivances) ■ Ammunition, smoke (water-activated contrivances), white phosphorus, with burster, expelling charge or propelling charge ■ Ammunition, smoke (water-activated contrivances), with or without white phosphorus or phosphides, with burster, expelling charge or propelling charge ■ Ammunition, smoke, white phosphorus, with burster, expelling charge or propelling charge, 1.2H, 1.3H ■ Ammunition, smoke, with or without burster, expelling charge or propelling charge, 1.2G, 1.3G, 1.4G, 8 ■ Amorces ■ Amorces (caps, toy) ■ Articles, pyrotechnic for technical purposes, 1.1G, 1.2G, 1.3G, 1.4G, 1.4S ■ Bombs, illuminating ■ Bombs, photo-flash, 1.1D, 1.1F, 1.2G, 1.3G ■ Bombs, smoke, non-explosive, with corrosive liquid, without initiating device, 8 ■ Bombs, target identification ■ Caps, toy ■ Cartridges, flash, 1.1G, 1.3G ■ Cartridges, illuminating ■ Cartridges, signal, 1.3G, 1.4G, 1.4S ■ Fireworks, 1.1G, 1.2G, 1.3G, 1.4G, 1.4S ■ Flares, aerial, 1.1G, 1.2G, 1.3G, 1.4G, 1.4S ■ Flares, aeroplane ■ Flares, distress, small ■ Flares, highway or railway ■ Flares, surface, 1.1G, 1.2G, 1.3G ■ Flares, water-activated ■ Flash powder, 1.1G, 1.3G ■ Fusees (railway or highway) ■ Fusees, railway or highway, explosive ■ Fuses, tracer ■ Grenades, illuminating ■ Grenades, smoke ■ Matches, trick ■ Projectiles, illuminating ■ Railway fusees ■ Railway torpedo ■ Signal devices, hand, 1.4G, 1.4S ■ Signal, highway ■ Signals, distress, ship, 1.1G, 1.3G ■ Signals, distress, ship, water-activated ■ Signals, railway track, explosive, 1.1G, 1.3G, 1.4G, 1.4S ■ Signals, smoke, 1.1G, 1.2G, 1.3G, 1.4G ■ Toy caps, 1.4S ■ Tracers for ammunition, 1.3G, 1.4G ■ Very signal cartridge

0015 ■ 0016 ■ 0037 ■ 0038 ■ 0039 ■ 0049 ■ 0050 ■ 0054 ■ 0092 ■ 0093 ■ 0094 ■ 0171 ■ 0191 ■ 0192 ■ 0193 ■ 0194 ■ 0195 ■ 0196 ■ 0197 ■ 0212 ■ 0245 ■ 0246 ■ 0254 ■ 0297 ■ 0299 ■ 0303 ■ 0305 ■ 0306 ■ 0312 ■ 0313 ■ 0333 ■ 0334 ■ 0335 ■ 0336 ■ 0337 ■ 0373 ■ 0403 ■ 0404 ■ 0405 ■ 0418 ■ 0419 ■ 0420 ■ 0421 ■ 0428 ■ 0429 ■ 0430 ■ 0431 ■ 0432 ■ 0487 ■ 0492 ■ 0493 ■ 2028

Pyrotechnic substances include those chemicals or mixtures that undergo combustion or explosive deflagration to produce heat, light, sound, gas, or smoke.

AMMUNITION

Pyrotechnic ammunition provides pyrotechnic effects for military or civil defense purposes by combining a pyrotechnic chemical with one or more of the following items:

- An expelling or propelling charge to project the round of ammunition toward its target.
- A burster which emits the chemical on explosion to the atmosphere or ignites the pyrotechnic chemical.
- An initiating device that detonates the burster or ignites the pyrotechnic chemical.

Illuminating Ammunition

Illuminating ammunition including *illuminating cartridges*, *illuminating grenades*, *illuminating bombs*, and other *illuminating projectiles* provides visible light or infrared radiation over a target area so that other weapons may be properly sighted or so that observers can see, particularly at night. Magnesium powder is usually the fuel for visible light; alkali nitrates produce infrared radiation on combustion. Visual sighting may be provided by overhead illumination or by ammunition fired behind a target to generate a silhouette. Illuminating bombs usually attach a device to a parachute allowing, for example, a typical 50mm pyrotechnic to provide around 250,000 candela for about 30 seconds as it falls. Illuminating grenades are often projected by a weapon and can contain around 1kg of illuminant providing enough light to illuminate a 200m diameter circle for about a minute (80,000 candela).

Photoflash bombs syn. target-identification bombs contain compositions that produce brief but intense light for medium altitude night photography. *Flash cartridges* are pyrotechnic cartridges that fulfil the same purpose for low altitude night photography.

Tracers

Tracers syn. tracer fuses are relatively slow burning low explosives that emit visible flame (and sometimes smoke) as they react. They are based on barium nitrate, powdered magnesium, and other metal salts which colour the flame; red is the most visible against most backgrounds. They are integral to some ammunition so that the path and strike of the projectile may be seen and trajectory adjustments made if necessary. In small arms fire, tracer compositions may be inserted into a void in a bullet which, ignited by the propellant, burns during flight. In larger rounds, tubes are filled with a tracer composition and inserted into the base of a projectile.

Smoke Ammunition

Smoke ammunition including *smoke bombs* and *smoke grenades* delivers smoke-producing chemicals to screen the movement of troops or equipment and for marking targets. White smoke is produced by the following chemicals as they explode or spontaneously combust on exposure to atmospheric moisture or oxygen:

- White phosphorus which ignites spontaneously and develops dense smoke very quickly. Unfortunately, the reaction is extremely exothermic so that much of the smoke is funnelled upward as the heat rises rather than hugging the ground as desired. The heat of reaction also presents great risk of fire in handling and storage.
- Titanium chloride, which burns more slowly and cooler than white phosphorus.
- A mixture of zinc and hexachloroethane.
- Corrosive liquids such as mixtures of sulphur trioxide and chlorosulphonic acid or a mixture of sulphur trioxide in fuming sulphuric acid.

Coloured smoke is usually a pyrotechnic mixture of chlorate salts and a fuel which burns and distributes dye particles into the air. Smoke ammunition includes a burster which opens the cartridge on arrival, or it includes a system which ignites the mixture in the container to allow the smoke to pass out through a series of perforations.

SIGNALS AND FLARES

Generally, the term *flare* is synonymous with **signal**. Pyrotechnic signals and flares burn to produce a single source of bright light for a few minutes to illuminate, mark, signal distress, or warn. In general, smoke and bright aerial flares (20,000 to 100,000 candela) are useful only during the day, while lower power flares (around 200 to 500 candela) are suitable at night. Those that are fired from a gun use *signal cartridges*, cases containing the pyro-technic, primer, and propellant. Signals and flares fall into four categories:

- Static flares that are hand-held or placed on the ground (*surface flares*).
- Floating flares that generate smoke from the water surface.
- *Aerial flares* (*syn. meteor* or *rocket flares*) that are propelled from a gun and signal as they ascend (to around 70m) and descend.
- Aerial flares that are propelled from a gun to around 400m and signal as they descend slowly under a parachute.

A simple example may include a *highway flare*, a pyrotechnic substance en-cased in a paper tube on one end of which is an igniting composition which is struck to initiate the pyrotechnic. On the other end is a spike which is stuck into the ground holding the flare vertical as it burns. Other flares include

- The *Very signal cartridge*, coloured signal flares fired from a short-barrel, large-calibre pistol named after its inventor, Edward W. Very of the U.S. Navy. The pyrotechnic substance is ignited by the propel-ling charge. They are often red, but coded green, white, and other col-ours are used.
- *Aeroplane flares*, which are fired from aircraft using a barrel and a propelling charge. At a safe distance from the aircraft, a time fuze ignites an expelling charge which ejects a flare attached to a parachute.
- *Railway track signal syn. railway fusees* or *railway torpedo*, warning signals placed on railway tracks that produce a loud report when crushed by the train.

FIREWORKS

Fireworks are articles that provide audible and visible pyrotechnic effects solely for entertainment purposes, although the term may be broadened to include civilian-use signals and flares. Projectile fireworks contain a charge of propellant, or they may be expelled from a mortar. For major displays, fireworks contain a bursting charge that explodes and ignites *stars*, small units of pyrotechnic chemicals compounded in a resin or shellac especially chosen and prepared for their colours. Star compositions include chlorinated isoprene (green), copper carbonate (blue), strontium carbonate (red), and sodium nitrate (yellow). Magnesium powder increases the brilliance of the display while aluminium flakes give luminous tails. Audible explosions may be enhanced with *flash powder*, chemicals based on chlorates and other oxidizers and metal powders that produce an intense flash as well as an audible report. Nonprojectile fireworks contain similar compositions.

Amorce is a French word that means cap, primer, detonator, or fuse, terms that differ from the English usage of amorce used to describe the *toy caps* used in toy pistols or thrown down sharply in play. They contain extremely small quantities (around 1mg) of shock sensitive explosive mixtures of potassium chlorate and arsenic monosulphide or silver fulminate. *Trick matches* are toys that are extremely difficult to blow out.

PYROTECHNIC ARTICLES
Many other *pyrotechnic articles* are in use including gas generators, heat generators, theatrical effects (simulated bullets, smoke, flash powder), flashbulbs, airbag inflators, and thermitic welding kits. There is not a clear division between these and certain explosive articles.

RELATED TERMS
Ammunition, see *Ammunition*, p.8
Articles, see *Explosive Articles*, p.69
Bombs, see *Ammunition*, p.8
Cartridges, see *Ammunition*, p.8
Contrivances, water-activated, see *Explosive Articles*, p.69
Corrosive liquid, see *Corrosives and Class 8*, p.47
Explosive, see *Explosives and Class 1*, p.74
Grenades, see *Ammunition*, p.8
Liquid, see *Terminology, Liquid*, p.241
Projectiles, see *Ammunition*, p.8
Water-activated; Water-activated contrivances, see *Explosive Articles*, p.69

REGULATORY DEFINITIONS
Ammunition, Illuminating with or without burster, expelling charge or propelling charge
Ammunition designed to produce a single source of intense light for lighting up an area. The term includes illuminating cartridges, grenades and projectiles; and illuminating and target identification bombs. The term excludes the following articles which are listed separately: Cartridges, Signal; Signal Devices, Hand; Signals, Distress; Flares, Aerial and Flares, Surface. UN App. B, ICAO A2, US 173.59

Ammunition, Illuminating with or without burster, expelling charge or propelling charge.
Ammunition designed to produce a single source of intense light for lighting up an area. The term includes: Illuminating cartridges; Grenades; Projectiles; Illuminating and target identification bombs. The term excludes Cartridges, signal; Signal devices, hand; Signals, distress; Flares, aerial and Flares, surface. These articles are listed separately. IATA App. A

Ammunition, Smoke Ammunition containing smoke-producing substance such as chlorosulphonic acid mixture, titanium tetrachloride or white phosphorus; or smoke-producing pyrotechnic composition based on hexachloroethane or red phosphorus. Except when the substance is an explosive per se, the ammunition also contains one or more of the following: a propelling charge with primer and igniter charge; a fuze with burster or expelling charge. The term includes grenades, smoke but excludes Signals, Smoke which are listed separately. The term includes: Ammunition, Smoke with or without burster, expelling charge or propelling charge; Ammunition, Smoke, White Phosphorus with burster, expelling charge or propelling. UN App. B, ICAO A2, IATA App. A

Ammunition, smoke. Ammunition containing a smoke-producing substance such as chlorosulphonic acid mixture (CSAM), titanium tetrachloride (FM), white phosphorus, or smoke-producing substance whose composition is based on hexachlorothannol (HC) or red phosphorus. Except when the substance is an explosive per se, the ammunition also contains one or more of the following: a propelling charge with primer and igniter charge, or a fuze with burster or expelling charge. The term includes: Ammunition, smoke, with or without burster, expelling charge or propelling charge; Ammunition, smoke, white phosphorus with burster, expelling charge or propelling charge. US 173.59

Articles, Pyrotechnic for technical purposes Articles which contain pyrotechnic substances and are used for technical purposes such as heat generation, gas generation, theatrical effects, etc. The term excludes the following articles which are listed separately: all ammunition: Cartridges, Signal; Cutters, Cable, Explosive; Fireworks; Flares, Aerial; Flares, Surface; Release Devices, Explosive; Rivets, Explosive; Signal Device, Hand; Signals, Distress; Signals, Railway Track, Explosive; Signals, Smoke. UN App. B, ICAO A2, US 173.59

Articles, Pyrotechnic for technical purposes Articles which contain pyrotechnic substances and are used for technical purposes such as heat generation, gas generation, theatrical effects, etc. The term excludes Ammunition (all); Cartridges, signal; Cutters, cable, explosive; Fireworks; Flares, aerial; Flares, surface; Release Devices, Explosive; Rivets, Explosive; Signal devices, hand; Signals, distress; Signals, railway track, explosive; Signals, smoke. These articles are listed separately. IATA App. A

Caps, Toy (Amorces). Articles consisting of a small quantity of an explosive substance between two strips or discs of paper or contained in a plastic cup or covered by varnishing or other means. ICAO A2

Caps, Toy (Amorces). Articles consisting of a small quantity of an explosive substance between two strips or discs of paper or contained in a plastic cup or covered by varnishing or other means. These caps should not contain more than 16 milligrams (0.25 grain) of the explosive mixture each. IATA App. A

Cartridges, Flash Articles consisting of a casing, a primer and flash powder, all assembled in one piece ready for firing. UN App. B, ICAO A2, IATA App. A

Cartridges, flash. Articles consisting of a casing, a primer and flash powder, all assembled in one piece for firing. US 173.59

Cartridges, Signal Articles designed to fire coloured flares or other signals from signal pistols, etc. UN App. B, ICAO A2, IATA App. A

Cartridges, signal. Articles designed to fire colored flares or other signals from signal pistols or devices. US 173.59

Fireworks Pyrotechnic articles designed for entertainment. UN App. B, ICAO A2, US 173.59

Fireworks. Fireworks are pyrotechnic articles designed for entertainment. IATA App. A

Flares Articles containing pyrotechnic substances which are designed for use to illuminate, identify, signal or warn. The term includes: Flares, Aerial; Flares, Surface. UN App. B, IATA App. A

Flares. Articles containing pyrotechnic substances which are designed to illuminate, identify, signal, or warn. The term includes: flares, aerial and flares, surface. US 173.59

Flash Powder Pyrotechnic substance which, when ignited, produces an intense light. UN App. B, ICAO A2, US 173.59, IATA App. A

Pyrotechnic substance is a substance or a mixture of substances designed to produce an effect by heat, light, sound, gas or smoke or a combination of these as the result of non-detonative self-sustaining exothermic chemical reactions. UN 2.1.1.3(b); ICAO 2-1.5

A pyrotechnic substance is a substance, or a mixture of substances, designed to produce an effect by heat, light, sound, gas or smoke, or a combination of these, as a result of non-detonative, self-sustaining, exothermic chemical reactions. IMO Class 1, 1.4.2

Pyrotechnic Substance. A mixture or compound designed to produce an effect by heat, light, sound, gas or smoke, or a combination of these, as the result of non-detonative, self-sustaining, exothermic, chemical reactions. IATA App. A

Signals Articles containing pyrotechnic substances designed to produce signals by means of sound, flame or smoke or any combinations thereof. The term includes: Signal Devices, Hand; Signals, Distress, ship; Signals, Railway Track, Explosive; Signals, Smoke. UN App. B, ICAO A2, US 173.59, IATA App. A

Tracers For Ammunition Sealed articles containing pyrotechnic substances, designed to reveal the trajectory of a projectile. UN App. B, ICAO A2, US 173.59, IATA App. A

References
27CFR55, 33CFR175, DOMS, DOMT, DOW, E2, E3, EOE, GOCE, JDOMT, KOC, MH14, SEO, SG, TIE, VSW

Radioactive Materials and Class 7

In each of the following entries *Radioactive Material* has been abbreviated to *RM*:

Fissile RM ▪ RM, excepted package, articles, 7 ▪ RM, excepted package, articles manufactured from depleted uranium, 7 ▪ RM, excepted package, articles manufactured from natural or depleted uranium or natural thorium, 7 ▪ RM, excepted package, articles manufactured from natural thorium, 7 ▪ RM, excepted package, articles manufactured from natural uranium, 7 ▪ RM, excepted package - articles manufactured from natural uranium or depleted uranium or natural thorium ▪ RM, excepted package - empty packaging, 7 ▪ RM, excepted package, instruments, 7 ▪ RM, excepted package - instruments or articles, 7 ▪ RM, excepted package - limited quantity of material, 7 ▪ RM, fissile, n.o.s., 7 ▪ RM, low specific activity (LSA), n.o.s., 7 ▪ RM, low specific activity (LSA-I) non fissile or fissile excepted, 7 ▪ RM, low specific activity (LSA-II) fissile, 7 ▪ RM, low specific activity (LSA-II) non fissile or fissile excepted, 7 ▪ RM, low specific activity (LSA-III) fissile, 7 ▪ RM, low specific activity (LSA-III) non fissile or fissile excepted, 7 ▪ RM, low specific activity, n.o.s., 7 ▪ RM, LSA, n.o.s., 7 ▪ RM, n.o.s., 7 ▪ RM, SCO, 7 ▪ RM, special form, n.o.s., 7 ▪ RM, special form, Type A package, non fissile or fissile excepted ▪ RM, surface contaminated object, 7 ▪ RM, surface contaminated objects (SCO), 7 ▪ RM, surface contaminated objects (SCO-I or SCO-II) fissile, 7 ▪ RM, surface contaminated objects (SCO-I or SCO-II) non fissile or fissile excepted, 7 ▪ RM, transported under special arrangement, fissile, 7 ▪ RM, transported under special arrangement, non fissile or fissile excepted, 7 ▪ RM, Type A package, fissile non-special form, 7 ▪ RM, Type A package, non-special form, non fissile or fissile excepted, 7 ▪ RM, Type A package, special form, fissile, 7 ▪ RM, Type B(M) package fissile, 7 ▪ RM, Type B(M) package non fissile or fissile excepted, 7 ▪ RM, Type B(U) package, fissile, 7 ▪ RM, Type B(U) package, non fissile or fissile excepted, 7 ▪ RM, Type C package, fissile, 7 ▪ RM, Type C package, non fissile or fissile excepted, 7 ▪ RM, uranium hexafluoride, fissile, 7 ▪ RM, uranium hexafluoride non fissile or fissile excepted, 7

2908 ▪ 2909 ▪ 2910 ▪ 2911 ▪ 2912 ▪ 2913 ▪ 2915 ▪ 2916 ▪ 2917 ▪ 2918 ▪ 2919 ▪ 2974 ▪ 2977 ▪ 2978 ▪ 2982 ▪ 3321 ▪ 3322 ▪ 3323 ▪ 3324 ▪ 3325 ▪ 3326 ▪ 3327 ▪ 3328 ▪ 3329 ▪ 3330 ▪ 3331 ▪ 3332 ▪ 3333

All materials are composed of the atoms of one or more chemical elements, each with its own characteristic number (the atomic number) of protons (positively charged particles) in its nucleus. Carbon, for example, has six protons, while nitrogen has seven. The chemical reactions and physical properties of materials are primarily dependent on the bonds between these atoms. Atomic nuclei also contain neutrons (particles with no charge), the number of which can vary with little or no change to the physical or chemical properties of the element. The sum of the protons and neutrons in a nucleus is the *atomic weight*. (Protons and neutrons are themselves made of subatomic particles.)

While the number of protons is fixed, the number of neutrons, and thus the atomic weight of an element, can vary; for example, carbon exists naturally with atomic weights of 11, 12, and 14 corresponding to five, six, or eight neutrons, respectively, in addition to its six protons. These three forms of carbon are its *isotopes*. Some elements have no isotopes while others may have a great many (e.g., caesium has 52 natural and synthetic isotopes). Generally, the greater the atomic number, the greater the number of possible isotopes.

RADIOACTIVITY

Not all of the 300 naturally occurring isotopes are stable. Unstable nuclei (protons and neutrons) spontaneously transform (decay) to achieve stability and are the *radioactive materials* *syn. radionuclides* or *radioisotopes*. Many other radioactive isotopes are made artificially by bombarding atoms with neutrons or charged particles in processes that occur in nuclear energy reactors and particle accelerators. As nuclei decay, they emit one of four types of radiation characteristic of the atom:

□ α (alpha) radiation, helium nuclei consisting of two protons and two neutrons with an atomic number of two, an atomic weight of four, and a positive charge. Alpha particles have an energy of around 4 to 8 MeV, travel only a few centimetres in air, and are easily stopped by metals or paper.

□ β (beta) radiation, positively or negatively charged particles (electrons and positrons) of around 0 to 4 MeV. Generally, they require about 1000 times more mass than alpha particles to stop them. A 1 MeV beta particle will travel about 4m in air, but only 5cm in water.

□ γ (gamma) radiation, electromagnetic radiation of extremely short wavelength[20] (between that of cosmic rays and x-rays), extremely high energy, and no mass. Being electrically neutral they penetrate and damage many materials, including living tissue, without being easily absorbed. A 1 MeV gamma ray travels over 70m in air and 10cm in water. They are best stopped by dense materials such as lead. Some nuclides like plutonium-239 emit radiation in the x-ray spectrum.[21]

□ Neutron radiation occurs only on fission (fission is the splitting of a nucleus) or by the bombardment (irradiation) of atomic nuclei with charged particles. Neutrons have mass but no charge, and like gamma radiation penetrate and damage many materials.

FISSILE MATERIAL

Neutron radiation is used to fission unstable nuclei. Uranium-235 (^{235}U) is unstable and when struck by a neutron will disintegrate to form several other elements (fission products) of extremely high energy, all of which are radioactive. In addition, the average ^{235}U nuclei will emit 2.5 neutrons and, if a sufficient *critical mass* (about 15kg for ^{235}U) of the material is present, a self-perpetuating chain reaction of tremendous potential force will propagate. This principle underlies atomic bombs and nuclear power generation. While

---

[20] 1 to 0.001 Angstroms.

[21] 100 to 0.1 Angstroms.

any nucleus (other than hydrogen-1) can in principle undergo fission, in transportation terms *fissile material* is taken to mean uranium-233 and -235, and plutonium-238, -239, and -241, the common fissile materials in commerce.

Natural uranium is found in pitchblende (metallic veins of uraninite and uranium oxide) and other ores as a mixture of isotopes: ^{234}U, 0.0055%; ^{235}U, 0.720%; and ^{238}U, 99.2745%. Uranium-233, a fissile isotope, is produced artificially by bombardment of *natural thorium* (essentially 100% thorium-232) with neutrons and is used as a nuclear fuel. Uranium-238 is nonfissile but will produce fissile plutonium-239 on neutron bombardment. Uranium-235 is readily fissionable and was separated from natural uranium for use in the first atom bombs. Enriched natural uranium, in which the relative concentration of ^{235}U is increased through chemical and physical means, is used in nuclear fuels. Uranium hexafluoride is widely used to separate uranium isotopes by gaseous diffusion (*depleted uranium* is the product left when ^{235}U has been mostly removed from natural uranium).

ACTIVITY AND UNITS

The more unstable the isotope, the more rapid the rate of decay. The larger the amount of radioactive material, the greater the radiation. The number of atomic disintegrations per second is termed the *activity*, the original unit of which was the *Curie* (Ci) named after Marie S. Curie, the Polish scientist who conducted pioneering work in radioactivity. It was defined as 3.7×10^{10} disintegrations/second, roughly that of 1g of natural radium. The international unit is the *becquerel* (Bq) after Henri Becquerel, a Frenchman who shared a Nobel prize with Marie Curie and her husband. It is defined as 1 disintegration/second.

As decay occurs, the remaining activity declines. The time it takes for a radionuclide to lose half its activity is its *half-life* ($t_{1/2}$), which may range from extremely short to extremely long periods. The half-lives of polonium-214 and uranium-238, for example, are 163.7μs and 4.46×10^9 years, respectively. As individual isotopes decay they can form new stable or unstable isotopes in a series of steps that eventually ends in a stable nucleus. The type of decay and the half-lives of the intermediaries in a decay series is characteristic of the isotope; e.g., radioactive thorium-232 undergoes the following decay steps to result in a stable isotope of lead:

ISOTOPE	$t_{1/2}$
^{232}Th	1.4×10^{10} years
^{228}Ra + α	5.76 years
^{228}Ac + β⁻	6.15h
^{228}Th + β⁻	1.913 years
^{224}Ra + α	3.66 days

ISOTOPE	$t_{1/2}$ (CONT.)
^{220}Rn + α	55.6s
^{216}Po + α	0.145s
^{212}Pb + α	10.64h
^{212}Bi + β^-	1.009h
^{212}Po + β^- and ^{208}Tl + α	0.298μs, 3.053min
^{208}Tl + α ^{208}Pb + β^-	3.053min, >2x10^{19}
^{208}Pb + β^-	>2x10^{19} years

The amount of radioactivity any individual receptor (e.g., humans) receives is the *dose*. There are a number of units used to quantify dose:

◻ The *roentgen* (R), after W.K. Roentgen, the German physicist who discovered x-rays. It was originally applied to gamma- and x-radiation, but extended to cover particle radiation. The roentgen is a measure of ionization strength.

◻ The *radiation absorbed dose* or *rad*, measuring the amount of energy absorbed by the irradiated material.

◻ The *roentgen equivalent man* or *rem* is a measure of dose equivalent; that is, a rad modified by a factor that takes into account the varying biological effects of different types of radiation.

◻ Also in use are the *gray* and the *sievert* (Sv). A gray is 100 rads while a sievert is the absorbed dose that produces the same effect as 1 gray of x- or gamma-rays.

TYPES OF RADIOACTIVE MATERIALS
Radioactive materials are used widely in all parts of the nuclear fuel industry (i.e., mining; milling; chemical conversion; isotope separation; fuel fabrication; reactor operation; spent fuel storage; waste management, including sludges, millings, cooling water filter media, absorbents, fuel rods, piping, resins, etc.), as well the nuclear weapons industry. In addition, radionuclides are used to sterilize wastewater and foods, in analytical crystallographic techniques and metallurgical testing, in industry to gauge thickness, in nuclear medicine in which biologically active compounds are made with isotopes to accumulate in specific organs where they radiate for beneficial effect, and in many other applications.

PACKAGING
Transportation regulations quantify the risk of radioactive materials by considering the type and degree of radiation, the critical mass of fissile materials, and the heat generated by the radiation.

Radionuclides are assigned activity limits, A_1 and A_2, which form the basis of packaging choice:

- A_1 for *special form radioactive materials*: indispersible radioactive solids or sealed capsules containing a radioactive material that presents a low risk of contamination if spilled.
- A_2 for other radioactive materials.

Iodine-131, for example, carries the A_1 and A_2 values of 3 and 0.7 TBq, respectively (a TBq is a terabecquerel or 10^{12} Bqs).

Type A, B, and C Packages

Radioactive material packagings are designed to deal with many factors: temperature variation, shielding, contamination, containment systems, pressure relief, absorbents, corrosion, moisture, heat buildup, water and snow immersion, cooling systems, critical mass, mechanical and structural integrity, etc. There are three main types:

- *Type A packages*, the basic types of packagings or freight containers. They hold activities up to A_1 for special form or A_2 for other radioactive material.
- *Type B(U) packages* and *Type B(M) packages* can hold activities greater than A_1 or A_2 as long as they meet additional specifications, tests, and approval by the competent transportation authorities. Requirements for Type B(U) packages are the more stringent of the two and require *unilateral* design approval, hence *B(U)*. Type B(M) packages, having fewer performance standards, require *multilateral* approval, hence *B(M)*, of the design and shipment by any states through which they pass.
- *Type C packages* have to meet the most rigorous standards.

Industrial Packages

Less dangerous radioactive materials permit less stringent packages. Specifically, *Industrial Package Types 1, 2, and 3* are in use for *low-specific activity (LSA)* materials and *surface-contaminated objects (SCO)*:

- LSA material is that which has by its nature a limited specific activity. It is divided into three groups: (1) *LSA-I* includes ores or naturally occurring radionuclides, unirradiated natural or depleted uranium, and other low activity radioactive materials; (2) *LSA-II* includes certain concentrations of tritium and solids, liquids, and gases with no more than 10,000 to 100,000 times less activity than A_2 per gram; and (3) *LSA-III* includes solid materials that are imbedded in a binder, relatively insoluble, and with not more than 500 times less activity than A_2 per gram.
- SCOs are not themselves radioactive, but have radioactive material distributed on their surfaces, including a variety of contaminated

wastes and residues. Depending on whether the radiation is fixed, nonfixed, or accessible, *SCO-I* includes that which ranges up to 4,000 to 40,000 Bq/cm^2, while *SCO-II* ranges up to around 80,000 to 800,000 Bq/cm^2.

Excepted Packages

Small quantities of fissile material are excepted (*fissile excepted*) from some fissile packaging requirements. They are not, however, excepted from the requirements appropriate to their other radioactive properties. Reduced requirements also apply to the following *excepted packages*:

- Limited quantities of materials with low activities.
- Instruments or articles with low activities (e.g., clocks, sun lamps, photoelectric cells, fire alarm sensors, radioactivity counters, x-ray tubes).
- Articles manufactured from natural thorium or depleted uranium (e.g., armour-piercing ammunition, alloys, light filaments).
- Empty packages with low remaining nonfixed contamination.

Special Arrangement

If a radioactive material does not meet with the packaging requirements, equivalent means must be found and the material must be transported under multilateral approval (*special arrangement*) by the states of origin, transit, and destination.

HAZARDS

Radiation includes charged particles, neutral particles, or electromagnetic emissions of extremely high energy. In living systems, the energy or charge is sufficient to remove electrons (ionize) from atoms in the path of the radiation. The resulting free radicals (ionized atoms and molecules) are charged, short-lived, and highly reactive particles that break down many organic molecules leaving mutated DNA and damaged proteins, amino acids, bone marrow, and body tissue. Skin burns, damaged cells, and a host of cancers are possible.

While some radionuclides may be quickly expelled others may be ingested, inhaled, or absorbed into the body where, depending on their half-lives and type of radioactivity, they may continue to emit harmful radiation for many years, if not life. Some may concentrate in certain organs; plutonium, for example, concentrates in blood-forming tissues in bone and in the liver, while plutonium oxide lodges in the lungs.

Radionuclides or fission products also generate heat. For example 100g of plutonium-239 will rise about 30 to 60°C above ambient temperature due to the heat of alpha-emissions while smaller samples may heat to incandescence. Other radionuclides give off gases as they decay or exhibit any of the other hazard classes (e.g., uranium hexafluoride is extremely corrosive and powdered thorium is explosive).

<u>RELATED TERMS</u>
None

<u>REGULATORY DEFINITIONS:</u>

<u>Class 7</u>
Radioactive material shall mean any material containing radionuclides where both the activity concentration and the total activity in the consignment exceed the values specified in paras 401-406. IAEA para. 236

Class 7 material (radioactive material) shall mean any material containing radionuclides where both the activity concentration and the total activity in the consignment exceed the values specified in paragraphs 401-406 of the Regulations for the Safe Transport of Radioactive Material, (1996) IAEA Safety Standards Series No. ST-1. UN 2.7.1

For the purposes of transport, any material with a specific activity greater than 70 kBq/kg (0.002 µCi/g) must be declared as a radioactive material. Nevertheless, a material coming within that definition may be present in such small quantity or incorporated in another material to such an extent that its hazardous nature is very much reduced and it may be excepted from certain packing and labelling requirements (see schedule 1, 2, 3 or 4 as appropriate). IMO Class 7, 1.1.5

Any material with a specific activity greater than 70 kBq/kg (2 nCi/g) is included in Class 7. ICAO 2-7.1

Radioactive material means any material having a specific activity greater than 70 Bq per gram (0.002 microcurie per gram) (see definition of "specific activity"). US 173.403

For the purposes of these Regulations, a radioactive material is any article or substance with a specific activity greater than 70 kBq/kg (0.002 µCi/g). IATA 3.7.1

<u>Other Definitions</u>
A_1 shall mean the activity value of special form radioactive material which is listed in Table I or derived in Section IV and is used to determine the activity limits for the requirements of these Regulations. A_2 shall mean the activity value of radioactive material, other than special form radioactive material, which is listed in Table I or derived from Section IV and is used to determine the activity limits for the requirements of these Regulations. IAEA para. 201

A_1 *and* A_2 *values for radionuclides* A_1 means the maximum activity of special form radioactive material permitted in a Type A package. A2 means the maximum activity of radioactive material other than special form radioactive material permitted in a Type A package. The values are listed in appendix 1 to this class. IMO Class 7, 2.1

A_1. The maximum activity of special form radioactive material permitted in a Type A package. A_1 values for those radionuclides commonly transported are listed in Table 2-9. ICAO 2-7.2

A_1 *and* A_2 *(Radioactive Material Only):* A_1: the maximum activity of special form radioactive material permitted in a Type A package. A_2: the maximum activity of radioactive material, other than special form radioactive material, permitted in a Type A package. IATA App. A

A_1 means the maximum activity of special form Class 7 (radioactive) material permitted in a Type A package. US 173.403

A_2 means the maximum activity of Class 7 (radioactive) material, other than special form, LSA or SCO, permitted in a Type A package. These values are either listed in Sec. 173.435 or derived in accordance with the procedure prescribed in Sec. 173.433. US 173.403

A_2. The maximum activity of radioactive material, other than special form radioactive material, permitted in a Type A package. A_2 values for those radionuclides commonly transported are listed in Table 2-9. ICAO 2-7.2

Articles manufactured from natural or depleted uranium or natural thorium Manufactured articles in which the sole radioactive material is unirradiated natural uranium, unirradiated depleted uranium or unirradiated natural thorium may be transported as an excepted package, provided that the outer surface of the uranium or thorium is enclosed in an inactive sheath made of metal or some other substantial material. ICAO 2-7.9.4

Contamination shall mean the presence of a radioactive substance on a surface in quantities in excess of 0.4 Bq/cm2 for beta and gamma emitters and low toxicity alpha emitters, or 0.04 Bq/cm2 for all other alpha emitters. IAEA para. 214

Contamination means the presence of radioactive material on a surface in quantities in excess of 0.4 Bq/cm^2 (10^{-5} μCi/cm^2) for beta and gamma emitters and low-toxicity alpha emitters, or 0.04 Bq/cm^2 (10^{-6} μCi/cm^2) for all other alpha emitters. Low-toxicity alpha emitters are: natural uranium; depleted uranium; uranium-235; uranium-238; natural thorium; thorium-232; thorium-228 and thorium-230 when contained in ores, or physical or chemical concentrates; or radionuclides with a half-life of less than 10 days. IMO Class 7, 2.8.1

Contamination. The presence of a radioactive substance on a surface in quantities in excess of 0.4 Bq/cm^2 (0.01 nCi/cm^2) for beta and gamma emitters and low toxicity alpha emitters, or 0.04 Bq/cm^2 (0.001 nCi/cm^2) for all other alpha emitters. This is either: *Fixed contamination* -- contamination other than non-fixed contamination; or *Non-fixed contamination* -- contamination that can be removed from a surface during normal handling. ICAO 2-7.2, IATA App. A

Fissile material shall mean uranium-233, uranium-235, plutonium-239, plutonium-241, or any combination of these radionuclides. Excepted from this definition is: (a) natural uranium or depleted uranium which is unirradiated, and (b) natural uranium or depleted uranium which has been irradiated in thermal reactors only. IAEA para. 222

Fissile materials means uranium-233, uranium-235, plutonium-238, plutonium-239, plutonium-241, or any combination of these radionuclides. Unirradiated natural uranium or unirradiated depleted uranium, and natural uranium or depleted uranium which has been irradiated in thermal reactors only, are not included in this definition. IMO Class 7, 2.12.2

Fissile material. The radionuclides uranium-233, uranium-235, plutonium-238, plutonium-239, plutonium-241, or any combination of these but excluding unirradiated natural uranium and depleted uranium, and natural uranium or depleted uranium which has been irradiated in thermal reactors only. ICAO 2-7.2.

Fissile material means plutonium-238, plutonium-239, plutonium-241, uranium-233, uranium-235, or any combination of these radionuclides. The definition does not apply to unirradiated natural uranium and depleted uranium, and natural uranium or depleted uranium that has been irradiated in a thermal reactor. Certain additional exceptions are provided in Sec. 173.453. US 173.403

Fissile Material. Uranium-233, uranium-235, plutonium-238, plutonium-239, plutonium-241 or any combination of these. Unirradiated natural and depleted uranium and natural uranium or depleted uranium which has been irradiated in thermal reactors only are not included under this definition. IATA App. A

Fissile means capable of undergoing fission, a process in which the atoms of the fissile radionuclide are split by neutron radiation into two approximately equal parts (fission products), with the attendant release of more neutrons and energy in the form of heat and ionizing radiation. IMO Class 7, 2.12.1

Fixed contamination shall mean contamination other than non-fixed contamination. IAEA para. 216

Fixed contamination means contamination other than non-fixed contamination. IMO Class 7, 2.8.3

Low dispersible radioactive material shall mean either a solid radioactive material or a solid radioactive material in a sealed capsule, that has limited dispersibility and is not in powder form. IAEA para. 225

207

Low-specific activity (LSA) material shall mean radioactive material which by its nature has a limited specific activity, or radioactive material for which limits of estimated average specific activity apply. External shielding materials surrounding the LSA material shall not be considered in determining the estimated average specific activity. LSA material shall be in one of three groups: (a) LSA-I (i) Uranium and thorium ores and concentrates of such ores, and other ores containing naturally occurring radionuclides which are intended to be processed for the use of these radionuclides; (ii) Solid unirradiated natural uranium or depleted uranium or natural thorium or their solid or liquid compounds or mixtures; (iii) Radioactive material for which the A_2 value is unlimited, excluding fissile material in quantities not excepted under para. 672; or (iv) Other radioactive material in which the activity is distributed throughout and the estimated average specific activity does not exceed 30 times the values for activity concentration specified in paras 401-406, excluding fissile material in quantities not excepted under para. 672. (b) LSA-II (i) Water with tritium concentration up to 0.8 TBq/L; or (ii) Other material in which the activity is distributed throughout and the estimated average specific activity does not exceed 10^{-4} A_2/g for solids and gases, and 10^{-5} A_2/g for liquids. (c) LSA-III Solids (e.g., consolidated wastes, activated materials), excluding powders, in which: (i) The radioactive material is distributed throughout a solid or a collection of solid objects, or is essentially uniformly distributed in a solid compact binding agent (such as concrete, bitumen, ceramic, etc.); (ii) The radioactive material is relatively insoluble, or it is intrinsically contained in a relatively insoluble matrix, so that, even under loss of packaging, the loss of radioactive material per package by leaching when placed in water for seven days would not exceed 0.1 A_2; and (iii) The estimated average specific activity of the solid, excluding any shielding material, does not exceed 2×10^{-3} A_2/g. IAEA para. 226

Low specific activity material means radioactive material which by its nature has a limited specific activity or radioactive material for which limits of estimated average specific activity apply. External shielding materials surrounding the LSA material should not be considered in determining the estimated average specific activity. Low specific activity material shall be in one of three groups: LSA-I, LSA-II and LSA-III: LSA-I (i) Ores containing naturally occurring radionuclides (e.g. uranium, thorium), and uranium or thorium concentrates of such ores; (ii) Solid unirradiated natural uranium or depleted uranium or natural thorium or their solid or liquid compounds or mixtures; or (iii) radioactive material, other than fissile material, for which the A_2 value is unlimited. LSA-II (i) Water with tritium concentration up to 0.8 TBq/l (20 Ci/l); or (ii) Other material in which the activity is distributed throughout and the estimated average specific activity does not exceed 10^{-4} A_2/g for solids and gases, and 10^{-5} A_2/g for liquids. LSA-III Solids (e.g. consolidated wastes, activated materials) in which: (i) The radioactive material is distributed throughout a solid or a collection of solid objects, or is essentially uniformly distributed in a solid compact binding agent (such as concrete, bitumen, ceramic, etc); (ii) The radioactive material is relatively insoluble, or it is intrinsically contained in a relatively insoluble matrix, so that, even under loss of packaging, the loss of radioactive material per package by leaching when placed in water for seven days would not exceed 0.1 A_2; and (iii) The estimated average specific activity of the solid, excluding any shielding material, does not exceed 2×10^{-3} A_2/g. For LSA material, see schedules 5, 6 and 7. IMO Class 7, 2.14

Low-specific activity (LSA) material. Radioactive material which by its nature has a limited specific activity, or radioactive material for which limits of estimated average specific activity apply (see 7.3). ICAO 2-7.2

LSA material is classified in one of three groups as follows: a) LSA-I: 1) ores containing naturally occurring radionuclides (e.g. uranium, thorium) and uranium or thorium concentrates of such ores: 2) solid, unirradiated natural uranium or depleted uranium or natural thorium or their solid or liquid compounds or mixtures; or 3) radioactive material, other than fissile material, for which the A_2 value is unlimited. b) LSA-II: i) water with tritium concentration up to 0.8 TBq/L (20 Ci/L); or 2) other material in which the activity is distributed throughout and the estimated average specific activity does not exceed 10^{-4} A_2/g for solids and gases, and 10^{-5} A_2/g for liquids. c) LSA-III - solids (e.g. consolidated wastes, activated materials), in which: 1) the radioactive material is distributed throughout a solid or a collection of solid objects, or is essentially uniformly distributed in a solid compact binding agent (such as concrete, bitumen, ceramic, etc.); 2) the radioactive material is relatively insoluble, or it is intrinsically contained in a relatively insoluble matrix, so that, even under loss of packaging, the loss of radioactive material per package by leaching when placed in water for seven days would not exceed 0.1 A_2; and 3) the estimated average specific activity of the solid, excluding any shielding material, does not exceed $2x10^{-3}$ A_2/g. ICAO 2-7.3.2

Low Specific Activity (LSA) material means Class 7 (radioactive) material with limited specific activity which satisfies the descriptions and limits set forth below. Shielding materials surrounding the LSA material may not be considered in determining the estimated average specific activity of the package contents. LSA material must be in one of three groups: (1) LSA-I. (i) Ores containing only naturally occurring radionuclides (e.g., uranium, thorium) and uranium or thorium concentrates of such ores; or (ii) Solid unirradiated natural uranium or depleted uranium or natural thorium or their solid or liquid compounds or mixtures; or (iii) Class 7 (radioactive) material, other than fissile material, for which the A_1 value is unlimited; or (iv) Mill tailings, contaminated earth, concrete, rubble, other debris, and activated material in which the Class 7 (radioactive) material is essentially uniformly distributed and the average specific activity does not exceed 10^{-6} A_2/g. (2) LSA-II. (i) Water with tritium concentration up to 0.8 TBq/liter (20.0 Ci/liter); or (ii) Material in which the Class 7 (radioactive) material is distributed throughout and the average specific activity does not exceed 10^{-4} A_2/g for solids and gases, and 10^{-5} A_2/g for liquids. (3) LSA-III. Solids (e.g., consolidated wastes, activated materials) that meet the requirements of Sec. 173.468 and which: (i) The Class 7 (radioactive) material is distributed throughout a solid or a collection of solid objects, or is essentially uniformly distributed in a solid compact binding agent (such as concrete, bitumen, ceramic, etc.); and (ii) The Class 7 (radioactive) material is relatively insoluble, or it is intrinsically contained in a relatively insoluble material, so that, even under loss of packaging, the loss of Class 7 (radioactive) material per package by leaching when placed in water for seven days would not exceed 0.1 A_2; and (iii) The average specific activity of the solid does not exceed $2x10^{-3}$ A_2/g. US 173.403

Low Specific Activity (LSA) Material. Radioactive material which by its nature has a limited specific activity, or radioactive material for which limits of estimated average activity apply. IATA App. A

LSA-I material is: (a) ores containing naturally occurring radionuclides, e.g. uranium, thorium, and uranium or thorium concentrates of such ores; (b) solid, unirradiated natural uranium or depleted uranium or natural thorium or their solid or liquid compounds or mixtures; or (c) radioactive material, other than fissile material, for which the A_2 value is unlimited. IATA 10.3.5.1.1

LSA-II material is: (a) water with tritium concentration up to 0.8 TBq/L (20 Ci/L); or (b) other material in which the activity is distributed throughout and the estimated average specific activity does not exceed 10^{-4} A_2/g for solid and gases, and 10^{-5} A_2/g for liquids. IATA 10.3.5.1.2

LSA-III material is Solids, e.g. consolidated wastes, activated materials, in which: (a) the radioactive material is distributed throughout a solid or collection of solid objects, or is essentially uniformly distributed in a solid compact binding agent (such as concrete, bitumen, ceramic, etc.); (b) the radioactive material is relatively insoluble, or it is intrinsically contained in a relatively insoluble matrix, so that, even under loss of packaging, the loss of radioactive material per package by leaching when placed in water for seven days would not exceed 0.1 A_2; and (c) the estimated average specific activity of the solid excluding any shielding material, does not exceed $2x10^{-3}$ A_2/g. IATA 10.3.5.1.3

Natural thorium means thorium with the naturally occurring distribution of thorium isotopes (essentially 100 percent by weight of thorium-232). US 173.403

Natural uranium shall mean chemically separated uranium containing the naturally occurring distribution of uranium isotopes (approximately 99.28% uranium-238, and 0.72% uranium-235 by mass). Depleted uranium shall mean uranium containing a lesser mass percentage of uranium-235 than in natural uranium. Enriched uranium shall mean uranium containing a greater mass percentage of uranium-235 than 0.72%. In all cases, a very small mass percentage of uranium-234 is present. IAEA para. 246

Natural uranium means chemically separated uranium containing the naturally occurring distribution of uranium isotopes (approximately 99.28% uranium-238, and 0.72% uranium-235 by mass). *Depleted uranium* means uranium containing a lesser mass percentage of uranium-235 than in natural uranium. *Enriched uranium* means uranium containing a greater mass percentage of uranium-235 than in natural uranium. In all cases, a very small mass percentage of uranium-234 is present. IMO Class 7, 2.30

Non-fixed contamination means contamination that can be removed from a surface during normal handling. IMO Class 7, 2.8.2

Non-fixed contamination shall mean contamination that can be removed from a surface during routine conditions of transport. IAEA para. 215

Non-fixed radioactive contamination means radioactive contamination that can be readily removed from a surface by wiping with an absorbent material. Non-fixed (removable) radioactive contamination is not significant if it does not exceed the limits specified in Sec. 173.443. US 173.403

Non-Fixed Radioactive Contamination (Radioactive Material Only). Radioactive contamination that can be removed from a surface by wiping with a dry smear. IATA App. A

Normal form Class 7 (radioactive) material means Class 7 (radioactive) material which has not been demonstrated to qualify as "special form Class 7 (radioactive) material." US 173.403

Radioactive contents means the radioactive material together with any contaminated solids, liquids, and gases within the packaging. IMO Class 7, 2.19

Radioactive contents. The radioactive material together with any contaminated solids, liquids and gases within the packaging. ICAO 2-7.2, IATA App. A

Radioactive contents means a Class 7 (radioactive) material, together with any contaminated liquids or gases within the package. US 173.403

Radioactive instrument and article means any manufactured instrument and article such as an instrument, clock, electronic tube or apparatus, or similar instrument and article having Class 7 (radioactive) material in gaseous or non-dispersible solid form as a component part. US 173.403

Radioactive material which by its nature has a limited specific activity, or radioactive material for which limits of estimated average specific activity apply, is termed *low specific activity*, or LSA, material. External shielding material surrounding the LSA material must not be considered in determining the estimated average specific activity. ICAO 2-7.3.1, IATA 10.3.5.1

Radioactive materials are grouped according to their form and/or characteristics. These include: Special Form; Low Specific Activity (LSA); Surface Contaminated Object (SCO); Fissile; Other Form. A radioactive material may meet the definition of one or more of the above. IATA 10.3.3

Radionuclide is a shortened form of "radioactive nuclide", practically synonymous with "radioisotope" or "radioactive isotope". A particular chemical element may consist of a number of nuclides, some of which may be radionuclides, characterized by the name of the chemical element followed by a number denoting the atomic mass of the nuclide in question, e.g. cobalt-60 and uranium-235. IMO Class 7, 2.20

Special arrangement shall mean those provisions, approved by the competent authority, under which consignments which do not satisfy all the applicable requirements of these Regulations may be transported. IAEA para. 238

Special arrangement means those provisions approved by the competent authority under which a consignment which does not satisfy all the applicable provisions of this class may be transported. For international shipments of this type, multilateral approval is required; see schedule 13. IMO Class 7, 2.21

Special arrangement. Those provisions approved by the competent authority under which a consignment which does not satisfy all the applicable requirements of these Instructions may be transported. For international shipments of this type multilateral approval is required (see Part 4;1.3.3.3). ICAO 2-7.2

Special form radioactive material shall mean either an indispersible solid radioactive material or a sealed capsule containing radioactive material. IAEA para. 239

Special form radioactive material means either indispersible solid radioactive material or a sealed capsule containing radioactive material. The sealed capsule shall be so constructed that it can be opened only by destroying the capsule. Special form radioactive material shall meet the following requirements: .1 it shall have at least one dimension not less than 5 mm; and .2 it shall comply with the requirements specified in IAEA paragraph 503. IMO Class 7, 2.22.1

Special form radioactive material is either an indispersible solid radioactive material or a sealed capsule containing radioactive material. Special form radioactive material must meet the following requirements: a) if in a sealed capsule, that capsule must be so constructed that it can only be opened by destroying it; b) it must have at least one dimension not less than 5mm; c) the design must have received unilateral approval. ICAO 2-7.4.1

Special form Class 7 (radioactive) material means Class 7 (radioactive) material which satisfies the following conditions: (1) It is either a single solid piece or is contained in a sealed capsule that can be opened only by destroying the capsule; (2) The piece or capsule has at least one dimension not less than 5 millimeters (0.2 inch); and (3) It satisfies the test requirements [text continues]. US 173.403

Special form radioactive material is either an indispersible solid radioactive material or a sealed capsule containing radioactive material. IATA 10.3.4.1

Special form radioactive material. Either an indispersible solid radioactive material or a sealed capsule containing radioactive material. ICAO 2-7.2

Specific activity of a radionuclide shall mean the activity per unit mass of that nuclide. The specific activity of a material shall mean the activity per unit mass or volume of the material in which the radionuclides are essentially uniformly distributed. IAEA para. 240

Specific activity means the activity of a radionuclide per unit mass of that nuclide. The specific activity of a material in which the radionuclide is essentially uniformly distributed is the activity per unit mass of the material. IMO Class 7, 2.23

Specific activity. The activity of the radionuclide per unit mass of that nuclide. The specific activity of a material in which the radionuclides are essentially uniformly distributed is the activity per unit mass of the material. ICAO 2-7.2

Specific activity of a radionuclide means the activity of the radionuclide per unit mass of that nuclide. The specific activity of a material in which the radionuclide is essentially uniformly distributed is the activity per unit mass of the material. US 173.403

Surface contaminated object (SCO) shall mean a solid object which is not itself radioactive but which has radioactive material distributed on its surfaces. SCO shall be in one of two groups: (a) SCO-I: A solid object on which: (i) the non-fixed contamination on the accessible surface averaged over 300 cm^2 (or the area of the surface if less than 300 cm^2) does not exceed 4 Bq/cm^2 for beta and gamma emitters and low toxicity alpha emitters, or 0.4 Bq/cm^2 for all other alpha emitters; and (ii) the fixed contamination on the accessible surface averaged over 300 cm^2 (or the area of the surface if less than 300 cm^2) does not exceed $4x10^4$ Bq/cm^2 for beta and gamma emitters and low toxicity alpha emitters, or $4x10^3$ Bq/cm^2 for all other alpha emitters; and (iii) the non-fixed contamination plus the fixed contamination on the inaccessible surface averaged over 300 cm^2 (or the area of the surface if less than 300 cm^2) does not exceed $4x10^4$ Bq/cm^2 for beta and gamma emitters and low toxicity alpha emitters, or $4x10^3$ Bq/cm^2 for all other alpha emitters. (b) SCO-II: A solid object on which either the fixed or non-fixed contamination on the surface exceeds the applicable limits specified for SCO-I in (a) above and on which: (i) the non-fixed contamination on the accessible surface averaged over 300 cm^2 (or the area of the surface if less than 300 cm^2) does not exceed 400 Bq/cm^2 for beta and gamma emitters and low toxicity alpha emitters, or 40 Bq/cm^2 for all other alpha emitters; and (ii) the fixed contamination on the accessible surface averaged over 300 cm^2 (or the area of the surface if less than 300 cm^2) does not exceed 8x10^5 Bq/cm^2 for beta and gamma emitters and low toxicity alpha emitters, or 8x10^4 Bq/cm^2 for all other alpha emitters; and (iii) the non-fixed contamination plus the fixed contamination on the inaccessible surface averaged over 300 cm^2 (or the area of the surface if less than 300 cm^2) does not exceed 8x10^5 Bq/cm^2 for beta and gamma emitters and low toxicity alpha emitters, or 8x10^4 Bq/cm^2 for all other alpha emitters. IAEA para. 241

Surface contaminated object (SCO) means a non-radioactive solid having activity distributed on its surfaces as specified in schedule 8. (a) SCO-I: A solid object on which: (i) the non-fixed contamination on the accessible surface, averaged over 300 cm^2 (or the area of the surface if less than 300 cm^2), does not exceed 4 Bq/cm^2 (10^{-4} μCi/cm^2) for beta and gamma emitters and low-toxicity alpha emitters, or 0.4 Bq/cm^2 (10^{-5} μCi/cm^2) for all other alpha emitters; and (ii) the fixed contamination on the accessible surface, averaged over 300 cm^2 (or the area of the surface if less than 300 cm^2), does not exceed 4x10^4 Bq/cm^2 (1 μCi/cm^2) for beta and gamma emitters and low-toxicity alpha emitters, or 4x10^3 Bq/cm^2(0.1 μCi/cm^2) for all other alpha emitters; and (iii) the non-fixed contamination plus the fixed contamination on the inaccessible surface averaged over 300 cm^2 (or the area of the surface if less than 300cm^2), does not exceed 4x10^4 Bq/cm^2 (1 μCi/cm^2) for beta and gamma emitters and low-toxicity alpha emitters, or 4x10^3 Bq/cm^2 (0.1 μCi/cm^2) for all other alpha emitters. (b) SCO-II: A solid object on which either the fixed or non-fixed contamination on the surface exceeds the applicable limits specified for SCO-I in (a) above and on which: (i) the non-fixed contamination on the accessible surface, averaged over 300 cm^2 (or the area of the surface if less than 300 cm^2), does not exceed 400 Bq/cm^2 (10^{-2} μCi/cm^2) for beta and gamma emitters and low-toxicity alpha emitters, or 40 Bq/cm^2 (10^{-3} μCi/cm^2) for all other alpha emitters; and (ii) the fixed contamination on the accessible surface, average over 300 cm^2 (or the area of the surface if less than 300cm^2), does not exceed 8x10^5 Bq/cm^2 (20 μCi/cm^2) for beta and gamma emitters and low-toxicity alpha emitters, or 8x10^4 Bq/cm^2 (2 μCi/cm^2) for all other alpha emitters; and (iii) the non-fixed contamination plus the fixed contamination on the inaccessible surface, averaged over 300 cm^2 (or the area of the surface if less than 300 cm^2), does not exceed 8x10^5 Bq/cm^2 (20 μCi/cm^2) for beta and gamma emitters and low-toxicity alpha emitters, or 8x10^4 Bq/cm^2 (2 μCi/cm^2) for all other alpha emitters. IMO Class 7, 2.24

Surface contaminated object (SCO). A solid object which is not itself radioactive but which has radioactive material distributed on its surfaces (see 7.5). ICAO 2-7.2

Surface contaminated object (SCO) means a solid object which is not itself radioactive but which has radioactive material distributed on its surfaces. SCO is classified in one of two groups as follows: a) SCO-I: A solid object on which: 1) the non-fixed contamination on the accessible surface averaged over 300 cm^2 (or the area of the surface if less than 300 cm^2) does not exceed 4 Bq/cm^2 (0.1 nCi/cm^2) for beta and gamma emitters and low toxicity alpha emitters, or 0.4 Bq/cm^2 (0.01 nCi/cm^2) for all other alpha emitters; and 2) the fixed contamination on the accessible surface averaged over 300 cm^2 (or the area of the surface if less than 300 cm^2) does not exceed 40 kBq/cm^2 (1 μCi/cm^2) for beta and gamma emitters and low toxicity alpha emitters, or 4 kBq/cm^2 (0.1 μCi/cm^2) for all other alpha emitters; and 3) the non-fixed contamination plus the fixed contamination on the inaccessible surface averaged over 300 cm^2 (or the area of the surface if less than 300 cm^2) does not exceed 40 kBq/cm^2 (1 μCi/cm^2) for beta and gamma emitters and low toxicity alpha emitters, or 4 kBq/cm^2 (0.1 μCi/cm^2) for all other alpha emitters. b) SCO-II: A solid object on which either the fixed or non-fixed contamination on the surface exceeds the applicable limits specified for SCO-I in a) above and on which: 1) the non-fixed contamination on the accessible surface averaged over 300 cm^2 (or the area of the surface if less than 300 cm^2) does not exceed 400 Bq/cm^2 (10 nCi/cm^2) for beta and gamma emitters and low toxicity alpha emitters, or 40 Bq/cm^2 (1 nCi/cm^2) for all other alpha emitters; and 2) the fixed contamination on the accessible surface averaged over 300 cm^2 (or the area of the surface if less than 300 cm^2) does not exceed 800 kBq/cm^2 (20 μCi/cm^2) for beta and gamma emitters and low toxicity alpha emitters, or 80 kBq/cm^2 (2 μCi/cm^2) for all other alpha emitters;

and 3) the non-fixed contamination plus the fixed contamination on the inaccessible surface averaged over 300 cm^2 (or the area of the surface if less than 300 cm^2) does not exceed 800 kBq/cm^2 (20 μCi/cm^2) for beta and gamma emitters and low toxicity alpha emitters, or 80 kBq/cm^2 (2 μCi/cm^2) for all other alpha emitters. ICAO 2-7.5

Surface Contaminated Object (SCO) means a solid object which is not itself radioactive but which has Class 7 (radioactive) material distributed on any of its surfaces. SCO must be in one of two groups with surface activity not exceeding the following limits: (1) SCO-I: A solid object on which: (i) The non-fixed contamination on the accessible surface averaged over 300 cm^2 (or the area of the surface if less than 300 cm^2) does not exceed 4 Bq/cm^2 (10^{-4} microcurie/cm^2) for beta and gamma and low toxicity alpha emitters, or 0.4 Bq/cm^2 (10^{-5} microcurie/cm^2) for alpha emitters; (ii) The fixed contamination on the accessible surface averaged over 300 cm^2 (or the area of the surface if less than 300 cm^2) does not exceed 4x10^4 Bq/cm^2 (1.0 microcurie/cm^2) for beta and gamma and low toxicity alpha emitters, or 4x10^3 Bq/cm^2 (0.1 microcurie/cm^2) for all other alpha emitters; and (iii) The non-fixed contamination plus the fixed contamination on the inaccessible surface averaged over 300 cm^2 (or the area of the surface if less than 300 cm^2) does not exceed 4x10^4 Bq/cm^2 (1 microcurie/cm^2) for beta and gamma and low toxicity alpha emitters, or 4x10^3 Bq/cm^2 (0.1 microcurie/cm^2) for all other alpha emitters. (2) SCO-II: A solid object on which the limits for SCO-I are exceeded and on which: (i) The non-fixed contamination on the accessible surface averaged over 300 cm^2 (or the area of the surface if less than 300 cm^2) does not exceed 400 Bq/cm^2 (10^{-2} microcurie/cm^2) for beta and gamma and low toxicity alpha emitters or 40 Bq/cm^2 (10^{-3} microcurie/cm^2) for all other alpha emitters; (ii) The fixed contamination on the accessible surface averaged over 300 cm^2 (or the area of the surface if less than 300 cm^2) does not exceed 8x10^5 Bq/cm^2 (20 microcurie/cm^2) for beta and gamma and low toxicity alpha emitters, or 8x10^4 Bq/cm^2 (2 microcuries/cm^2) for all other alpha emitters; and (iii) The non-fixed contamination plus the fixed contamination on the inaccessible surface averaged over 300 cm^2 (or the area of the surface if less than 300 cm^2) does not exceed 8x10^5 Bq/cm^2 (20 microcuries/cm) for beta and gamma and low toxicity alpha emitters, or 8x10^4 Bq/cm^2 (2 microcuries/cm^2) for all other alpha emitters. US 173.403

Surface contaminated object (SCO) means a solid object which is not itself radioactive but which has radioactive material distributed on its surfaces. SCO is classified in one of two groups. SCO-I; SCO-II. IATA 10.3.6.1

SCO-I A solid object on which: (a) the non-fixed contamination on the accessible surface averaged over 300 cm^2 (or the area of the surface if less than 300 cm^2) does not exceed 4 Bq/cm^2 (0.1 nCi/cm^2) for beta and gamma emitters and low toxicity alpha emitters, or 0.4 Bq/cm2 (0.01 nCi/cm^2) for all other alpha emitters; (b) the fixed contamination on the accessible surface averaged over 300 cm^2 (or the area of the surface if less than 300 cm^2) does not exceed 40 kBq/cm^2 (1 μCi/cm^2) for beta and gamma emitters and low toxicity alpha emitters, or 4 kBq/cm^2 (0.1 μCi/cm^2) for all other alpha emitters; and (c) the non-fixed contamination plus the fixed contamination on the inaccessible surface averaged over 300 cm^2 (or the area of the surface if less than 300 cm^2) does not exceed 40 kBq/cm^2 (1 μCi/cm^2) for beta and gamma emitters and low toxicity alpha emitters, or 4 kBq/cm^2 (0.1 μCi/cm^2) for all other alpha emitters. IATA 10.3.6.1.1

SCO-II A solid object on which either the fixed or non-fixed contamination on the surface exceeds the applicable limited specified for SCO-I in 10.3.6.1.1 and on which: (a) the non-fixed contamination on the accessible surface averaged over 300 cm^2 (or the area of the surface if less than 300 cm^2) does not exceed 400 Bq/cm^2 (10 nCi/cm^2) for beta and gamma emitters and low toxicity alpha emitters, or 40 Bq/cm^2 (1 nCi/cm^2) for all other alpha emitters; (b) the fixed contamination on the accessible surface averaged over 300 cm^2 (or the area of the surface if less than 300 cm^2) does not exceed 800 kBq/cm^2 (20 μCi/cm^2) for beta and gamma emitters and low toxicity alpha emitters, or 80 kBq/cm^2 (2 μCi/cm^2) for all other alpha emitters; and (c) the non-fixed contamination plus the fixed contamination on the accessible surface averaged over 300 cm^2 (or the area of the surface if less than 300 cm^2) does not exceed 800 kBq/cm^2 (20 μCi/cm^2)

for beta and gamma emitters and low toxicity alpha emitters, or 80 kBq/cm^2 (2 μCi/cm^2) for all other alpha emitters. IATA 10.3.6.1.2

Unirradiated thorium shall mean thorium containing not more than 10^{-7} g of uranium-233 per gram of thorium-232. IAEA para. 244

Unirradiated thorium means thorium containing not more than 10^{-7} g of uranium-233 per gram of thorium-232. IMO Class 7, 2.28; US 173.403

Unirradiated thorium. Thorium containing not more than 10^{-7} g of uranium-233 per gram of thorium-232. ICAO 2-7.2, IATA App. A

Unirradiated uranium shall mean uranium containing not more than 2x10^3 Bq of plutonium per gram of uranium-235, not more than 9x10^6 Bq of fission products per gram of uranium-235 and not more than 5x10^{-3} g of uranium-236 per gram of uranium-235. IAEA para. 245

Unirradiated uranium means uranium containing not more than 10^{-6} g of plutonium per gram of uranium-235 and not more than 9 MBq (0.2 mCi) of fission products per gram of uranium-235. IMO Class 7, 2.20

Unirradiated uranium. Uranium containing not more than 10^{-6} g of plutonium per gram of uranium-235 and not more than 9 MBq (0.2 mCi) of fission products per gram of uranium-235. ICAO 2-7.2, IATA App. A

Unirradiated uranium means uranium containing not more than 10^{-6} grams plutonium per gram of uranium-235 and a fission product activity of not more than 9 MBq (0.24 millicuries) of fission products per gram of uranium-235. US 173.403

Uranium -- natural, depleted or enriched. Natural uranium means chemically separated uranium containing the naturally occurring distribution of uranium isotopes (approximately 99.28 per cent uranium-238 and 0.72 per cent uranium-235 by mass). Depleted uranium means uranium containing a less mass percentage of uranium-235 than in natural uranium. Enriched uranium means uranium containing a greater mass percentage of uranium-235 than in natural uranium. In all cases, a very small mass percentage of uranium-234 is present. ICAO 2-7.2

Uranium--natural, depleted or enriched means the following: (1) "Natural uranium" means uranium with the naturally occurring distribution of uranium isotopes (approximately 0.711 weight percent uranium-235, and the remainder essentially uranium-238). (2) "Depleted uranium" means uranium containing less uranium-235 than the naturally occurring distribution of uranium isotopes. (3) "Enriched uranium" means uranium containing more uranium-235 than the naturally occurring distribution of uranium isotopes. US 173.403

REFERENCES
DGR, HCC, HMM, HOC, IMDG, KOE, RFT, THS

Refrigerants and Halocarbons

Freon ▪ R11 ▪ R12 ▪ R12B1 ▪ R12B2 ▪ R13 ▪ R13B1 ▪ R14 ▪ R21 ▪ R22 ▪ R23 ▪ R32 ▪ R40 ▪ R41 ▪ R114 ▪ R115 ▪ R116 ▪ R124 ▪ R125 ▪ R133a ▪ R134a ▪ R142b ▪ R143 ▪ R143a ▪ R152a ▪ R161 ▪ R218 ▪ R227 ▪ R404A ▪ R407A ▪ R407B ▪ R407C ▪ R500 ▪ R502 ▪ R503 ▪ R1113 ▪ R1132a ▪ R1216 ▪ R1318 ▪ RC318 ▪ Refrigerant gas, n.o.s., 2.1, 2.2 ▪ Refrigerant gas, n.o.s., (e.g., non-flammable halocarbons), 2.2

1009 ▪ 1018 ▪ 1020 ▪ 1021 ▪ 1022 ▪ 1028 ▪ 1029 ▪ 1030 ▪ 1063 ▪ 1078 ▪ 1082 ▪ 1858 ▪ 1941 ▪ 1954 ▪ 1958 ▪ 1959 ▪ 1973 ▪ 1974 ▪ 1976 ▪ 1982 ▪ 1983 ▪ 1984 ▪ 2035 ▪ 2193 ▪ 2422 ▪ 2424 ▪ 2453 ▪ 2454 ▪ 2517 ▪ 2599 ▪ 2602 ▪ 3159 ▪ 3220 ▪ 3252 ▪ 3296 ▪ 3337 ▪ 3338 ▪ 3339 ▪ 3340

Certain gases and low boiling point liquids are used as *refrigerant gases* to absorb heat in refrigeration equipment by undergoing a change in state from compressed liquid to gas, which changes the temperature of their environment. An ideal refrigerant for a mechanical system is noncorrosive, nonflammable, nontoxic, and free of water. Ammonia, ethyl chloride, methylene ether, carbon dioxide, and the *halocarbons* are common refrigerants.

Halocarbons describe organic compounds (those based on carbon chains attached to hydrogens) in which a number of hydrogens have been replaced by a combination of the halogens (fluorine, chlorine, bromine, iodine, and, theoretically, the rare astatine). A number of fluorinated, chlorinated, and brominated halocarbons, particularly those based on chains of only a few carbons, have gained widespread use as blowing agents, propellants, and refrigerants. Unfortunately, chlorofluorocarbons (*CFCs*), halocarbons based on chlorine and fluorine, when released, rise and are broken apart by ultraviolet radiation from the sun in the upper atmosphere where they react to destroy the protective ozone layer. As a result, ozone-depleting substances are the subject of many national and international laws intended to phase out or ban their use. *Freon*[22] is a trademark that collectively describes certain CFCs.

NOMENCLATURE

An abbreviated nomenclature for halocarbons has been developed based on the chemistry of carbon atoms which have four bonds with which to attach to other atoms. When bonding with another carbon to make a carbon link, both carbons will use one available bond, leaving each with three; double and triple bonds are possible between carbons in which each uses two or three available bonds, respectively. For example, in ethane each of the two carbons uses one bond to connect with the other and each of the remaining six bonds connects with hydrogen, making $H_3C\text{-}CH_3$, or C_2H_6. In ethylene there

---

[22] Trademark, Du Pont de Nemours, E. I. & Company.

is a double bond between the two carbons, leaving each with only two bonds to join with a total of four hydrogens, making $H_2C=CH_2$, or C_2H_4.

While imperfect, the halocarbon nomenclature system enables the simpler halocarbon structures to be relatively well described in a few characters: a letter indicating the chlorofluorocarbon's use (although R, for refrigerant, is used ubiquitously) followed by *ABCD*:

- A is the number of double bonds in the compound.
- B is the number of carbon atoms minus 1.
- C is the number of hydrogen atoms in the molecule plus 1.
- D is the number of fluorine atoms in the molecule.
- Where the number of hydrogen and fluorine atoms is less than that required to satisfy the four-bonds-per-carbon rule, chlorine atoms make up the sum.
- If A is zero, as in the case with methane- and ethane-based molecules, for example, it is omitted.
- If B is zero, as in the case of methane, it is omitted.

Using this system, the common names of some of these compounds become clear:

- Refrigerant 0013 or R13 indicates no double bonds, one carbon, no hydrogens, three fluorines, and one chlorine: chlorotrifluoromethane ($CClF_3$).
- R114 indicates no double bonds, two carbons, no hydrogens, four fluorines, and two chlorines: 1,2-dichloro-1,1,2,2-tetrafluoroethane ($CClF_2CClF_2$).
- R1318 indicates one double bond, four carbons, no hydrogens, and eight fluorines: octafluorobutene (C_4F_8).

A few exceptions modify the system:

- RC318 indicates the presence of a cyclic (hence C) carbon ring, in this case octafluorocyclobutane (C_4F_8).
- R12B1 is read *R12 with one bromine*. The above rules can then be followed with the knowledge that a bromine atom is included and R12B1 becomes bromochlorodifluoromethane ($CBrClF_2$).
- The letters a and b distinguish positions of halogens around the carbon atoms.

Mixtures of refrigerants are divided into azeotropic mixtures (the 500-series) and nonazeotropic mixtures (i.e., zeotropes) which are assigned the 400-series. These numbers are not related to the principal naming system and can cause confusion:

- R404A is a mixture of R125, R143a, and R134a in a 44:52:4% ratio.
- R407A, R32:R125:R134a in a 20:40:20% ratio.

216

- R407B, R32:R125:R134a in a 10:70:20% ratio.
- R407C, R32:R125:R134a in a 23:25:52% ratio.

RELATED TERMS

Azetropic mixture, see *Terminology, Azeotropic*, p.232

Compressed gas, see *Gases and Class 2*, p.104

Dispersant gas, n.o.s., see *Gases and Class 2*, p.104

Gas, see *Terminology, Gas*, p.239

Non-flammable gas, see *Gases and Class 2*, p.104

Refrigerating machines, see *Refrigerating Machines*, p.218

REGULATORY DEFINITIONS

Refrigerant gases are gases used as cooling substances in the refrigerant process, e.g. in refrigerating machines (UN No.2857 in this class). IMO Class 2, 1.11

Refrigerant gas or Dispersant gas. The terms Refrigerant gas or Dispersant gas apply to all non-poisonous refrigerant gases, dispersant gases (fluorocarbons) listed in Secs. 172.101, 173.304(a)(2), 173.314(c), 173.315(a)(1) and 173.315(h), and mixtures thereof, or any other compressed gas having a vapor pressure not exceeding 1792 kPa (260 psi) at 54°C (130°F), and restricted for use as a refrigerant, dispersant or blowing agent. US 173.115(j)

REFERENCES

HCC, KOE, MH14, THS, VNS

Refrigerating Machines

Refrigerating machines, 2.1, 2.2, 3

1954 ▪ 1993 ▪ 2857 ▪ 8023

Refrigerating machines include domestic refrigerators, air conditioners, and dehumidifiers; commercial systems for supermarkets, restaurants, and cold storage; industrial and research gas liquefaction and chemical processing; and mobile systems for trucks, railcars, and other vehicles. Refrigeration involves the mechanical compression of a refrigerant to a liquid which is then discharged through a valve. As the pressure on the compressed refrigerant drops, part of it expands to a vapour or gaseous state cooling the remaining liquid. The cooled liquid is passed through a heat exchange device which absorbs heat from an external source and vaporizes the refrigerant again. The refrigerant is cycled back into the condenser in a closed and sealed loop. Refrigerating machines use three general classes of refrigerants: halocarbons, hydrocarbons, and inorganic fluids.

RELATED TERMS

Gas, see *Terminology, Gas*, p.239
Liquefied gas, see *Gases and Class 2*, p.104
Non-flammable gas, see *Gases and Class 2*, p.104
Non-toxic gas, see *Gases and Class 2*, p.104

Refrigerants, see *Refrigerants and Halocarbons*, p.215
Solution, see *Terminology, Solutions*, p.247
Toxic gas, see *Gases and Class 2*, p.104

REGULATORY DEFINITIONS

Refrigerating machines include air conditioning units and machines or other appliances which have been designed for the specific purpose of keeping food or other items at low temperature in an internal compartment. [Text continues.] IATA 4.4 A26

REFERENCES

HCC, KOE, TPD

Self-Propelled Vehicles

Automobile ■ Automobile, motorcycle, tractor, or other self-propelled vehicle, engine, or other mechanical apparatus ■ Engines, internal combustion (flammable gas powered) including where fitted in machinery or vehicles, 9 ■ Engines, internal combustion (flammable liquid powered) including where fitted in machinery or vehicles, 9 ■ Engines, internal combustion, including when fitted in machinery or vehicles, 9 ■ Engine starting fluid ■ Engine starting fluid, with flammable gas, 2.1 ■ Motorcycle ■ Self-propelled vehicle ■ Tractors ■ Vehicle (flammable gas powered) including where containing an internal combustion engine, 9 ■ Vehicle (flammable liquid powered) including where containing an internal combustion engine, 9 ■ Vehicles, self-propelled ■ Vehicles, self-propelled including internal combustion engines or other apparatus containing an internal combustion engine or electric storage battery

3166

Self-propelled vehicles are those *automobiles*, *motorcycles*, aircraft, boats, snowmobiles, trucks, *tractors*, jet skis, lawn mowers, golfcarts, etc., that convert their own energy supply into motive power used for propulsion. Batteries do so by converting electrochemical energy; *engines* do so by burning fuels to release their chemical energy. The majority of this equipment uses internal combustion engines such as a jet, diesel, or gasoline engines that consume flammable gases or liquids. Machinery may also employ engines to do other work such as turning crankshafts or generating electricity. External combustion engines include the steam engine which burns fuel to heat water, which on conversion to steam provides motive force.

In transportation, vehicles and machinery contain integral parts that present hazards:

- Fuel in tanks, injectors, carburettors, gas regulators, and fuel system components.
- Batteries, usually wet electric storage batteries, but also liquid sodium and fuel cells.
- Compressed gas cylinders as fuel tanks, shock absorbers, tire inflation products, etc.
- Fire extinguishers.
- Air bags and modules.
- Ferro-magnetic material.

ENGINE STARTING FLUID
Engine starting fluids include ethers or other extremely volatile hydrocarbons used in gasoline and diesel engines to help in cold-weather starting.

RELATED TERMS
Aerosol, see *Aerosols*, p.3
Battery-powered equipment, see *Batteries*, p.27
Battery-powered vehicle, see *Batteries*, p.27

Combustion, see *Terminology, Combustion*, p.233
Flammable gas, see *Gases and Class 2*, p.104

Flammable liquid, see *Flammable Liquids and Class 3*, p.96

Fluid, see *Terminology, Fluid*, p.238

Self-propelled vehicle, see also *Batteries*, p.27

REGULATORY DEFINITIONS

Fuel tank means a tank other than a cargo tank, used to transport flammable or combustible liquid, or compressed gas for the purpose of supplying fuel for propulsion of the transport vehicle to which it is attached, or for the operation of other equipment on the transport vehicle. US 171.8

Motor vehicle includes a vehicle, machine, tractor, trailer, or semitrailer, or any combination thereof, propelled or drawn by mechanical power and used upon the highways in the transportation of passengers or property. It does not include a vehicle, locomotive, or car operated exclusively on a rail or rails, or a trolley bus operated by electric power derived from a fixed overhead wire, furnishing local passenger transportation similar to street-railway service. US 171.8

REFERENCES

49CFR, MEOS

Solid Bulk Materials

Calcined pyrites (pyritic ash, fly ash) ■ Charcoal ■ Coal ■ Direct reduced iron ■ Ferrophosphorus (including briquettes) ■ Ferrosilicon, with 25% to 30% silicon, or 90% or more silicon (including briquettes) ■ Fluorspar (calcium fluoride) ■ Lime (unslaked) (calcium oxide, quicklime, dolomitic quicklime) ■ Magnesia (unslaked) (Lightburned magnesia, calcined magnesite, caustic calcined magnesite) ■ Metal sulphide concentrates ■ Peat moss ■ Petroleum coke, calcined or uncalcined ■ Pitch prill, prilled coal tar, pencil pitch ■ Sawdust ■ Silicomanganese with a silicon content of 25% or more ■ Tankage (garbage tankage (containing 8% or more moisture)), (rough ammonia tankage (containing 7% or more moisture)), (tankage fertilizer (containing 8% or more moisture)) ■ Vanadium ore ■ Woodchips ■ Wood pulp pellets

The *Code of Safe Practice for Solid Bulk Cargoes* (COS) recognizes that certain solid materials transported in large volumes present particular hazards due to their extreme bulk that are not present when shipped in smaller quantities or when packaged. These hazards are maximized when transported in the cargo space of ships without any intermediate containment.

STRUCTURAL DAMAGE AND STABILITY LOSS
Certain cargo can shift during transportation if improperly stored, or it may cause structural damage to the ship. These include stone chips and pebbles, some ores and minerals, sands, etc. Some may harden to solids on exposure to moisture.

LIQUEFACTION
Through a combination of existing moisture content, moisture buildup, and small particle size, cargoes may liquefy and become viscous fluids. Vibration and the motion of transportation will aid mixing and cause instability, particularly in bad weather, possibly to the point of capsizing the ship. Materials prone to liquefaction include mineral concentrates, other fine particulates, and cargoes already high in moisture such as fish and peat.

CARGOES POSSESSING CHEMICAL HAZARDS
Many bulk solids undergo chemical reactions often triggered by oxidation or reaction with water that can lead to oxygen reduction, emission of toxic or flammable gases, self-heating, or corrosive reaction products. Other materials release irritating, toxic, or explosive concentrations of dusts. Many of these materials already meet one or more dangerous goods hazard classes including sulphur (division 4.1), copra (division 4.2), aluminium ferrosilicon powder (division 4.3), calcium nitrate (division 5.1), and castor beans (class 9). Other solid bulk materials present similar risks, but are not defined as dangerous goods when shipped by sea. Rather, these substances are defined as *materials hazardous only in bulk* (MHB):

▫ *Charcoal* (16291-96-6) is subject to spontaneous ignition and self-heating when oxidation is catalyzed by moisture. Oxygen depletion can result.

- *Coal* emits methane and hydrogen gas and carbon monoxide on oxidation. Self-ignition is also possible.
- *Direct reduced iron*, produced from the reduction of iron oxide, can react with air and moisture to produce hydrogen gas and heat.
- *Ferrophosphorus* is an iron and phosphorus alloy (18 to 25% phosphorus) used to adjust phosphorus concentration in steel.
- *Ferrosilicon* is an iron-silicon alloy (20 to 95% silicon) used to adjust silicon concentration in steel. Between 30 and 90% silicon, the alloy is flammable and evolves gases in moisture.
- *Fluorspar* (7789-75-5) is a mineral source of calcium fluoride. Its dust is an irritant.
- *Lime* is calcium oxide (1305-78-8). *Quicklime* is unslaked lime. *Dolomitic quicklime* is a carbonate of calcium and magnesium. On contact with moisture (slaking), these substances generate heat as they react to produce calcium and magnesium hydroxides, solutions of which are alkaline skin irritants.
- *Magnesia* is magnesium oxide (1309-48-4). It is produced by calcining magnesite (magnesium carbonate) to varying degrees: *caustic calcined magnesite* retains around 2 to 10% carbon dioxide, *lightburned magnesia* is high-purity magnesium oxide. On contact with moisture from the air or body, magnesia forms magnesium hydroxide generating heat and a caustic residue.
- *Metal sulphides*. Many metals (e.g., iron, copper, lead, nickel) are found as mineral ores in sulphide form (pyrites) of which iron sulphide is most common. These are concentrated by physical and chemical means to produce an ore ready for economical extraction. Some may be subject to oxidation to generate heat, deplete oxygen, and emit toxic fumes. *Calcined pyrites* are used in the production of sulphuric acid and sulphur dioxide. In the presence of moisture, the calcined residue or dust can generate acids.
- *Peat* is a high carbon content residue resulting from the decomposition of plants by the action of water. It can contain up to 85 or 90% water, but on drying to around 40% water it serves as a fuel or source of hydrocarbons (including methane gas which it also evolves as it is dried). Peat oxidizes in the atmosphere to the point of self-heating (evolving carbon dioxide) or spontaneous combustion. It can also generate explosive concentrations of dust.
- *Petroleum coke* (64741-79-3) presents a spontaneous combustion hazard and can form explosive concentrations of dust.
- *Pitch prill, prilled coal tar*, and *pencil pitch* are combustible solids and dust irritants.
- *Silicomanganese* is an alloy based on manganese, silicon, and carbon along with some impurities. On contact with moisture it can generate hydrogen, phosphine, and arsine gases.

- *Tankage* can spontaneously ignite, generate ammonia-type gases, and harbour infectious substances.
- *Vanadium ore,* of which there are over 65 sources, ranges from inert micas (Roscoelite) to lead-containing vanadinite and radioactive poisons (Carnotite). They can generate toxic dusts.
- Wood products such as *sawdust, woodchips,* and *wood pulp pellets* oxidize leading to self-heating, depletion of oxygen, and the buildup of carbon dioxide.

RELATED TERMS

Calcined, see *Terminology, Calcined,* p.233
Charcoal, see also *Carbon,* p.36
Coal, see also *Coal,* p.44
Metal sulphides, see also *Metals, Inorganic Compounds,* p.156
Pencil pitch, see *Bituminous Products,* p.32
Petroleum Coke, see also *Petroleum,* p.183
Pitch prill, see *Bituminous Products,* p.32
Prilled coal tar, see *Coal,* p.44

Sawdust, see also *Fibres and Fibrous Products,* p.90
Tankage, see also *Fertilizer,* p.88
Unslaked, see *Terminology, Slaked,* p.247
Woodchips, see also *Fibres and Fibrous Products,* p.90
Wood pulp pellets, see also *Fibres and Fibrous Products,* p.90

REGULATORY DEFINITIONS

A solid bulk cargo is a material, other than a liquid or gas, which is loaded directly into a cargo space of a ship without any intermediate form of containment; this includes a material loaded in a barge on a barge-carrying ship. IMO Gen. Intro. 24.1.2

REFERENCES

COS, HCC, IMDG, KOE, WNN

Solvents

Plastic solvent, n.o.s. ■ Solvents, flammable, n.o.s. ■ Solvents, flammable, toxic, n.o.s.

Solvents are those chemicals that make up the bulk of a solution in which a solute has been dissolved by molecular or ionic forces. Solutions include gases in liquids; liquids in liquids; or gases, liquids, and solids in solids (mixtures of gases are said to be miscible). Common solutions include seawater, steel, carbonated beverages, and paints. Solvents work according to complex principles resulting in a solution which is more stable than its constituents. They fall into two main categories:

    □   Polar solvents, including water, alcohol, and acids, in which the uneven distribution of electrical charge across the solvent molecule encourages other polar compounds and ionic compounds, like salts, to go into solution.

    □   Non-polar solvents like the hydrocarbons that form molecular solutions.

As well as being the basis of many fundamental biochemical and hydrogeologic processes, solvents in industry fulfil the following functions:

    □   Dissolution, e.g., degreasing, *plastic solvents* (those used to dissolve plastics).

    □   Extraction, e.g., removal of essential oils from plants, leaching metal ores.

    □   Softening, e.g., detackification.

    □   Suspension/dispersion, e.g., coatings, aerosols, pigmenting.

    □   Viscosity control, e.g., plastics, plastic solvents used to thin adhesives.

    □   Chemical intermediaries, e.g., chemical synthesis.

    □   Manufacturing and processing to improve workability, e.g., mould release.

    □   Heat transfer fluids, e.g., chemical manufacturing.

    □   Inert reaction media.

Most nonhydrocarbon solvents, whether inorganic or organic, are pure substances. They include water, ethanol, and glycol ether. Most hydrocarbon solvents are mixtures including the complex mixtures of hydrocarbons present in petroleum distillates or the carefully engineered solvent blends used in automotive paints. These blends and mixtures are chosen to produce the desired solvency, evaporation rates, flash point, and other factors applicable to any process. A significant number of organic solvents are flammable.

From a transportation perspective, many solvents are on the lists of regulated materials in transportation as individual chemicals (e.g., toluene) or mixtures (e.g., petroleum distillates). The transportation term *solvents*, then, is usually applied to hydrocarbon or other organic solvents, often mixtures, that exhibit flammability or toxicity and that are not described under other entries.

<u>RELATED TERMS</u>

Flammable liquid, n.o.s., see *Flammable Liquids and Class 3*, p.96

Plastic solvent, see *Polymers and Resins*, p.189

Solvent, see *Terminology, Solutions*, p.247

Toxic, see *Toxic Substances and Division 6.1*, p.255

<u>REGULATORY DEFINITIONS</u>

Plastic Solvent, n.o.s. A name commonly used for mixtures of liquids employed for dissolving plastics or for thinning plastic cements. In general, they may contain flammable liquids, such as acetone, amyl acetate, or some of the alcohols or ketones. The classification is determined by the flash point. ICAO A2

Plastic Solvent. Is a name commonly used for mixtures of liquids employed for dissolving plastics or for thinning plastic cements. In general, they may contain flammable liquids, such as acetone, amyl acetate, or some of the alcohols or ketones. The classification is determined by the flash point. IATA App. A

Solvents. Substances capable of dissolving other substances to form a uniformly dispersed mixture or solution. Examples of organic solvent groups are esters, ethers, ketones, amines and nitrated and chlorinated hydrocarbons. Many solvents are flammable and toxic to varying degrees. ICAO A2

Solvents. Substances capable of dissolving other substances to form a uniformly dispersed mixture or solution. Examples of organic solvent groups are esters, ethers, ketones, amines and nitrated and chlorinated hydrocarbons. Many solvents are flammable and toxic to varying degrees. IATA App. A

<u>REFERENCES</u>

HCC, ISH, KOE, TPD

Spontaneously Combustible Materials and Division 4.2

Articles, pyrophoric, 1.2L ■ Fish meal (fish scrap), unstabilized, 4.2 ■ Fish meal, stabilized, 9 ■ Fish meal, unstabilized, 4.2 ■ Fish scrap, stabilized, 9 ■ Fish scrap, unstabilized, 4.2 ■ Pyrophoric articles ■ Pyrophoric liquid, inorganic, n.o.s., 4.2 ■ Pyrophoric liquid, organic, n.o.s., 4.2 ■ Pyrophoric metal, n.o.s., 4.2 ■ Pyrophoric solid, inorganic, n.o.s., 4.2 ■ Pyrophoric solid, organic, n.o.s., 4.2 ■ Self-heating liquid, inorganic, n.o.s., 4.2, 6.1, 8 ■ Self-heating liquid, organic, n.o.s., 4.2, 6.1, 8 ■ Self-heating solid, inorganic, n.o.s., 4.2, 6.1, 8 ■ Self-heating solid, n.o.s., 4.2, 5.1 ■ Self-heating solid, organic, n.o.s., 4.2, 6.1, 8 ■ Substances liable to spontaneous combustion, n.o.s.

0380 ■ 1374 ■ 1383 ■ 2216 ■ 2845 ■ 2846 ■ 3088 ■ 3126 ■ 3127 ■ 3128 ■ 3184 ■ 3185 ■ 3186 ■ 3187 ■ 3188 ■ 3190 ■ 3191 ■ 3192 ■ 3194 ■ 3200

Substances liable to spontaneous combustion are those materials that, though not exposed to any particular source of heat or ignition source, still combust. This can occur

- When the oxidation of a sensitive material (e.g., phosphorus) by oxygen in the atmosphere reaches the material's autoignition temperature.
- If oxidation takes place when catalyzed by moisture, as in the case of wet cotton, paper, or sodium.
- When autoignition temperatures are reached as a consequence of internal exothermic reactions like the polymerization of some drying oils.
- When autoignition temperatures are reached as substances like sewage and compost are decomposed by bacterial action.

Whether a material is *pyrophoric* or *self-heating* is a matter of degree. Usually, pyrophoric materials are those that ignite quickly due to atmospheric oxidation. Self-heating materials are those that generate heat over a longer period. The property is partially dependent on the relative surface area available for reaction; hence, the finer the particle size, the greater the tendency for spontaneous combustion. The tendency to oxidation or moisture catalyzation can be reduced by stabilizing the materials with compatible coverings, including inert gases, desiccants, oils, etc. As oxidation progresses, dangerous reductions of oxygen can occur in confined spaces, even if the material is not visibly aflame.

Pyrophoric articles include any device containing a chemical that spontaneously combusts, such as some pyrotechnics, smoke devices, explosives, etc.

Fish meal syn. fish scrap is a protein-rich, dried fertilizer or animal feed processed from the inedible portion of fish by dry or wet rendering. It may be spontaneously combustible due to oxidation catalyzed by moisture, auto-oxidation of remaining fish oils, bacterial buildup, or a combination of all three. Fish meal is stabilized with antioxidants such as ethoxyquinone or butylated hydroxytoluene (BHT).

RELATED TERMS

Corrosive, see *Corrosives and Class 8*, p.47
Inorganic, see *Terminology, Inorganic*, p.241
Liquid, see *Terminology, Liquid*, p.241
Metal, see *Terminology, Metals*, p.242
Organic, see *Terminology, Organic*, p.244
Oxidizing, see *Oxidizers and Class 5*, p.170

Solid, see *Terminology, Solid*, p.247
Stabilized, see *Terminology, Stabilized*, p.248
Toxic, see *Toxic Substances and Division 6.1*, p.255
Unstabilized, see *Terminology, Unstabilized*, p.252

REGULATORY DEFINITIONS

Articles, Pyrophoric Articles which contain a pyrophoric substance (capable of spontaneous ignition when exposed to air) and an explosive substance or component. The term excludes articles containing white phosphorus. UN App. B, ICAO A2, US 173.59, IATA App. A

Substances liable to spontaneous combustion Substances which are liable to spontaneous heating under normal conditions encountered in transport, or to heating up in contact with air, and being then liable to catch fire. UN 2.4.1.1(b); ICAO 2-4

Pyrophoric substances, which are substances, including mixtures and solutions (liquid or solid), which even in small quantities ignite within five minutes of coming in contact with air. These are the Division 4.2 substances [and] are the most liable to spontaneous combustion. UN 2.4.3.1.1(a)

Self-heating substances are substances, other than pyrophoric substances, which in contact with air without energy supply are liable to self-heating. These substances will ignite only when in large amounts (kilograms) and after long periods of time (hours or days) and are called self-heating substances. UN 2.4.3.1.1(b)

Self-heating of substances, leading to spontaneous combustion, is caused by reaction of the substance with oxygen (in the air) and the heat developed not being conducted away rapidly enough to the surroundings. Spontaneous combustion occurs when the rate of heat production exceeds the rate of heat loss and the auto-ignition temperature is reached. UN 2.4.3.1.2, IMO Class 4.2, 1.3

The substances in this class are either liquids or solids which are liable to spontaneous heating under normal conditions encountered in transport, or to heating up in contact with air, and being then liable to catch fire. IMO Class 4.2, 1.1

This class comprises: .1 pyrophoric substances; and .2 self-heating substances. IMO Class 4.2, 1.2

Two types of substances can be distinguished with spontaneous combustion properties: .1 substances, including mixtures and solutions (liquid or solid), which even in small quantities ignite within 5 minutes of coming into contact with air. These substances are the most liable to spontaneous combustion and are called pyrophoric substances; and .2 other substances which in contact with air without energy supply are liable to self-heating. These substances will ignite only when in large amounts (kilograms) and after long periods of time (hours or days) and are called self-heating substances. IMO Class 4.2, 1.4

Some substances may also give off toxic gases if involved in a fire. IMO Class 4.2, 1.5

The following types of substances are classified in Division 4.2: a) pyrophoric substances; and b) self-heating substances. ICAO 2-4.2.1

Self-heating of substances, leading to spontaneous combustion, is caused by reaction of the substance with oxygen (in the air) and the heat developed not being conducted away sufficiently rapidly to the surroundings. Spontaneous combustion occurs when the rate of heat production exceeds the rate of heat loss and the auto-ignition temperature is reached. Two types of substances can be distinguished with spontaneous combustion properties: a) substances, including mixtures and solutions (liquid or solid), which even in small quantities ignite within 5 minutes of coming into contact with air. These substances are the most liable to spontaneous combustion and are called pyrophoric substances; b) other substances which in contact with air without energy supply are liable to self-heating. These substances will ignite only when in large amounts (kilograms) and after long periods of time (hours or days) and are called self-heating substances. ICAO 2-4.2.2

Certain substances in contact with water emit flammable gases which can form explosive mixtures with air. Such mixtures are easily ignited by all ordinary sources of ignition, for example, naked lights, sparking handtools or unprotected light bulbs. The resulting blast wave and flames may endanger people and the environment. The test method in Part 8;1.3 must be used to determine whether the reaction of a substance with water leads to the development of a dangerous amount of gases which may be flammable. It must not be applied to pyrophoric substances. ICAO 2-4.3.2

Pyrophoric Liquid/Solid, Organic/Inorganic. A substance that may ignite in air at or below room temperature in the absence of added heat, shock or friction. ICAO A2

Division 4.2 (Spontaneously Combustible Material). For the purposes of this subchapter, spontaneously combustible material (Division 4.2) means--(1) A pyrophoric material. A pyrophoric material is a liquid or solid that, even in small quantities and without an external ignition source, can ignite within five (5) minutes after coming in contact with air when tested according to UN Manual of Tests and Criteria. (2) A self-heating material. A self-heating material is a material that, when in contact with air and without an energy supply, is liable to self-heat. A material of this type which exhibits spontaneous ignition or if the temperature of the sample exceeds 200°C (392°F) during the 24-hour test period when tested in accordance with UN Manual of Tests and Criteria, is classed as a Division 4.2 material. US 173.124(b)

Division 4.2 - Substances liable to spontaneous combustion. Substances which are liable to spontaneous heating under normal conditions encountered in transport, or to heating up in contact with air, and being then liable to catch fire. The following types of substances are classified in Division 4.2: pyrophoric substances; and self-heating substances. IATA 3.4.2.1

Pyrophoric substances (liquid or solid) including mixtures and solutions are substances which, even in small quantities, ignite within 5 minutes of coming in contact with air. These substances are the most liable to spontaneous combustion. IATA 3.4.2.2.1

Self-heating of a substance, leading to spontaneous combustion, is caused by a reaction of that substance with oxygen (in the air) if the heat developed is not conducted away sufficiently rapidly to the surroundings. Spontaneous combustion occurs when the rate of heat production exceeds the rate of heat loss and the auto-ignition temperature is reached. Two types of substances can be distinguished with spontaneous combustion properties. IATA 3.4.2.2

Self-heating substances are substances which in contact with air without an additional energy supply are liable to self-heating. These substances will ignite only in large amounts (kilograms) and after long periods of time (hours or days). IATA 3.4.2.2.2

Pryophoric Liquid/Solid, Organic/Inorganic. A substance that may ignite in air at or below room temperature in the absence of added heat, shock or friction. All are decidedly combustible and all fume strongly on exposure to air to produce fumes that are somewhat irritating and may be somewhat toxic. IATA App. A

REFERENCES
DOSATT, IMDG, MH14

Terminology

Activated	231	Homologues	239
Amorphous	231	Hydrocarbons	240
Anhydrous	231	Immiscible	240
Aqueous	231	Hypergolic	240
Atomic Number	231	Inert	240
Atomic Weight	231	Inhibited	241
Azeotropic	232	Inorganic	241
Boiling Point	232	Isomers	241
Butter	232	Liquid	241
By Mass	232	Metals	242
By Volume	233	Miscible	243
By Weight	233	Mixture	243
Calcined	233	Molecular Weight	243
Combustion	233	Natural	243
Complex	234	Nonactivated	244
Compound	234	Nonvolatile	244
Concentrates	234	Oil	244
Derivative	234	Organic	244
Desensitized	234	Oxidation	245
Dimer	235	Oxidation Number	245
Dispersion	235	Phlegmatized	245
Dry	235	Plasticized	246
Elements	235	Powder	246
Flammable Range	237	Reaction	246
Flash Point	237	Reduction	247
Fire Point	238	Salts	247
Flowers	238	Sensitized	247
Fluid	238	Slaked	247
Fuel	238	Solid	247
Fulminating	238	Solutions	247
Fuming	238	Sponge	248
Gas	239	Stabilized	248
Gel	239	Structural Notation	248
Glacial	239	Substituted	251
Heavy	239	Symmetrical	252

Terminology

Synthetic 252
Tinning Flux 252
Uncalcined 252
Uninhibited 252
Unphlegmatized 252

Unstabilized 252
Unsymmetrical 252
Volatile....................................... 252
Wetted.. 253

ACTIVATED

Activation is the process by which a substance is treated with heat, radiation, or other means so that its ability to react or undergo physical and chemical change is enhanced. For example, carbon is activated by superheated steam to greatly increase its available internal surface area and ability to absorb. *Nonactivated* materials are those that have not undergone such treatment; the term is used to differentiate them from their activated counterparts.

AMORPHOUS

Amorphous describes all liquids and some solids that are noncrystalline; that is, they have no ordered lattice structure at the molecular level. Amorphous solids include rubber, glass, carbon black, and phosphorus.

ANHYDROUS

Anhydrous describes substances that do not contain water in the form of a hydrate or water of crystallization. It is used to differentiate these substances in the hydrous form in which they are often found. For example, anhydrous ammonia, a gas at standard temperature and pressure, readily dissolves in water to form its hydrate, ammonium hydroxide.

AQUEOUS

Aqueous refers to solutions in which the solvent is water. In some cases, the substance can only exist in aqueous solution, such as chloric acid. Where the word *aqueous* or another qualifier is absent, a liquid *solution* is assumed to be water-based. See *Terminology, Solutions*, p.247.

ATOMIC NUMBER

Atomic number is the number of protons present in the nucleus of an element. The atomic number rises by whole number through the elements from 1 (hydrogen) to 112 (ununbium). See *Terminology, Atomic Weight*, p.231; *Terminology, Elements*, p.235.

ATOMIC WEIGHT

Each element has an *atomic weight*, a measure of the number of protons (the atomic number) and neutrons that make up its nucleus. Average atomic weights are used when an element exists naturally as different isotopes (elements with differing numbers of neutrons in the nucleus). For example, chlorine (atomic number 17) has two stable isotopes: chlorine-35 and chlorine-37 which occur naturally in a 75.4:24.6% ratio. Thus, the average atomic weight is 35.5:

$$(35 \times 75.4\%) + (37 \times 24.6\%) = 35.5$$

Carbon (which is 99% carbon-12) has an atomic weight of 12.0 while hydrogen (mostly hydrogen-1, but with some hydrogen-2) has an atomic weight of 1.01.

Note that atomic weights have no units. For that, the quantity of a *mole* must be considered. A mole is defined as the number of atomic entities present in 12g of carbon. This sum, known as Avogadro's number, is astonishingly large: 6.02×10^{23}. Using the concept of atomic weight and Avogadro's number, weights may be assigned to substances. Thus, one mole of methane (CH_4) weighs approximately 16g:

$$1 \text{ carbon (12g)} + 4 \text{ hydrogens (1g each)} = 16g$$

See *Terminology, Atomic Number*, p.231; *Terminology, Elements*, p.235; *Terminology, Molecular Weight*, p.243

AZEOTROPIC
A liquid mixture of two or more substances that boils at a constant minimum or maximum boiling point lower or higher than that of its constituents is called *azeotropic*. The liquid and the vapour produced on boiling have the same composition and, in this, the mixture acts like a single substance. Chlorotrifluoromethane and trifluoromethane forms such a mixture.

BOILING POINT
Boiling point is the temperature at which liquids become gases. Technically, this occurs when the vapour pressure of the liquid is the same as the atmospheric pressure of its environment. One result of this is that boiling point changes with elevation; e.g., at the top of a mountain where the lessened force of gravity has lowered the concentration of air (air pressure), water boils at a lower temperatures than if it were at sealevel.

BUTTER
Certain metallic chlorides such as antimony trichloride are hygroscopic, meaning they readily absorb moisture from the air, and can have a viscous, or even looser consistency, like *butter*.

BY MASS
When specifying a concentration of one substance in another, the terms *by mass*, *by volume*, and *by weight* are sometimes used to specify important differences. For example, $100cm^3$ of a 24% solution of ethanol (density of $0.816g/cm^3$) in water (density of $1.00g/cm^3$) *by volume* contains 19.6g ($24cm^3$ x $0.816g/cm^3$) ethanol, whereas a 24% solution *by weight* contains 24g of ethanol. Uncertainty exists when the basis of measurement is not provided.

By mass is often used synonymously with *by weight*, but the term *mass* refers to a quantity of matter regardless of the effect of gravity and other forces; *weight*, however, is dependent on those forces. For example, while a lead mass will weigh less in the water than on dry ground and less again on the surface of the moon, its mass remains constant.

Weight. The force at which a body is attracted towards the earth and is equal to the mass multiplied by the acceleration due to gravity. For practical purposes, mass and weight are used
interchangeably in these Regulations. IATA App. A

BY VOLUME
See *Terminology, By Mass*, p.232.

BY WEIGHT
See *Terminology, By Mass*, p.232.

CALCINED
Calcined materials are those which have been processed by incineration or
burning which drives off volatiles and oxidizes the substances to varying
degrees, often leaving a product better disposed for subsequent chemical
processing. Calcination is often used in metallurgy to prepare ore for extraction, as in the case when pyrites (e.g., iron sulphide ore) is calcined to iron
oxide. Petroleum coke is calcined to drive off volatiles, mostly hydrogen and
methane.

COMBUSTION
Combustion is the process by which gases, liquids, and solids burn, or more
specifically undergo exothermic oxidation: a reaction that evolves heat and
forms an oxide. Commonly, combustion involves an organic compound or
fuel in the presence of oxygen from the air. A typical reaction is that between ethanol and oxygen to produce carbon dioxide, water, and heat:

$$C_2H_5OH \text{ (liquid)} + 3O_2 \text{ (gas)} \rightarrow 2CO_2 \text{ (gas)} + 3H_2O \text{ (liquid)} + heat$$

The basic constituents of the above reaction are a fuel and an oxidizer. Combustion can proceed when these are present even if oxygen or a carbon-based
fuel is not; for example, hydrogen (fuel) and chlorine (oxidizer) combust to
form hydrogen chloride:

$$Cl_2 + H_2 \rightarrow 2HCl + heat$$

Combustion, however, requires an energy input to initiate the reaction.
Mixing ethanol and oxygen together at normal temperatures will not result in
combustion unless triggered by enough energy to break enough chemical
bonds to start the reaction. Combustion is initiated through various possible
routes:

▫ Pyrophoric materials will combust spontaneously, often started by the
heat generated from oxidation of the combustible material by atmospheric oxygen or moisture. Sodium combusts this way as it generates
hydrogen gas and sodium hydroxide in contact with moisture from the
air.

▫ From exposure to temperatures above the autoignition temperature of
a material. Autoignition temperature is the lowest temperature neces-

sary to initiate self-sustained combustion without an ignition source; for example, the autoignition temperature of ethanol is 793°F.

□ From contact with an ignition source, such as spark or flame, when the material is at or above its flash point or fire point.

Those materials that ignite easily and burn rapidly are said to be flammable materials; combustible materials are said to be those that will burn, but not quite so readily. *Inflammable* is a synonym of *flammable*. *Nonflammable*, its antonym, describes materials that do not combust or do so with difficulty. See *Terminology, Reaction*, p.246.

REGULATORY DEFINITIONS

Spontaneous Ignition Temperature. The lowest temperature at which a substance will ignite spontaneously without an external source of ignition. IATA App. A

COMPLEX

A *complex* is a compound formed when a metal ion shares an available electron and bonds with nonmetallic ions or an organic molecule. An example is the boron trifluoride acetic acid complex in which boron trifluoride and acetic acid are joined by the boron (a metal) and the oxygen in the acetic acid's hydroxyl group (-OH).

COMPOUND

A *compound* is any chemical substance, organic or inorganic, made up of atoms of more than one element. Thus antimony compounds include any of the many chemical combinations of antimony and other elements, including antimony bromide, antimony pentoxide, and antimony sulphate. The most elements thus far identified in a single compound is ten.

CONCENTRATES

A *concentrate* is a mixture or derivation of a mixture in which one or more components has undergone a concentration increase. In metallurgy, mineral ores are concentrated in several stages prior to refining. Metal sulphide ore may be concentrated by hydrometallurgical leaching and physical separation before being pyrometallurically extracted.

DERIVATIVE

Classification of compounds is sometimes best made by describing them as derivatives of their parent compound. For example, there are many salt derivatives of picric acid, valued for their explosive properties.

DESENSITIZED

Certain materials including some explosives, oxidizers, and organic peroxides are subject to violent reaction on exposure to heat, friction, impact, or certain chemicals. This risk may be reduced by the addition of a compatible *desensitizer* (e.g., water, organic liquids or solids, or inorganic solids) during transportation. *Sensitized* materials are those that have a reactive substance added to them to make them more sensitive. For example, relatively insensitive explosives are sensitized with primary explosives or fuels. See *Terminology, Inhibited,* p.241; *Terminology, Phlegmatized,* p.245; *Terminology, Stablized,* p.248; *Terminology, Wetted,* p.253.

DIMER

Dimers consist of molecules made up of two identical smaller molecules (*monomers*). For example, acrolein dimer $(C_3H_4O)_2$, is made up of two acrolein molecules (C_3H_4O). *Trimers* and *tetramers* are made up of three and four monomers, respectively. Any molecule with five or more monomers is a *polymer*.

DISPERSION

Dispersions are finely divided particles suspended in some liquid, often flammable, to inhibit oxidation, reaction with moisture, or to accurately control reaction rates; e.g., sodium may be dispersed in toluene, xylene, naphtha, or kerosene.

DRY

The simple application of the term *dry* indicates the absence of water. However, dry materials include

- Liquids, like ethanol, that are free of residual water.
- Solids that contain a nonwater liquid.
- Fibres and seed cakes that contain oils.
- Solids handled with desiccants to capture moisture and leave chemicals free-flowing.

ELEMENTS

There are 112 known *elements*, which represent the basic building blocks of all substances. The first 92, ending in uranium, are naturally occurring, the other 20 are artificial. Each element is symbolized by one or two letters (although multiple spellings are possible). Strung together, these symbols represent molecules; e.g., C_2H_5OH represents ethanol, a chain of two carbons: one bonded to three hydrogens, the other to two hydrogens and a hydroxyl group (-OH); NaCl stands for sodium chloride in which the sodium ion (Na^+) has bonded with the chloride ion (Cl^-).

The elements have chemical and physical relationships that allow them to be plotted by atomic number in the following representation known as the *Periodic Table*:

Periodic Table of the Elements

IA	IIA											IIIA	IVA	VA	VIA	VIIA	VIIIA
1 H 1.01																	2 He 4.00
3 Li 6.94	4 Be 9.01											5 B 10.8	6 C 12.0	7 N 14.0	8 O 16.0	9 F 19.0	10 Ne 20.2
11 Na 23.0	12 Mg 24.3											13 Al 27.0	14 Si 28.1	15 P 31.0	16 S 32.1	17 Cl 35.5	18 Ar 40.0
19 K 39.1	20 Ca 40.1	21 Sc 45.0	22 Ti 47.9	23 V 50.9	24 Cr 52.0	25 Mn 54.9	26 Fe 55.8	27 Co 58.9	28 Ni 58.7	29 Cu 63.5	30 Zn 65.4	31 Ga 69.7	32 Ge 72.6	33 As 74.9	34 Se 79.0	35 Br 79.9	36 Kr 83.8
37 Rb 85.5	38 Sr 87.6	39 Y 88.9	40 Zr 91.2	41 Nb 92.9	42 Mo 95.9	43 Tc 98	44 Ru 101	45 Rh 103	46 Pd 106	47 Ag 108	48 Cd 112	49 In 115	50 Sn 119	51 Sb 122	52 Te 128	53 I 127	54 Xe 131
55 Cs 133	56 Ba 137	57 La 139	72 Hf 178	73 Ta 181	74 W 184	75 Re 186	76 Os 190	77 Ir 192	78 Pt 195	79 Au 197	80 Hg 201	81 Tl 204	82 Pb 207	83 Bi 209	84 Po 209	85 At 210	86 Rn 222
87 Fr 223	88 Ra 226	89 Ac 227	104 Rf 261	105 Db 262	106 Sg 263	107 Bh 262	108 Hs 265	109 Mt 266	110 Unn 269	111 Unu 272	112 Unb 277						

alkali metals IA — alkaline earth metals IIA — transition metals — halogens VIIA — rare gases VIIIA — other metals

Atomic Number / Symbol / Atomic Weight

Lanthanides

58 Ce 140	59 Pr 141	60 Nd 144	61 Pm 145	62 Sm 150	63 Eu 152	64 Gd 157	65 Tb 159	66 Dy 163	67 Ho 165	68 Er 167	69 Tm 169	70 Yb 173	71 Lu 175

Actinides

90 Th 232	91 Pa 231	92 U 238	93 Np 237	94 Pu 244	95 Am 243	96 Cm 247	97 Bk 247	98 Cf 251	99 Es 252	100 Fm 257	101 Md 258	102 No 259	103 Lr 262

Ac, Actinium
Ag, Silver
Al, Aluminium or Aluminum
Am, Americium
Ar, Argon
As, Arsenic
At, Astatine
Au, Gold
B, Boron
Ba, Barium
Be, Beryllium
Bh, Bohrium
Bi, Bismuth
Bk, Berkelium
Br, Bromine
C, Carbon
Ca, Calcium
Cd, Cadmium
Ce, Cerium
Cf, Californium
Cl, Chlorine
Cm, Curium
Co, Cobalt
Cr, Chromium
Cs, Caesium or Cesium
Cu, Copper
Db, Dubnium
Dy, Dysprosium
Er, Erbium
Es, Einsteinium
Eu, Europium
F, Fluorine
Fe, Iron
Fm, Fermium
Fr, Francium
Ga, Gallium
Gd, Gadolinium
Ge, Germanium
H, Hydrogen
He, Helium
Hf, Hafnium
Hg, Mercury
Ho, Holmium
Hs, Hassium
I, Iodine
In, Indium
Ir, Iridium
K, Potassium
Kr, Krypton
La, Lanthanum
Li, Lithium
Lr, Lawrencium
Lu, Lutetium
Md, Mendelevium
Mg, Magnesium
Mn, Manganese
Mo, Molybdenum
Mt, Meitnerium
N, Nitrogen
Na, Sodium
Nb, Niobium or Columbium
Nd, Neodymium
Ne, Neon
Ni, Nickel
No, Nobelium
Np, Neptunium
O, Oxygen
Os, Osmium
P, Phosphorus
Pa, Protactinium
Pb, Lead
Pd, Palladium
Pm, Promethium
Po, Polonium
Pr, Praseodymium
Pt, Platinum
Pu, Plutonium
Ra, Radium
Rb, Rubidium
Re, Rhenium
Rf, Rutherfordium

Rh, Rhodium	Si, Silicon	Th, Thorium	V, Vanadium
Rn, Radon	Sm, Samarium	Ti, Titanium	W, Tungsten
Ru, Ruthenium	Sn, Tin	Tl, Thallium	Xe, Xenon
S, Sulphur or Sulfur	Sr, Strontium	Tm, Thulium	Y, Yttrium
Sb, Antimony	Ta, Tantalum	U, Uranium	Yb, Ytterbium
Sc, Scandium	Tb, Terbium	Unb, Ununbium	Zn, Zinc
Se, Selenium	Tc, Technetium	Unn, Ununnilium	Zr, Zirconium
Sg, Seaborgium	Te, Tellurium	Unu, Unununium	

FLAMMABLE RANGE

Flammable liquids are those whose vapours form sufficient concentrations in air to ignite on exposure to an ignition source. Vapour concentration is dependent on the vapour pressure of the material and the ambient temperature and pressure. The lowest concentration at which ignition occurs is the lower limit of the fuel-to-air ratio. As a rule of thumb, the lower limit is often a fuel-air ratio of around 50% of the theoretical concentration for complete combustion (i.e., sufficient oxygen in the air to fully oxidize the fuel). The upper limit of flammability is often around 300% of the ideal ratio; higher concentrations have insufficient oxygen to support ignition. The lower and upper concentrations delimit the *flammable range* of a material which widens with increasing ambient temperature until at some point the upper flammability limit is delimited by the autoignition temperature. For example, at ambient temperatures and pressures, the flammable range of carbon monoxide in air is between 12.5 and 74% carbon monoxide. See *Terminology, Combustion*, p.233; *Terminology, Volatile*, p.252.

REGULATORY DEFINITIONS

Flammable range. The term flammable range means the difference between the minimum and maximum volume percentages of the material in air that forms a flammable mixture. US 173.115(h)

FLASH POINT

Flash point is the temperature at which vapours evolved from a flammable liquid in air ignite on exposure to an ignition source. Flash point is measured in two ways: (1) in a closed-cup tester in which the liquid is heated in a mostly closed environment that allows the vapours to concentrate; or (2) in an open-cup tester which allows vapours to dissipate and, because the temperature must be higher to accumulate sufficient vapours to ignite, results in a higher recorded flash point.

Flammable liquids may not continue to burn after they have been ignited at their flash point. The temperature at which flammable liquids evolve vapours quickly enough to support sustained and continuous combustion is the *fire point*. A typical sample of No. 1 fuel oil may have a closed-cup flash point of 54°C, an open-cup flash point of 59°C, and a fire point of 63°C. See *Terminology, Combustion*, p.233; *Terminology, Volatile*, p.252.

Fire Point. The lowest temperature at which a liquid evolves vapour in sufficient concentration that when it is ignited in air, the liquid will continue to burn. It is usually close to the flash point. IATA App. A

The flashpoint of a flammable liquid is the lowest temperature of the liquid at which its vapour forms an ignitable mixture with air. It gives a measure of the risk of formation of explosive or ignitable mixtures when the liquid escapes from its packing. A flammable liquid cannot be ignited so long as its temperature remains below the flashpoint. IMO Gen. Intro. 6.1

Flash point. The lowest temperature of a liquid at which flammable vapour is given off in a test vessel in sufficient concentration to be ignited in air when exposed momentarily to a source of ignition. ICAO 1-3.1

Flash point means the minimum temperature at which a liquid gives off vapor within a test vessel in sufficient concentration to form an ignitable mixture with air near the surface of the liquid. [Text continues.] US 173.120(c)(1)

Flash Point. Is defined as the lowest temperature at which flammable vapour is given off a liquid in a test vessel in sufficient concentration to be ignited in air when exposed momentarily to a source of ignition. This does not mean the temperature at which a liquid ignites spontaneously (See "Spontaneous ignition temperature"). IATA App. A

FIRE POINT
See *Terminology, Flash Point*, p.237.

FLOWERS
Certain solid materials sublimate, that is, they pass directly from solid to vapour without a liquid phase. *Flowers* is an obsolete term describing the fine powder collected by condensing these vapours, hence flowers of sulphur.

FLUID
Fluids are those substances, including gases, liquids, and finely-divided solids, that change shape and flow uniformly on application of an external force. The term is not synonymous with *liquid* although it is sometimes used as such.

FUEL
Fuels are those substances which undergo controlled chemical and nuclear reactions to evolve useful energy for heat or power. Chemical fuels are principally carbonaceous organic substances, such as petroleum, coal, and wood, which undergo combustion. Liquefied gases such as hydrogen and ammonia are also used. Nuclear fuels, plutonium and uranium, release nuclear energy on fission.

FULMINATING
Fulminating chemicals are those which *fulminate*, an old term meaning to flash like lightning and produce a loud report. In short, they explode. Contemporary definitions include sensitive explosives containing carbon-nitrogen-oxygen groups often derived from fulminic acid, such as fulminating mercury, gold, platinum, or silver.

FUMING

Certain highly active, concentrated liquids evolve smoke-like vapours (*fumes*) on contact with air. For example, arsenic trichloride fumes in contact with moisture in the air as do 85% or higher concentrations of nitric acid.

GAS

A *gas* is a physical state of matter generally characterized by low density and viscosity and the abilities to greatly expand and contract with changes in temperature and pressure, mix completely with other gases, and occupy any container uniformly. These features are particularly striking when contrasted to the other states of matter, liquid and solid. See *Terminology, Liquid*, p.241, *Terminology, Solid*, p.247.

GEL

Gels are structures formed by high molecular weight substances in long chains or fibres. These absorb many times their own weight in water or other chemicals and swell to a viscous or semisolid jelly or gel when cooled as the chains and fibres become entangled. On heating, gels return to their liquid state. Gels include soaps, certain proteins, starches, sugars, etc. Gelatine, obtained from collagen (a fibrous protein in the connective tissues of animal and human skin, muscles, and tendons) absorbs up to ten times its weight in water to form a gel. Nitrocellulose forms a gelatinous substance with nitroglycerine, thereby making an explosive less sensitive to shock.

GLACIAL

The term *glacial*, meaning *icy* or to *produce ice*, indicates the tendency of a material of high concentration to crystallize at room temperature, as does acetic acid, the glacial form of which is 99.8% pure or more. The term does not refer to water crystals in this context.

HEAVY

Heavy in nuclear chemistry refers to the relative mass of one isotope to another of the same element. The vast majority of hydrogen, the lightest element, has a single proton making up its atomic nucleus. There are two other isotopes of hydrogen (i.e., heavy hydrogen): the rarer deuterium (one proton and one neutron) and the artificially produced tritium (one proton and two neutrons). Isotopes are species of an element with the same atomic number (number of protons) but with varying numbers of neutrons. See *Terminology, Atomic Number*, p.231; *Terminology, Atomic Weight*, p.231; *Terminology, Metals*, p.242.

HOMOLOGUES

Homologues are members of a homologous series in which each successive member (all of which are organic) has one more carbon in the chain. For example, the C_2 through C_{12} homologues of alkylphenol, range from ethylphenol ($C_2H_5C_6H_4OH$) to the mixed isomers of dodecylphenol ($C_{12}H_{25}$-C_6H_4OH) and include many alkylphenols in between.

HYDROCARBONS

The *hydrocarbons* are a vast range of organic compounds based on chains of carbon atoms with hydrogen attached to any remaining free bonds (each carbon atom has four available bonds). There are two broad groups:

- *Aliphatic,* those with straight or branched carbon chains including (1) *paraffins* (alkanes) those with single bonds between the carbons; (2) *olefins* (alkenes) those with double bonds between one or more carbons; (3) *acetylenes* (alkynes) those with triple bonds between one or more carbons.
- *Cyclic,* those with rings of carbons including (1) *alicyclic*, rings made up of aliphatic compounds; (2) *aromatic*, molecules made up of one or more six-carbon rings in which the carbons share the bonds between them.

Highly complex mixtures of hydrocarbons are found naturally in petroleum, natural gas, shale oil, asphalt, coal, and other deposits. They form the basis of our energy supply as fuels and much of the chemical industry. They range from the simplest, methane gas, to liquids, waxes, and hard resinous solids. They undergo many reactions as hydrocarbon groups attach themselves to other atoms or species. Some of these groups have evolved specific names:

- *Alkyl* groups are paraffins in which one of the hydrogens is removed leaving an available carbon bond. The simplest in the series in the methyl group (-CH_3), but an infinite number exist characterized by the formula C_nH_{2n+1} where *n* is the number of carbons.
- The *allyl* group is a chain of three carbons with a double bond between two of the carbons ($H_2C=CHCH_2$-). Allyl compounds include allyl iodide and allyl isothiocyanate.
- The *aryl* group (C_6H_5-) is that remaining after a hydrogen has been removed from a benzene ring.

IMMISCIBLE
See *Terminology, Miscible*, p.243.

HYPERGOLIC
Hypergolic liquids are fuels that ignite spontaneously on contact with liquid oxidizers, particularly for use as rocket fuels; e.g., nitric acid (oxidizer) with aniline, triethylamine, hydrazine, or liquid oxygen with lithium.

INERT
Inert materials have little or no chemical affinity or activity and, therefore, undergo few if any reactions. Inert gases include the rare gases, although nitrogen and carbon dioxide are also used as inert atmospheres to protect

highly reactive substances from contact with atmospheric oxygen; clay, asbestos, talc, and sand are inert solids.

INHIBITED

Inhibitors are the opposites of catalysts. They are designed to retard or stop unwanted chemical reactions such as decomposition or oxidation. Antioxidants inhibit oxidation of foods, rubber, and other organic materials. For example, acrolein undergoes polymerization very readily unless it is inhibited, usually with hydroquinone. *Uninhibited* chemicals are those that might otherwise have had an inhibitor added or which have had the stablizer removed. See *Terminology, Stablized*, p.248.

INORGANIC

See *Terminology, Organic*, p.244.

ISOMERS

Isomers describe a collection of two or more molecules that have the same number and type of atoms but differ in their arrangement or symmetrical configuration. For example, propanol (C_3H_7OH), an alcohol based on a chain of three carbons, has two isomers: one with the hydroxyl group (-OH) on the end of the carbon chain (n-propanol *syn.* 1-propanol or propyl alcohol) and the other with the hydroxyl group attached to the middle carbon (isopropanol *syn.* 2-propanol or isopropyl alcohol). Isomers may have discernibly different physical and chemical properties; e.g., n-propanol and its isomer have boiling points of 97.2°C and 82.4°C, respectively. See *Terminology, Symmetrical*, p.252; *Terminology, Structural Notation*, p.248.

LIQUID

Liquids are substances in the amorphous state of matter between gases and solids in which the molecules exist in no set relationship to one another and are able to move and flow within the limits set by their intermolecular forces. Liquids are more concentrated than gases but are usually less dense than solids. Water is an exception; it is denser than ice. See *Terminology, Gas*, p.239; *Terminology, Solid*, p.247.

REGULATORY DEFINITIONS

Liquids are, unless there is an explicit or implicit indication to the contrary in these Regulations, dangerous goods with a melting point or initial melting point of 20°C or lower at a pressure of 101.3 kPa. [Text continues.] UN Part 1

Liquid means a material, other than an elevated temperature material, with a melting point or initial melting point of 20°C (68°F) or lower at a standard pressure of 101.3 kPa (14.7 psi). A viscous material for which a specific melting point cannot be determined must be subjected to the procedures specified in ASTM D 4359 "Standard Test Method for Determining Whether a Material is Liquid or Solid". US 171.8

Liquid means a material that has a vertical flow of over two inches (50mm) within a three-minute period, or a material having one gram or more liquid separation, when determined in accordance with the procedures specified in ASTM D 4359-84, "Standard Test Method for Determining Whether a Material is a Liquid or a Solid," 1990 edition, which is incorporated by reference. US 130.5

Liquid dangerous goods. Unless otherwise provided for in these Instructions, dangerous goods with a melting point or initial melting point of 20°C or lower at a pressure of 101.3 kPa must be considered as liquids. [Text continues.] ICAO 1-3.1

Liquid Dangerous Goods. Unless otherwise provided in these Regulations, dangerous goods with a melting point or an initial melting point of 20°C or lower at a pressure of 101.3 kPa must be considered as liquids. [Text continues.] IATA App. A

METALS

Metals are elements generally characterized by forming positive ions in solution, whose oxides form hydroxides rather than acids in water, can conduct electricity and heat, have high physical strength, and can be formed and worked. More than three quarters of all elements demonstrate these properties, although the exact nature of some transuranic elements (those with atomic numbers above uranium) is only presumed. See *Terminology, Elements*, p.235.

Alkali and Alkaline Earth Metals

Alkali metals are the most basic and reactive of metallic elements comprising *Group IA* of the periodic table. Generally, they are soft, readily fused and volatilized, and react vigorously, particularly in contact with water, acids, or oxidizing materials. Their reactivity is such that they are never found in elemental form in nature. *Alkaline earth metals* are those which occupy the next periodic series, *Group IIA*. They form alkaline oxides or *earths* and are generally less reactive than the alkali metals although they are also not found naturally in elemental form. From a transportation perspective, these metals exhibit a range of flammable and reactive hazards.

Heavy Metal

Heavy metals are those in the series of metals that have a greater atomic weight than sodium (atomic weight 23.0). Heavy metals include lead, mercury, aluminium, chromium, etc. Heavy metals can be involved in many reactions; e.g., 2,4-dinitroresorcinol is used to create heavy metal complexes.

Ferrous Metals

Ferrous metals are alloys based on iron. Iron is distinguished in being the only metal whose hardness and strength can be increased through tempering: heat treating and quenching. It is widely used as cast iron and a range of steel alloys. Ferrous compounds are iron(II) compounds.

Metalloids

Some elements exhibit properties that lie between definite nonmetals and definite metals. These elements, historically called *metalloids*, mimic metals in that they are solid and semiconductors. They include arsenic, antimony, boron, carbon, germanium, polonium, phosphorus, selenium, silicon, and tellurium.

Rare-Earth Metals

Strictly, the *rare-earth metals* are oxides of the lanthanide elements (lanthanide plus the 14 metals in the lanthanide series). They are chemically related and difficult to separate. The lanthanides are also members of the *rare metals*, a larger group of the less commonly occurring metals. Some rare metals (e.g., plutonium and promethium) occur only as a result of nuclear fission.

MISCIBLE

The ability of one liquid or gas to dissolve into another is *miscibility*. In the case of liquids and liquid-gas mixtures, miscibility is determined by the chemical similarity of the constituents or the types and strengths of bonds between them. For example, alcoholic beverages are based on miscible mixtures of water and ethanol, whereas oil and water are immiscible. Some liquid mixtures are not fully miscible at all concentrations. Gases are miscible with each other in all proportions.

REGULATORY DEFINITIONS

Miscible. The ability of a liquid (or gas) to dissolve uniformly in another liquid (or gas). Miscibility depends on the chemical nature of the substances involved and in some cases, liquids may only be partially miscible. Liquids which do not mix at all are said to be *immiscible*. IATA App. A

MIXTURE

Mixtures are coalitions of more or less uniformly distributed chemical substances that can usually can be separated by physical means. See *Terminology, Solutions*, p.247.

REGULATORY DEFINITIONS

Mixture means a material composed of more than one chemical compound or element. US 171.8

MOLECULAR WEIGHT

Molecular weight is the sum of the atomic weights of the constituent elements of a molecule. Thus the molecular weight of styrene (C_8H_8) is approximately

(8 x atomic weight of carbon: 12) + (8 x atomic weight of hydrogen: 1) = 104

Polystyrene, the polymer of up to 3000 repeating units of styrene has a molecular weight over 300,000. See *Terminology, Atomic Weight*, p.231.

NATURAL

A vast number of chemical substances and elements occur *naturally* in plants, minerals, and the atmosphere. These can be refined, physically or chemically manipulated, or put to use directly. In contrast, *synthetic* chemicals are those fabricated by chemical or physical reactions, although at some point the raw materials must have occurred naturally. Synthetics may be the chemical twin of a compound found in nature or entirely unique. Hence, *natural gasoline* might be distinguished from the synthetic hydrocarbon fuel derived from coal in the Fischer-Tropsch process.

Non-Activated
See *Terminology, Activated*, p.231.

Non-Volatile
See *Terminology, Volatile*, p.252.

Oil

Oil is a term applied to a wide variety of substances related more by their consistency than their chemical properties and may be generally categorized by chemical group:

- Glycerides of fatty acids (principally oleic, palmitic, stearic, and linolenic acids) derived from animals, seeds, and nuts. The more hydrogens the molecule contains (i.e., the greater the degree of saturation), the thicker it becomes. Examples include lard, fish oil, castor oil, linseed oil, and corn oil.
- Hydrocarbon mixtures of hundreds of chemical compounds represented by the mineral oils, those contained in geological deposits, e.g., petroleum, kerogen, natural asphalt, etc.
- Terpenes contained in the essential oils derived from stems, leaves, and flowers of plants and trees.

Regulatory Definitions

Animal fat means a non-petroleum oil, fat, or grease derived from animals, not specifically identified elsewhere in this part. US 130.5

Oil means oil of any kind or in any form, including, but not limited to, petroleum, fuel oil, sludge, oil refuse, and oil mixed with wastes other than dredged spoil. US 130.5

Other non-petroleum oil means a non-petroleum oil of any kind that is not an animal fat or vegetable oil. US 130.5

Petroleum oil means any oil extracted or derived from geological hydrocarbon deposits, including fractions thereof. US 130.5

Vegetable oil means a non-petroleum oil or fat derived from plant seeds, nuts, kernels or fruits, not specifically identified elsewhere in this part. US 130.5

Organic

Historically, *organic* described those compounds with a living animal or plant origin based on hydrocarbons and their derivatives which distinguished them from *inorganic* compounds derived from minerals. The distinction between these two major branches of chemical compounds still exists, although with the rise of synthetic chemistry in which millions of compounds have been speciated, organic has come to mean compounds based on hydrocarbons and their derivatives, while inorganic means the chemistry of the elements including all those substances not based on carbon, other than a few simple compounds like the carbon oxides, carbon disulphide, carbonyls, carbonates, cyanides, and carbides. The dividing line between the two remains indistinct.

REGULATORY DEFINITIONS

Inorganic. Inorganic compounds are all elements, alloys, and compounds which do not contain a carbon-carbon linkage. Some examples of inorganic substances are carbides, carbon disulphide; such toxic materials as phosgene or chlorine and inorganic acids such as hydrochloric, hydrofluoric, and sulphuric acids. IATA App. A

Organic. Organic substances are compounds which contain a carbon-carbon linkage. There are a few exceptions but by methodical reasons they are counted to belong to the organic compounds (e.g., Methane CH_4), for example carbon oxides, the carbides, carbon disulphide, etc.; ternary compounds such as metallic cyanides, metallic carbonyls, phosgene ($COCl_2$), carbonyl sulphide (COS), etc.; and the metallic carbonates, such as calcium carbonate and sodium carbonate. Some examples of organic substances are hydrocarbons, alcohols, ethers, aldehydes, ketones, organic solvents and organic acids. IATA App. A

OXIDATION

The term *oxidation* is used to describe the chemical combination of oxygen with another chemical, as in the process of combustion. Chemically, oxidation involves reactions in which electrons are transferred from one species to another even if oxygen is not involved. As one species is oxidized it loses electrons; the other species is *reduced* and gains the lost electrons. Corrosion of iron or iron-containing alloys such as steel is a common *redox* (reduction-oxidation) reaction in which oxygen from the air, catalyzed by moisture, reacts with iron to form rust (iron oxide). This being of a lower density and of less strength than iron, gradually degrades the structural integrity of the substrate.

OXIDATION NUMBER

Through oxidation of one element or molecule and reduction of another, ionic bonds are formed as the resulting ions electrically attract each other (i.e., a positive charge attracts a negative charge). For example, in arsenic bromide ($AsBr_3$), elemental arsenic has lost three electrons to exist in the As^{3+} state (electrons carry a negative charge) while each of three bromine atoms has picked up one electron making three Br^- species, and the formula $AsBr_3$ is electrically balanced. Similar transfers are the basis of a vast number of chemical reactions.

The *oxidation number* is the number of electrons that must be transferred to either oxidize or reduce the new species back to its elemental state. In the case of $AsBr_3$ the numbers are +3 for arsenic and -1 for bromine. *Oxidation states* are assigned to those elements that have been oxidized. In the case of arsenic in this example, it exists in the oxidation state of III, written *arsenic(III)* or *As(III)*. Certain elements can exist in multiple oxidation states, for example, arsenic(III) oxide and arsenic(V) oxide have the formulae As_2O_3 and As_2O_5, respectively. See *Terminology, Oxidation*, p.245.

PHLEGMATIZED

A *phlegmatizer* is a chemical desensitizer that is often characterized as a thick, viscous, mucous-like fluid. The term implies that the phlegmatized material will be indolent, apathetic, or sluggish. Phlegmatizers may particularly render the substance less susceptible to heat and shock by providing an insulating effect. For example, nitroglycerine is phlegmatized with lactose or glucose, and gunpowder is phlegmatized with dibutyl phthalate, camphor, and dinitrotoluene.

Unphlegmatized chemicals are those that might otherwise have had a phlegmatizer added or which have had the phlegmatizer removed. See *Terminology, Desensitized*, p.234.

REGULATORY DEFINITIONS

Phlegmatiser. A solid or liquid which is added to a substance such as an explosive or an organic peroxide in order to reduce its sensitivity to heat and impact, thereby assisting in ensuring safety during transport. IATA App. A

PLASTICIZED

Plasticizers are added to increase the workability, flexibility, or distensibility of a material. Usually, the plasticizer dissolves into the material. They are often high molecular weight organic liquids or low-melting point solids such as derivatives of citric, oleic, phthalic, sebacic, and stearic acids.

POWDER

Powders are dry solids in a fine state of subdivision prepared by comminution, precipitation, combustion, or sublimation. They are generally considered to be of 1mm diameter or less. Those powders that become easily airborne may be termed *dusts*.

REACTION

Chemical *reactions* include any number of changes that take place within a molecule or between two or more chemicals or elements. Reactions progress by the breaking and making of chemical bonds and result in one or more new products. The laws of thermodynamics dictate that chemical reactions result overall in a more energetically stable state. Thus, some chemicals react *exothermically* and give off energy in the form of heat, light, and sound to result in more energetically stable products. This energy gradient provides the motive force of the reaction. As energy is lost to the environment, the degree of disorder has increased fulfilling the thermodynamic law of entropy.

Endothermic reactions, however, take energy from the environment (heat, light, shock, noise). This may result in products that are more energetically unstable, but the overall energy of the system, including the environment from which the energy has been drawn, is in a more stable state.

Some reactions are reversible; others may not go to completion but hover in an equilibrium of reactants and products, depending on external conditions.

Nuclear reactions involve fission (splitting) of a nucleus or fusion (joining) of two or more nuclei during which matter is converted into energy.

REDUCTION
See *Terminology, Oxidation*, p.245.

SALTS
Salts result from reactions in which the hydrogen of an acid is replaced by a metal or equivalent ion. In the reaction of hydrochloric acid with sodium hydroxide, sodium chloride (common salt) and water are formed:

$$NaOH + HCl \rightarrow NaCl + H_2O$$

The possible range of salts is enormous. Metal salts of organic acids include sodium benzoate derived from reaction of benzoic acid and sodium bicarbonate solution:

$$C_6H_5COOH + NaHCO_3 \rightarrow C_6H_5COONa + CO_2 + H_2O$$

SENSITIZED
See *Terminology, Desensitized*, p.234.

SLAKED
Slaking is the process of hydrating materials, particularly lime, by treatment with water in which the lime generates heat and crumbles into powder. The result is slaked lime, or calcium hydroxide. Lime will also slake through exposure to moisture in the air, although at a slower rate.

SOLID
Solids are generally the densest form of matter in which the atoms or molecules are packed in close formation, often in a lattice or crystalline structure. See *Terminology, Gas*, p.239; *Terminology, Liquid*, p.241.

REGULATORY DEFINITIONS
Solids are dangerous goods, other than gases, that do not meet the definition of liquids in this paragraph. UN Part 1

Solid means a material which is not a gas or a liquid. US 171.8

Solid dangerous goods. Dangerous goods, other than gases, that do not meet the definition of Liquid dangerous goods. ICAO 1-3.1, IATA App. A

SOLUTIONS
The true meaning of *solution* applies to uniform molecular or ionic mixtures of substances, either of which may be a solid, liquid, or gas. The material that is dissolved into the other is called a *solute*; the receiving substance is called a *solvent* (e.g., sugar is the solute and water the solvent in a sugar-water solution). Steel is an example of a solid-solid solution of carbon, iron, and other substances. The tendency of one substance to dissolve into another is called *solubility*. If one material does not form a solution with another it is *insoluble*; saturated solutions occur when the solvent can dissolve no more of

the solute, which remains in its original form. Mixtures of gases are said to be miscible in one another rather than being in solution. See *Terminology, Miscible*, p.243; *Terminology, Mixture*, p.243.

SPONGE

Sponge is a term applied to the porous particulate form of a metal generated by the reduction of its oxide, or to the crude elemental forms obtained by the reduction of a metal ore. Titanium sponge is that obtained from the reduction of titanium tetrachloride with carbon.

STABILIZED

Stabilizers is a very general term that includes inhibitors, antioxidants, and emulsifiers which keep or retard a substance from changing its chemical form or nature. For example, acetone cyanohydrin readily decomposes to hydrocyanic acid and acetone unless it is stabilized. Unstabilized substances are those that might otherwise have had a stabilizer added or which have had the stabilizer removed. See *Terminology, Inhibited*, p.241.

STRUCTURAL NOTATION

The lists of regulated dangerous goods in transportation include many chemical prefixes that codify the structure of molecules, usually organic. Over time, many conventions have arisen that result in the acceptance of today's somewhat inconsistent scheme. On occasion, these prefixes may be placed in the centre of a chemical name, preceding a functional group or atom, or, very rarely, at the end of a name.

PREFIX	EXPLANATION
' (the prime mark)	The prime mark, , is used to distinguish one atom from another. In the case of 4,4'-diaminodiphenylmethane ($H_2NC_6H_4CH_2C_6H_4$. NH_2), two phenyl groups ($-C_6H_4$) are attached to a methyl group ($-CH_2$). At the 4-position on each phenyl group (i.e., the carbon directly opposite the bond to the methyl group) is an amino group ($-NH_2$).
alpha-, a-, α-	*alpha-* denotes the position of a functional group in an organic compound as being on the first carbon; e.g., in the case of α-methylbenzyl alcohol ($C_6H_5CH(CH_3)OH$) the hydroxyl group ($-OH$) is attached to the first carbon, the alpha carbon, extending from the benzene ring ($-C_6H_5$).
	Similarly, *beta, b-,* or β- indicates that the functional group is on the second carbon; e.g., in β-methyl acrolein ($CH_3C_2H_2CHO$) the methyl group ($-CH_3$) is attached to the second carbon, the β-carbon, in the acrolein group, the first being attached to the aldehyde group ($-CHO$).

PREFIX	EXPLANATION (CONT.)
	gamma- or γ- denotes functional groups in organic compounds as being on the third carbon; e.g., γ-methallyl chloride (C_4H_8Cl).
	omega- or ω-, the last letter in the Greek alphabet, denotes that the succeeding functional group is placed on the end of carbon chain as for ω-bromoacetophenone.
beta-, b-, β-	See *alpha-*
bi-, bis-	See *mono-*
cis-	In organic compounds with a carbon-carbon double bond, free rotation about the axis of the bond is not possible. There are two different spatial arrangements of any functional groups on either side of the bond. That in which the groups are on the same side is the *cis-* isomer; that in which the groups are on opposite sides is the *trans-* isomer.
cyclo-	*cyclo-* indicates that an aliphatic molecule contains a ring structure. Cyclobutane (C_4H_8), for example, is a ring of four carbons; cycloheptane (C_7H_{14}) is a ring of seven carbons.
D-	*D-* for *dextro* indicates the righthanded version of an enantiomer, an optical isomer; *L-* for *levo*, indicates the lefthanded version. In the molecule D(-)alpha phenylglycine chloride hydrochloride, one of the carbons is attached to four different function groups positioned at each of the four corners of a pyramidal shape about the carbon. Close inspection will show two possible enantiomers, mirror images for the relative configurations of the functional groups. Pairs of enantiomers may have differing chemical and physical properties.
di-	See *mono-*
gamma-, γ-	See *alpha-*
iso-	In the chains of carbon atoms that are the basis of organic chemistry, four or more carbons can be organized into a single chain or branched formation. Isomers differing in this fashion are denoted with the prefix *n-* (for *normal*) for the straight version and *iso-* (for *same*) for the branched version. For example, n-butylamine ($C_4H_9NH_2$) has a single chain of four carbons, while its isomer, isobutylamine has three carbons in a chain, with the fourth carbon joined to the middle carbon.
L-	See *D-*
meta-, m-	See *ortho-*
mono-	When a functional group name is preceded by the term *mono-* it indicates the presence of only a single group. For example, monoethylamine ($C_2H_5NH_2$) describes a single ethyl group ($-C_2H_5$) attached to the amino group ($-NH_2$). This is in contrast to diethylamine where there are two ethyl groups

PREFIX	EXPLANATION (CONT.)

((C$_2$H$_5$)$_2$NH); as a consequence, the *mono-* designation is often not used.

The prefixes *bi-* and *bis-* also mean two, but *di-* is the preferred option since *bi-* can also describe the presence of hydrogen in some molecules such as in sodium bicarbonate (NaHCO$_3$) as differentiated from sodium carbonate (Na$_2$CO$_3$). The prefixes *tri-* or *tris-* for three; *tetra-* for four; *penta-* for five, etc.; act similarly.

n-, normal See *iso-*

N- Nitrogen (symbol N) is a common element, particularly in organic compounds. The *N-* prefix indicates that the succeeding group is attached to a nitrogen in the molecule. For example, in *N-*methylbutylamine (C$_4$H$_9$NHCH$_3$) both the methyl group (-CH$_3$) and the butyl group (-C$_4$H$_9$) are attached to nitrogen in the amino group (-NH). Multiple *N*s indicate that the succeeding functional groups are each attached to the nitrogen. Thus, in *N,N-*dimethylformamide (HCON(CH$_3$)$_2$), both methyl groups are attached to the amino group.

numerals Numerals designate the positions of chemical groups or atoms in an organic molecule; thus, 1,1,1-trichloroethane has all three chlorine atoms (-Cl) attached to the same carbon in the ethane chain (Cl$_3$C-CH$_3$) and differs from its isomer, 1,1,2-trichloroethane (Cl$_2$HC-CH$_2$Cl). Numerals can also occur in the middle of names, as in 1,1,-dichloro-1-nitroethane in which the ethyl group of two carbons (-C-C-) has all three functional groups (two chlorines (-Cl) and one nitro (-NO$_2$)) connected to the first carbon.

Numerals are not always used conventionally. For example, 1,3-dichloropropanol-2 (ClCH$_2$CH$_2$OCH$_2$Cl) is a chain of three carbons with a chlorine at each end (carbons 1 and 3), and a hydroxyl (-OH) group on the middle carbon (carbon 2). Following the conventions described earlier it should be named 1,3-dichloro-2-propanol.

omega-, ω- See *alpha-*

ortho-, o- When there are two groups attached to a benzene ring of six carbons (C$_6$H$_6$), the terms *ortho-* *(meaning straight ahead)*, *meta-* *(meaning beyond)*, and *para-* (meaning *opposite*) are used to denote their relative positions. To use the system, imagine that one of the groups is fixed to one of the six carbons, given the number 1. When the second group is attached to the adjacent carbon (carbon 2), it is in the *o-* or *ortho-* position; carbon 3 is the *m-* or *meta-* position; and the next carbon (carbon 4), which is opposite to carbon 1 on the ring, is the *p-* or *para-* position. Note that moving to carbon 5 is equivalent to the *meta-* position, and carbon 6 is identical to the *ortho-* position.

PREFIX   EXPLANATION (CONT.)

For example, there are three isomers of aminophenol ($C_6H_4NH_2OH$) in which the amino group (-NH_2) and the hydroxyl group (-OH) can occupy three relative positions: o-aminophenol, m-aminophenol, and p-aminophenol. Each isomer has different properties.

para-, p-   See *ortho-*

penta-   See *mono-*

per   This prefix *per* describes one of three situations: (1) when a compound represents the maximum possible degree of substitution; thus, when methane (CH_4) has had each of its hydrogens replaced by chlorine, it becomes perchloromethane (CCl_4), otherwise known as tetrachloromethane; (2) to indicate an element in its highest oxidation state; e.g., chlorine in perchloric acid ($HClO_4$) exists as chlorine(VII); (3) to indicate the presence of the peroxy group (-O-O-) in a compound; e.g., peroxyacetic acid (CH_3.COOOH).

secondary-, sec-s   Used in reference to monohydric alcohols, amines, and a few related organic compounds, the terms *primary*, *secondary*, and *tertiary* describe molecular structure by indicating the number of alkyl groups (e.g., -CH_3 or -C_2H_5) to which a carbon with a functional group is attached. For example, methanol (CH_3OH) consists of a central carbon surrounded by three hydrogens and one hydroxyl group. If one of the hydrogens were replaced by a methyl group (-CH_3), the alcohol is *primary*, the central carbon is attached to one alkyl group. If a second hydrogen is similarly replaced, the alcohol is *secondary*, the central carbon is attached to two alkyl groups. Finally, if the last hydrogen is replaced, the alcohol is *tertiary*, the central carbon is attached to three alkyl groups. (The four monohydric alcohols cited are methanol, ethanol, sec-propyl alcohol (isopropanol), and tert-butyl alcohol.)

sym-   See *Terminology, Symmetrical*, p.252.

tertiary-, tert-, t-   See *secondary*

tetra-   See *mono-*

trans-   See *cis-*

tri-, tris-   See *mono-*

uns-   See *Terminology, Symmetrical*, p.252.

SUBSTITUTED

Substitution is an extremely common type of chemical reaction in which one element or functional group replaces another; rarely, however, is the product referred to as *substituted*. In the reaction of hydrogen chloride with ethanol, the hydroxyl group (-OH) is substituted by the chloride ion (Cl-) to produce ethyl chloride and water:

$$HCl + C_2H_5OH \rightarrow C_2H_5Cl + H_2O$$

SYMMETRICAL

Symmetry denotes the *symmetrical* arrangement of constituents with respect to the carbon chain or a functional group in a molecule. In the case of di-methylhydrazine $((CH_3)_2N_2H_2)$ two isomers are possible: 1,1-dimethylhydrazine, in which both methyl groups (-CH_3) are arranged on one side of the N-N bond, and the second in which the methyl groups are arranged on either side of the N-N bond. The former is unsymmetric and the latter symmetric. These varying arrangements may result in different physical and chemical properties; e.g., unsymmetrical dimethylhydrazine shows a degree of corrosivity that the symmetrical option does not. See *Terminology, Isomers*, p.241.

SYNTHETIC
See *Terminology, Natural*, p.243.

TINNING FLUX
Tinning is the process of applying a protective coating of tin to another metal. Flux is a material that is used to lower the melting temperature of the materials or to aid in their fusion in this and related soldering processes. Zinc chloride is a common tinning flux.

UNCALCINED
See *Terminology, Calcined*, p.233.

UNINHIBITED
See *Terminology, Inhibited*, p.241.

UNPHLEGMATIZED
See *Terminology, Phlegmatized*, p.245.

UNSTABILIZED
See *Terminology, Stabilized*, p.248.

UNSYMMETRICAL
See *Terminology, Symmetrical*, p.252.

VOLATILE
Depending on temperature and pressure, all materials in the solid and liquid state have a vapour pressure, that is, some of their surface atoms or molecules exist as a vapour or gas above the material. Thus under certain conditions a liquid evaporates and some solids sublimate. This is most familiar to us as we smell coffee or perfumes; the fragrant molecules leave the liquid and enter the surrounding atmosphere. Mothballs are an example of the tendency of solids, like naphthalene, to enter the gaseous state. *Volatility* is the meas-

ure of the tendency of the atoms or molecules of a solid or liquid to pass into the gaseous or vapour state. Volatile chemicals, like volatile organic chemicals, are those that exhibit this tendency strongly under ambient conditions and evaporate fairly quickly, particularly as temperatures approach the boiling point. Nonvolatile chemicals have relatively little vapour pressure under ambient conditions.

REGULATORY DEFINITIONS

Volatility refers to the relative rate of evaporation of materials to assume the vapor state. US 171.8

WETTED

Many substances will react with the moisture or oxygen (or other gases) in the air. As a consequence, they may be *wetted*: covered with water, alcohol, oil, or other material that separates the reactive substance from the atmosphere. The dilution afforded by this process may also serve to desensitize the mixture. Wetting agents are sometimes added to water to reduce its surface tension and help it spread and penetrate the substance. Many explosives are desensitized by being wetted. See *Terminology, Desensitized*, p.234.

REFERENCES

CAT, FOM, GOCT, HCC, HOC, HPT, IMDG, JDOMT, KOE, LCA, MEOS, MH14, REI, TEA, THS, TOE, TPD, VOA, WNN

Tires

Tires syn. tyres are the outer circumference of a wheel which makes contact with a surface. Metal tires are used on rail and other vehicles. Solid rubber tires are used in factories where they carry high loads to resist abrasion and puncture. Where the greatest need for protection from shock is required (e.g., in automobiles, aircraft, and other vehicles) *pneumatic* (compressed air-filled) tires are used. Tire assemblies consist of the tire and the wheel to which they are attached. Assemblies may also include rubber inner tubes which hold the air. On inflation, the edge of the tire is pressed against the rim of the wheel and forms an airtight seam or an inner tube is filled. Tires are manufactured to rated inflation pressures appropriate to the weight of the vehicle and driving conditions.

HAZARD
Uninflated tires present no hazard. Inflated tires contain compressed gas and can rupture, particularly if damaged or otherwise unserviceable.

RELATED TERMS
None

REGULATORY DEFINITIONS
None

REFERENCES
THE

Poisonous ▪ Solids containing toxic liquid, n.o.s., 6.1 ▪ Toxic liquid, inorganic, n.o.s., 6.1, 8 ▪ Toxic liquid, n.o.s., 4.3, 5.1, 6.1 ▪ Toxic liquid, organic, n.o.s., 3, 6.1, 8 ▪ Toxic solid, inorganic, n.o.s., 6.1, 8 ▪ Toxic solid, n.o.s., 4.2, 4.3, 5.1, 6.1 ▪ Toxic solid, organic, n.o.s., 4.1, 6.1, 8

2810 ▪ 2811 ▪ 2927 ▪ 2928 ▪ 2929 ▪ 2930 ▪ 3086 ▪ 3122 ▪ 3123 ▪ 3124 ▪ 3125 ▪ 3243 ▪ 3287 ▪ 3288 ▪ 3289 ▪ 3290

Toxic or *poisonous* substances are those which injure biological systems causing abnormal and unwanted biochemical or biophysical activity. Toxic chemicals damage tissue, impair the nervous system, or cause serious illness or death in a variety of ways:

- Inflammation, which produces an increased blood flow as the body works to heal itself.
- Degeneration, a broad range of cell damage and malfunction.
- Necrosis, cell death.
- Hypersensitization in which antigens trigger an allergic reaction.
- Immunosuppression in which the immune system is depressed and less able to fight infection.
- Neoplasis in which cell growth and control malfunctions lead to abnormal masses of cells which, if malignant, produce cancers (these are the *carcinogens*).
- Mutagenesis in which heritable DNA damage is caused.
- Enzyme inhibition which disrupts the normal functioning of enzymes.
- Biochemical uncoupling in which excessive heat can be generated.
- Lethal synthesis in which chemicals mimic biochemicals and inhibit normal reactions.
- Teratogenesis which produces effects on developing foetuses.
- Endocrine disruption in which chemicals mimic or block the effect of oestrogen.

DOSE AND EXPOSURE

Toxic effects can only occur on direct exposure to the substance through inhalation of vapours, gases, and mists; ingestion of solids and liquids; or absorption through broken and unbroken skin, eyes, and mucous membranes (branches of science are devoted to the study of fate and transport of toxics in the environment and workplace to determine how they reach a receptor). The degree of toxicity also depends on the species involved, the number of exposures, the magnitude of exposure, the timing of the dose (the effect of diurnal and seasonal factors), and the formulation of the toxic substance, including possible impurities. Synergistic effects also play their part; for example, smokers exposed to asbestos fibres are ten times more likely to develop lung cancer than asbestos workers who do not smoke, and five times as likely as smokers that are not exposed to asbestos. Due to the many unknowns associated with synergistic effects, risks are assumed to be at least additive.

Response to a toxic dose varies from *acute* to *chronic*. An acute effect is generally characterized by a single, rather high exposure with a rapid onset of symptoms and a quick resolution to the crises (e.g., ingesting some cyanide compounds will lead rapidly to death). Chronic responses are those associated with low-level, repeated exposures over long periods with a gradual onset of symptoms. Some chronic effects are caused by the bioaccumulation of toxics (i.e., accumulation in the body tissues), like PCBs or by latency periods in which symptoms are not realized until long after exposure, as is the case with asbestosis. There is a wide spectrum of responses between acute and chronic affects.

QUANTIFYING TOXICITY

The effects of many toxics are well documented: the carcinogenic effects of cigarette smoke, the liver damage caused by alcohol, and the poisonous effects of strychnine, for example. Long-term studies are necessary to determine chronic toxicity, but the effects may be approximated by giving relatively high controlled doses to a controlled test population (usually animals) and analysing the results. The data are then extrapolated and modified to a semblance of the equivalent human response. *Lethal dose* tests are those in which animals are made to ingest, inhale, or absorb specified concentrations of a substance over a controlled period. The concentration necessary to kill half the population (50%) is called the LD_{50} (lethal dose, 50) for oral and dermal toxicity, and LC_{50} (lethal concentration, 50) for inhalation toxicity.

POISONOUS BY INHALATION

In the United States, the Department of Transportation designates materials that are *poisonous by inhalation* as a subcategory requiring special handling and communication precautions. Liquids,[23] which at normal temperatures and pressures form air dispersions of molecules (i.e., a vapour or fume) in sufficient concentration and toxicity, fall into this category; e.g., bromine trifluoride evolves a toxic vapour. Depending on their inhalation toxicity and volatility, poisonous by inhalation liquids and solids are assigned to *Hazard Zone A* or *B*, *A* being the more toxic. (Solids are unlikely to have a vapour pressure high enough to be an inhalation hazard by these criteria.)

RELATED TERMS

Compounds, see *Terminology, Compound,* p.234

Corrosive, see *Corrosives and Class 8*, p.47

Flammable see *Flammable Liquids and Class 3*, p.96

Flammable see *Flammable Solids and Division 4.1*, p.99

Inhalation hazard, Zone A, Zone B, see also *Gases and Class 2*, p.104

Inorganic, see *Terminology, Inorganic*, p.241

---

[23] Toxic gases are also poison inhalation hazards, see *Gases and Class 2*, p.104.

Liquid, see *Terminology, Liquid,* p.241
Organic, see *Terminology, Organic,* p.244
Oxidizing, see *Oxidizers and Class 5,* p.170
Self-heating, see *Spontaneously Combustible Materials and Division 4.2,* p.226
Solids, see *Terminology, Solid,* p.247
Toxins, liquid, extracted from living sources,

n.o.s., *Infectious Substances and Division 6.2,* p.115
Toxins, solid, extracted from living sources, n.o.s., *Infectious Substances and Division 6.2,* p.115
Water-reactive, see *Dangerous When Wet Materials and Division 4.3,* p.58

REGULATORY DEFINITIONS

Toxic substances These are substances liable either to cause death or serious injury or to harm human health [if] swallowed or inhaled or by skin contact. UN 2.6.1(a); IMO Class 6, 1.1.1

Division 6.1-Toxic substances. Substances liable either to cause death or injury or to harm human health if swallowed, if inhaled or by skin contact. ICAO 6.1

For the purpose of this subchapter, *poisonous material (Division 6.1)* means a material, other than a gas, which is known to be so toxic to humans as to afford a hazard to health during transportation, or which, in the absence of adequate data on human toxicity: (1) Is presumed to be toxic to humans because it falls within any one of the following categories when tested on laboratory animals (whenever possible, animal test data that has been reported in the chemical literature should be used): (i) Oral Toxicity. A liquid with an LD_{50} for acute oral toxicity of not more than 500 mg/kg or a solid with an LD_{50} for acute oral toxicity of not more than 200 mg/kg. (ii) Dermal Toxicity. A material with an LD_{50} for acute dermal toxicity of not more than 1000 mg/kg. (iii) Inhalation Toxicity. (A) A dust or mist with an LC_{50} for acute toxicity on inhalation of not more than 10 mg/L; or (B) A material with a saturated vapor concentration in air at 20°C (68°F) of more than one-fifth of the LC_{50} for acute toxicity on inhalation of vapors and with an LC_{50} for acute toxicity on inhalation of vapors of not more than 5000 ml/m^3; or (2) Is an irritating material, with properties similar to tear gas, which causes extreme irritation, especially in confined spaces. US 173.132(a)

Division 6.1 Toxic substances are substances which are liable to cause death or injury or to harm human health if swallowed, inhaled or contacted by the skin. IATA 3.6.1.1

Hazard Zone means one of four levels of hazard (Hazard Zones A through D) assigned to gases, as specified in 173.116(a) of this subchapter, and one of two levels of hazards (Hazard Zones A and B) assigned to liquids that are poisonous by inhalation, as specified in 173.133(a) of this subchapter. A hazard zone is based on the LC_{50} value for acute inhalation toxicity of gases and vapours, as specified in 173.133(a). US 171.8

Material poisonous by inhalation means: (1) A gas meeting the defining criteria in 173.115(c) of this subchapter and assigned to Hazard Zone A, B, C, or D in accordance with 173.116(a) of this subchapter; (2) A liquid (other than as a mist) meeting the defining criteria in 173.132(a)(1)(iii) of this subchapter and assigned to Hazard Zone A or B in accordance with 173.133(a) of this subchapter; or (3) Any material identified as an inhalation hazard by a special provision in Column 7 of the 172.101 Table. US 171.8

REFERENCES
HCC, HMM, KOE, THS

Units of Measure

The following SI (International System of Units) and non-SI units are used in the transportation regulations. The SI system is based on fundamental standards, derived units, and prefixes.

<u>S.I. FUNDAMENTAL STANDARDS</u>

length	metre	m
mass	gram	g
time	second	s
electric current	ampere	A
temperature	kelvin	K
luminous intensity	candela	cd
amount of substance	mole	mol

<u>S.I. PREFIXES</u>

SYMBOL	PREFIX	(NON-S.I. NAME)	MULTIPLYING FACTOR
E	exa	quintillion[1]	1,000,000,000,000,000,000 or 10^{18}
P	peta	quadrillion[2]	1,000,000,000,000,000 or 10^{15}
T	tera	trillion[3]	1,000,000,000,000 or 10^{12}
G	giga	billion[4]	1,000,000,000 or 10^{9}
M	mega	million	1,000,000 or 10^{6}
k	kilo	thousand	1,000 or 10^{3}
h	hecto	hundred	100 or 10^{2}
da	deca	ten	10 or 10^{1}
d	deci	tenth	0.1 or 10^{-1}
c	centi	hundredth	0.01 or 10^{-2}
m	milli	thousandth	0.001 or 10^{-3}
μ	micro	millionth	0.000,001 or 10^{-6}
n	nano	billionth	0.000,000,001 or 10^{-9}
p	pico	trillionth	0.000,000,000,001 or 10^{-12}
f	femto	quadrillionth	0.000,000,000,000,001 or 10^{-15}
a	atto	quintillionth	0.000,000,000,000,000,001 or 10^{-18}

[1]Some nations may use *trillion*.
[2]Some nations may use *billiard*.
[3]Some nations may use *billion*.
[4]Some nations may use *milliard*.

S.I AND NON-S.I. DERIVED AND COMPOUND UNITS

The following derived and compound units are used in the transportation regulations.

AMOUNT OF SUBSTANCE

M............ 1 mole/litre
Mol......... mole (S.I. standard)

Conversions
1 mole = 6.0×10^{23} atoms or molecules

AREA

cm^2 .... $centimetre^2$
ft^2 .... $foot^2$
in^2 .... $inch^2$
km^2 .... $kilometre^2$
m^2 .... $metre^2$
m^2/g .... $metre^2/gram$ (surface area)

Conversions
$1 m^2 = 10.765 ft^2$

CONCENTRATION

mg/kg ..... milligrams/kilogram
mg/l .... milligrams/litre
ml/m^3 ...... $millilitres/metre^3$
mol/l .... moles/litre
pH .... log reciprocal hydrogen ion conc.
ppm .... parts per million

DENSITY

g/cm^3 ...... $grams/centimetre^3$
lb/gal ..... pounds/gallon
kg/l .... kilogram/litre
kg/m^3 ...... $kilogram/metre^3$

Conversions
1 lb/gal (U.S) = $119.83 kg/m^3$

ELECTRICITY AND MAGNETISM

A .... ampere (S.I. standard)
A/m .... amperes/metre (magnetic field strength)
C .... Coulomb (quantity of electricity)
G .... gauss
Ω .... ohm (electric resistance)
Ω/m .... ohm/metre
S .... siemens (conductance)
T .... tesla (magnetic flux density)

Conversions
1 mho = 1 S
$1G = 1 \times 10^{-4} T$

ENERGY/POWER

BTU .... British Thermal Unit
cal .... calorie
eV .... electron volt
gcal .... gram calorie
J.............. joule
J/g .... joules/gram (specific energy)
kgm .... kilogram metre
kWh .... kilowatt hour
MeV .... megaelectron volt
W .... Watt
W/m^2 ...... $watts/metre^2$

Conversions
1 BTU = 1054.4 J
1 cal = 4.190 J
1 eV = 1.602×10^{-19} J
1 gcal = 0.001 cal
1 J = 1 N·m
1 kgm = 9.807 J
1 kWh = 3.6 MJ
1 W = 1 J/s

FORCE/STRENGTH
kgf .... kilogram force
N .... newton (force)

Conversions
1 kgf = 9.807 N
1 dyne = 1×10^{-5} N
1 N = 1 kg·m/s^2

FREQUENCY
Hz .... Hertz (frequency)

Conversions
1 Hz = 1 cycle/second

LENGTH
Å   Angstrom
cm .... centimetre
ft .... foot
in .... in
km .... kilometre
m .... metre (S.I. standard)
μm .... micron
mm .... millimetre

Conversions
1 Å = 10^{-10}m
1 ft = 12 in
1 m = 3.281 ft

LIGHT
cd .... candela (S.I. standard)
        or candlepower
lm .... lumen

Conversions
1 lm = 1/683 W at 540 THz

MASS/WEIGHT
g .... gram (S.I. standard)
gr .... grain
kg .... kilogram
lb .... pound avoirdupois
mg .... milligram
oz .... ounce avoirdupois
t .... metric ton

Conversions
1 grain = 0.0648 g
1 lb = 453.59 g
1 lb = 16 oz
1 t = 1,000 kg

PRESSURE/STRESS
atmos ...... atmospheres
bar .... bar
g/m^2 .... grams/metre^2
kg/cm^2 .... kilograms/centimetre^2
kgf/cm^2 ... kilogram-
        force/centimetre^2
kPa .... kilopascal
lb/ft^2 .... pounds/foot^2
lb/in^2 .... pounds/inch^2
mmHg .... millimetre of mercury
        (torr)
N/m^2 .... newton/m^2 (stress)
N/m^2 .... newtons/metre^2
N/mm^2 .... newton/millimetre^2
        (stress)
N/mm^2 .... newtons/millimetre^2
Pa .... pascal (pressure)
psi .... pounds/inch^2
psia .... pounds/inch^2 absolute
psig .... pounds/inch^2 gauge
t/cm^2 .... metric tons/cm^2

Conversions
1 bar = 1×10^5 Pa
1 kgf/cm^2 = 98.07 kPa
1 psi = 6.895 kPa
1 atmos = 101.325 kPa
1 mmHg = 133.322 Pa
1 Pa = 1 N/m^2

RADIOACTIVITY
Bq........... becquerel (activity of a
radionuclide)
Bq/cm^2 ... becquerel/cm^2
Bq/l .... becquerel/litre
C/kg........ coulomb/kilogram
(exposure)
Ci........... curie
Ci/g .... curie/gram
Ci/l .... curie/litre
Gy .... gray (absorbed dose)
kBq/kg.... kilobecquerel/kilogram
$\mu Ci/cm^2$ .. microcurie/centimetre^2
nCi/g ...... nanocurie/gram
R............. roentgen (exposure)
R/h.......... roentgen/hour
rad .... radiation absorbed dose
rem .... roentgen equivalent
man
rem/h ...... rem/hour
S/m .... siemens/metre
Sv .... Sievert (dose equiva-
lent)
Sv/h .... sieverts/hour

Conversions
1 Bq = 1 disintegration/second
$1\ Ci = 3.7 \times 10^{10}\ Bq$
1 Sv = 100 rem
1 Gy = 100 rad
1 mho/cm = 100 S/m

SURFACE TENSION
dynes/cm^2.... dynes/centimetre^2

Conversions
1 bar = 106 dynes/cm^2
1 dyne/cm^2 = 0.1 Pa

TEMPERATURE
$^{\circ}$C .... degree Celsius
$^{\circ}$F .... degree Fahrenheit
K .... kelvin (S.I. standard)

Conversions
$^{\circ}$C = K + 273.15
$^{\circ}$C = ($^{\circ}$F - 32)/1.8

TIME
h .... hour
s.............. second (S.I. standard)

Conversions
1 h = 3600 s

TOXICITY
LC_{50} .... lethal concentration,
50%
LD_{50} .... lethal dose, 50%

VISCOSITY
kgs/m^2 .... kilogram sec-
onds/metre^2
mm^2/s ..... millimetre^2/second
(kinetic)
m^2/s .... metres^2/second (kinetic)
P............. poise (dynamic)
Pa·s .... Pascal seconds
(dynamic)
St............ Stokes (kinetic)

Conversions
1 P = 0.1 Pa·s
1 kgs/m^2 = 9.807 Pa·s
1 St = 10^{-4} m^2/s
1 Pa·s = 1 Ns/m^2

VOLUME
cm^3 .... centimetre^3
ft^3 .... foot^3
gal .... gallon
l or L ...... litre
m^3 .... metre^3
ml........... millilitre
pt............ pint
qt............ quart

Units of Measure

Conversions
1 gal = 8 pt
1 qt = 2 pt
1 pt = 20 oz (US 16 oz)
1 oz = 2.9574×10^{-5} m^3

1 ft^3 = 6.229 gal (7.481 US)
1 ft^3 = 0.028 m^3
1 m^3 = 223 gal (266 US)
1 gal = 4.54 l (3.785 US)

RELATED TERMS
None

REGULATORY DEFINITIONS
P.s.i.a. or psia means pounds per square inch absolute. US 171.8
P.s.i.g. or psig means pounds per square inch gauge. US 171.8

REFERENCES
APD, EAC, DGR, HOC, IMDG, RFT, ROT, TAS, THS, TWE

References

7CFR319 *Foreign Quarantine Notices*, Part 319, Title 7-Agriculture, Code of Federal Regulations; Office of the Federal Register, U.S. National Archives and Records Administration: Washington

10CFR436 *Federal Energy Management and Planning Programs*, Part 436, Title 10-Energy, Code of Federal Regulations; Office of the Federal Register, U.S. National Archives and Records Administration: Washington

15CFR700 *Defense Priorities and Allocation System*, Part 700, Title 15-Commerce and Foreign Trade, Code of Federal Regulations; Office of the Federal Register, U.S. National Archives and Records Administration: Washington

16CFR1210 *Safety Standard for Cigarette Lighters*, Part 16, Title 16-Commercial Practices, Code of Federal Regulations; Office of the Federal Register, U.S. National Archives and Records Administration: Washington

21CFR182 *Substances Generally Recognized as Safe*, Part 182, Title 21-Food and Drugs, Code of Federal Regulations; Office of the Federal Register, U.S. National Archives and Records Administration: Washington

21CFR868 *Anesthesiology Devices*, Part 868, Title 21-Food and Drugs, Code of Federal Regulations; Office of the Federal Register, U.S. National Archives and Records Administration: Washington

21CFR872 *Dental Devices*, Part 872, Title 21-Food and Drugs, Code of Federal Regulations; Office of the Federal Register, U.S. National Archives and Records Administration: Washington

27CFR21 *Formulas for Denatured Alcohol and Rum*, Part 21, Title 27-Alcohol, Tobacco Products and Firearms, Code of Federal Regulations; Office of the Federal Register, U.S. National Archives and Records Administration: Washington

27CFR55 *Commerce in Explosives*, Part 27, Title 27-Alcohol, Tobacco Products and Firearms, Code of Federal Regulations; Office of the Federal Register, U.S. National Archives and Records Administration: Washington

27CFR270 *Manufacture of Tobacco Products*, Part 270, Title 27-Alcohol, Tobacco Products and Firearms, Code of Federal Regulations; Office of the Federal Register, U.S. National Archives and Records Administration: Washington

29CFR520 *Employment Under Special Certificate or Messengers, Learners (including Student Learners), and Apprentices*, Part 520, Title 29-Labor, Code of Federal Regulations; Office of the Federal Register, U.S. National Archives and Records Administration: Washington

29CFR1910 *Occupational Safety and Health Standards*, Part 1910, Title 29-Labor, Code of Federal Regulations; Office of the Federal Register, U.S. National Archives and Records Administration: Washington

32CFR552 *Regulations Affecting Military Reservations*, Part 552, Title 32-National Defense, Code of Federal Regulations; Office of the Federal Register, U.S. National Archives and Records Administration: Washington

33CFR175 *Equipment Requirements*, Part 175, Title 33-Navigation and Navigable Waters, Code of Federal Regulations; Office of the Federal Register, U.S. National Archives and Records Administration: Washington

40CFR60 *Standards of Performance for New Stationary Sources*, Part 60, Title 40-Protection of Environment, Code of Federal Regulations; Office of the Federal Register, U.S. National Archives and Records Administration: Washington

40CFR63   *National Emission Standards for Hazardous Air Pollutants for Source Categories*, Part 63, Title 40-Protection of Environment, Code of Federal Regulations; Office of the Federal Register, U.S. National Archives and Records Administration: Washington

40CFR261   *Identification and Listing of Hazardous Waste*, Part 261, Title 40-Protection of Environment, Code of Federal Regulations; Office of the Federal Register, U.S. National Archives and Records Administration: Washington

40CFR439   *Pharmaceutical Manufacturing Point Source Categories*, Part 439, Title 40-Protection of Environment, Code of Federal Regulations; Office of the Federal Register, U.S. National Archives and Records Administration: Washington

40CFR455   *Pesticide Chemicals*, Part 455, Title 40-Protection of Environment, Code of Federal Regulations; Office of the Federal Register, U.S. National Archives and Records Administration: Washington

42CFR72   *Interstate Shipment of Etiologic Agents*, Part 72, Title 42-Public Health, Code of Federal Regulations; Office of the Federal Register, U.S. National Archives and Records Administration: Washington

46CFR160   *Lifesaving Equipment*, Part 160, Title 46-Shipping, Code of Federal Regulations; Office of the Federal Register, U.S. National Archives and Records Administration: Washington

49CFR   Title 49-Transportation, Code of Federal Regulations; Office of the Federal Register, U.S. National Archives and Records Administration, Revised as of Oct 1, 1998 as amended by Federal Registers through January 31, 1999

ACD   *A Consumer's Dictionary of Cosmetic Ingredients*, 4th Edition; Winter, R.; Crown Trade Paperbacks: New York, 1994

ACM   *Articles Containing Mercury*, Transportation of Hazardous Waste, personal communication; Teichman, Donald W., Jr.

ADM   *A Dictionary of Mining, Mineral, and Related Terms;* Thrush, Paul W.; U.S. Department of the Interior: Washington, 1968

ADO   *A Dictionary of Machining;* Simons, Eric N.; Frederick Muller: London, 1972

AEO   *An Encyclopaedia of the Iron & Steel Industry;* 2nd Edition; Osborne, A.K.; The Technical Press: London 1967

AH   *Adhesives Handbook;* Shields, J., B.Sc.; 2nd Edition.; Newnes-Butterworths: London, 1976

AMR   *ASM Metals Reference Book, A Handbook of Data about Metals and Metalworking;* American Society for Metals: Metals Park, 1981

AND   *A New Dictionary of Physics*; Gray, H.J., Isaacs, A.; Longman Group: London, 1975

ANS   *American National Standard for Primary Batteries - Lithium Primary Batteries - Specifications;* ANSI C18.3M 91; American National Standards Institute: New York, 1991

APD   *A Physicists Desk Reference*, The Second Edition of Physics Vade Mecum; Anderson, Herbert L., ed.; American Institute of Physics: New York, 1989

BCO   *Basel Convention on the Control of Transboundary Movements of Hazardous Wastes and Their Disposal, with Annexes;* United Nations: Basel, 1989

BH   *Blasters' Handbook*, 16th Edition; E.I. du Pont de Nemours: Wilmington, 1980

BHA   *Battery Hazards and Accident Prevention;* Levy, S.C., and Bro, P.; Plenum Press: New York, 1994

BMD   *Black's Medical Dictionary*, 37th Edition.; Macpherson, Gordon, ed.; Barnes & Noble Books: Lanham, 1992

BMS *Bristol-Myers Squibb Company*, Pharmaceutical Group, personal communication; Kearney, James; January 6, 1998

BRB *Battery Reference Book*, 2nd Edition; Crompton, T.R.; Society of Automotive Engineers, Inc.; Reed Educational and Professional Publishing Ltd, 1996

BRZ *Big River Zinc Corp.*, personal communication; James, Steve; March 23, 1999

BTH *Battery Technology Handbook;* Kiehne, H.A., ed.; Marcel Dekker: New York, 1989

CAB *Coals and Bitumens and Related Fossil Carbonaceous Substances;* Tomkeieff, S.I.; Pergamon Press: London, 1954

CAC *Cottonseed and Cottonseed Products;* Bailey, A.E., ed.; Interscience Publishers: New York, 1948

CAT *Chemistry and Technology of Lime and Limestone;* Boynton, Robert S.; Interscience Publishers: New York, 1966

CBI *Carbon Black - Its Manufacture, Properties, and Uses;* Neal, R.O., and Perrott, G. St. J.; Bureau of Mines, Department of the Interior: Washington, 1922

CCO *Combined Catalogs of Sporting Goods Manufacturers, 1929;* Johnson, Cal, ed.; Sporting Goods Journal, National Sports Publications, 1929

CEO *Concise Encyclopedia of Polymer Science and Engineering;* Kroschwitzi, J.I., ed.; John Wiley & Sons: New York, 1990

CERCLA *Comprehensive Environmental Response, Compensation, and Liability Act* of 1980, (Superfund) as amended, 42 USC 9601 et seq, United States Code

CFA *Crash, Fire and Rescue Handbook* EA-250-2; Bellomo, Charles and Lynch, John; IAP: Casper, 1982

CIM *Celluloid, Its Manufacture Applications and Substitutes;* Masselon, Roberts and Cillard; Charles Griffin: London, 1912

CON *Chemistry of Natural Protein Fibers;* Asquith, R.S., ed.; Plenum Press: New York, 1977

COS *Code of Safe Practice for Solid Bulk Cargoes (BC Code)*, Supplement to the International Maritime Dangerous Goods Code, International Maritime Organization: London 1997

COTB *Convention of the Prohibition of the Development, Production and Stockpiling of Bacteriological (Biological) and Toxin Weapons and on Their Destruction;* United Nations: 1971. See also, *Protocol for the Prohibition of the Use in War of Asphyxiating, Poisonous or Other Gases and of Bacteriological Methods of Warfare*; Geneva, 1925 (the Geneva Convention)

COTC *Convention on the Prohibition of the Development, Production, Stockpiling and Use of Chemical Weapons and on Their Destruction;* United Nations: Paris, 1993

CP *Chemical Products*, Vol. 2., Handbook of Paint and Coating Raw Materials; Ash, Michael and Irene; Gower Publishing: Aldershot, 1996

CPS *Creosote Pilings Spark Controversy in California;* Wastler, A.R.; Journal of Commerce and Commercial, January 25, 1994

CSA *Castor, Sesame and Safflower;* Weiss, E.A.; Barnes & Noble: New York, 1971

CSS *Charcoal: Small Scale Production and Use;* Boutette, M.; Friedr. Vieweg & Sohn: Wiesbaden, Germany, 1984

CTA *Coal-Tar and Water-Gas Tar Creosotes: Their Properties and Methods of Testing;* Bulletin No. 1036, October 20, 1922; United States Department of Agriculture: Washington D.C.

References

CTD   *Chemical Tradename Dictionary*, Ash, Michael & Irene; VCH Publishers: New York, 1993

D   *Demolitions*, Military Engineering, Volume IV, Part I, 1942, (Provisional); The War Office (His Majesty's), September 23, 1942

DC   *Detonating Cord*, product literature; The Ensign Bickford Company; 5/94 Rev., Stock #. 29302

DFA   *Definitions for Asbestos and Other Health-Related Silicates;* Levadie, B., ed.; American Society for Testing and Materials, Special Technical Publication 834, 1984

DGR   *Dangerous Goods Regulations*, 40th Edition; International Air Transport Association (IATA): Montreal, 1998

DIM   *Dorlang's Illustrated Medical Dictionary*, 28th Edition; W.B. Saunders; Philadelphia, 1988

DOCN   *Dictionary of Chemical Names and Synonyms;* Howard, Philip H. and Neal, Michael; Lewis Publishers: Boca Raton

DOCO   *Dead or Creosote Oil*, Letter from the Chairman of the United States Tariff Commission transmitting in response to Senate Resolution No.470 (Seventy-First Congress) a report on dead or creosote oil, March 24, 1932, Washington D.C.

DODT   *Department of Defense Dictionary of Military Terms;* The Joint Chiefs of Staff; Arco: New York, 1988

DOM   *Dictionary of Metallurgy;* Birchon, D.; George Newnes: London, 1965

DOMA   *Dictionary of Metallurgical and Chemical Machinery, Appliances and Material*, 2nd Edition; Metallurgical and Chemical Engineering: New York, 1910

DOMS   *Dictionary of Military Science*, Shafritz, Jay M., Shafritz, Todd J.A., and Robertson, David B.; Facts On File: New York, 1989

DOMST   *Dictionary of Machine Shop Term;* Telford, Arthur C.; American Technical Society: Chicago, 1947

DOMT   *Dictionary of Military Terms;* Dupuy, Johnson and Hayes, 1986

DOS   *Dictionary of Science and Technology;* Morris, Christopher; Academic Press: San Diego, 1992

DOSATT   *Dictionary of Scientific and Technical Terms*, 5th Edition; Parker, Sybil, ed.; McGraw-Hill: New York, 1994

DOST   *Dictionary of Space Technology;* Williamson, Mark; Adam Hilger, 1990

DOTU   *Dictionary of Terms Used in the Hides, Skins, and Leather Trade;* Agriculture Handbook, No. 465; Thomas, Charles C.; U.S. Department of Agriculture: Springfield, 1966

DOW   *Dictionary of Weapons and Military Terms;* Quick, John; McGraw-Hill: New York, 1973

E1   *Explosives;* La Motte, Arthur; International Textbook: Great Britain, 1938

E2   *Explosives, Matches and Fireworks;* Reilly, Joseph; Gurney and Jackson: London, 1938

E3   *Explosives;* Meyer, Rudoff and Kohler, Josef; VCH Verlagsgesellschaft mbH, 1993

EAC   *European Agreement concerning the international carriage of dangerous goods by road (ADR) and protocol of signature*; Inland Transport Committee, Economic Commission for Europe; United Nations: New York, 1998

EOA   *Encyclopedia of Agricultural Science;* Arutzen, C.J. and Ritter, E.M.; Academic Press: San Diego, 1994

EOE — *Encyclopedia of Explosives and Related Terms, Volume 1;* Feforoff, Basel T.; Picatinny Arsenal: Dover, 1960

EOI — *Encyclopaedia of Inorganic Chemistry,* King, R. Bruce, ed.; John Wiley & Son: Chichester: 1994

EOTF — *Encyclopedia of Textiles, Fibers, and Nonwoven Fabrics;* Grayson, Martin; John Wiley & Sons: New York, 1984

EOTM — *Evaluation of the M-16 Rifle as a Line-Throwing Gun;* Pierce, W.T.; Office of Research & Development; U.S. Coast Guard: Washington, 721203/009

EVT — *Electric Vehicles: Technology, Performance and Potential;* International Energy Agency/Organisation for Economic Co-Operation and Development: Paris, 1993

FAF — *Fuels and Fuel Technology, A Summarized Manual,* 2nd Edition; Francis, Wilfrid and Peters, Martin C.; Pergamon Press: Oxford, 1980

FDO — *Fairchild's Dictionary of Textiles,* 7th Edition; Tortora, Phyllis G., ed.; Fairchild Publications: New York, 1996

FOM — *Fuel Oil Manual;* Schmidt, Paul F.; Industrial Press: New York, 1985

FSA — *Fire Suppression and Detection Systems, 2nd Edition;* Bryan, John L.; Macmillian: New York, 1982

FTR — *Finding the Rx for Managing Medical Wastes;* Congress of the United States, Office of Technology Assessment

GAE — *Glossary, Aircraft Engine Starting and Auxiliary Power Systems;* Aerospace Recommended Practice, ARP 906A; Society of Automotive Engineers, 1986

GM — *General Metals;* Feirer, John L.; McGraw-Hill: New York, 1952

GOA — *Glossary of Automotive Inflatable Restraint Systems;* Surface Vehicle Information Report, J1538; Society of Automotive Engineers, Rev. 1995-04

GOCE — *Glossary of Commercial Explosives Industry Terms;* Publication No, 12; Institute of Makers of Explosives, 1985

GOCT — *Glossary of Chemical Terms;* Hampel, Clifford A.; Van Nostrand Reinhold: 1982

GOT — *Glossary of Terms Relating to Rubber and Rubber-Like Material;* American Society for Testing and Materials, 1956

HCC — *Hawley's Condensed Chemical Dictionary,* 11th Edition; Sax, N. Irving and Lewis, Richard J., Sr.; Van Nostrand Reinhold: New York, 1987

HCD — *Hackn's Chemical Dictionary,* 4th Edition; Grant, Julius; McGraw-Hill: New York, 1969

HG — *Hand Grenades;* Ainslie, Graham M.; John Wiley & Sons: New York, 1917

HMM — *Hazardous Materials Management;* Carson, Tom and Cox, Doye, eds.; Institute of Hazardous Materials Management, Rockville, 1992

HMW — *Hazardous Materials/Waste Handling for the Emergency Responder;* York, Kenneth J. and Grey, Gerald L.; Fire Engineering: New York, 1989

HOA — *Handbook of Aviation Fuel Properties;* CRC Rep. No. 530; Coordinating Research Council: Atlanta, 1983

HOB — *Handbook of Batteries,* 2nd Edition; Linden, David, ed.; McGraw-Hill: New York, 1995

HOC — *Handbook of Chemistry & Physics,* 79th Edition, 1998-1999; Lide, David R., ed.; CRC Press: Boca Raton: 1998

HOE — *Handbook of Electric Blasting,* (Revised 1985); Field Technical Operations; Atlas Powder Co., 1985

References

HOHM   *Handbook of Hazardous Materials;* Corn, Morton, ed.; Academic Press: San Diego, 1993

HOO   *Handbook of Oil Industry Terms and Phrases;* Langenkamp, R.D.; Pennwell Books: Tulsa, 1981

HOT   *Handbook of Toxinology;* Shier, W. Thomas and Mebs, Dietrich; Marcel Dekker: New York, 1990

HOTC   *Handbook of Terms Commonly Used in the Steel Industry;* Handbook 12, Iron Age

HPT   *Hybrid Propulsion Technologies for Urban Bus Transit,* UMTA-MA-06-0120-84-4; Gunderson, R. and Wychorski, H.; U.S. Department of Transportation: Washington, November, 1984

HSR   *Hazardous Substances Resource Guide;* Pohanish, Richard P. and Green, Stanley A.; Gale Research: Detroit, 1993

ICI   *International Cosmetic Ingredient Dictionary,* 5th Edition, Volume 1; Wenniger, J.A., and McEwen Jr., G.N., eds.; The Cosmetic Toiletry, and Fragrance Association: Washington, 1993

ICT   *Industrial Chemicals - Their Characteristics and Development;* Agam, G.; Elsevier: New York, 1994

IMDG   *International Maritime Dangerous Goods Code,* including Amendment 29-98, 11 to 20 May, 1998; International Maritime Organization (IMO): London, 1997

ISD   *IEEE Standard Dictionary of Electrical and Electronics Terms,* ANSI/IEEE Std 100-1988; Jay, Frank, ed.; The Institute of Electrical and Electronic Engineers: New York, 1988

ISH   *Industrial Solvents Handbook, 4th Edition;* Flick, Ernest W.; Noyes Data Corporation: Park Ridge, 1991

ISO   *ISO 14000 Guide,* The New International Environmental Management Standards; Cascio, Joseph, Woodside, Gayle, and Mitchell, Philip; McGraw-Hill: New York

ITT   *Introduction to the Technology of Explosives,* Cooper, Paul W. and Kurowski, Stanley R., VCH Publishers: New York, 1996

IW   *Incendiary Weapons;* Stockholm International Peace Research Institute; The MIT Press: Cambridge, 1975

JDOMS   *Jane's Directory of Military Small Arms Ammunition;* Hogg, Ian V.; Jane's Publishing: London, 1985

JDOMT   *Jane's Dictionary of Military Terms;* Hayward, P.H.C.; Macdonald & Jane's: London, 1975

KOC   *Kirk-Othmer Concise Encyclopedia of Chemical Technology;* John Wiley & Sons: New York, 1984

KOE   *Kirk-Othmer Encyclopedia of Chemical Technology,* 4th Edition; John Wiley & Sons: New York, 1995

LAC   *Lithium, A Conference Focusing on the Transportation and Disposal of Lithium Batteries,* June 23-26, 1987; New York, 1987

LBM   *Laboratory Biosafety Manual;* World Health Organization: Geneva, 1993

LCA   *Limes, Cements and Mortars;* Jones, David T.; International Textbook; Scranton, 1970

LDD   *Leather Dressing, Dyeing, and Finishing;* Woodroffe, D.; Quality Books: Teignmouth, 1953

LOSA   *Letter of Submittal to The President,* Annex III (Regulations for the Prevention of Pollution of Harmful Substances Carried by Sea in Packaged Forms or in Freight

Containers, Portable Tanks or Road and Rail Tank Wagons); U.S. Department of State: Washington, November 4, 1989

LOSB   *Letter of Submittal to The President,* Basel Convention on the Control of Trans-boundary Movements of Hazardous Wastes and Their Disposal; U.S. Department of State: Washington, May 31, 1991

MARPOL   *The 1973 International Convention for the Prevention of Pollution from Ships,* as modified and incorporated by the 1978 Protocol relating thereto (MARPOL 73/78), 1973

MEOE   *McGraw-Hill Encyclopedia of Environmental Sciences,* 2nd Edition; Parker, S.P., ed.; McGraw Hill: New York, 1980

MEOP   *Macmillan Encyclopedia of Physics;* Ryder, J.S., ed.; Simon & Shuster Macmillan, 1996

MEOS   *McGraw-Hill Encyclopedia of Science & Technology,* 8th Edition; McGraw-Hill: New York, 1997

MFP   *Mineral Facts & Problems, 1970;* U.S Department of Interior, Bureau of Mines: Washington, 1970

MH   *Metals Handbook;* American Society for Metals: Cleveland, 1948

MH12   *Materials Handbook,* 12th Edition; Brady, G.S. and Clauser, H.R.; McGraw-Hill: New York 1986

MH14   *Materials Handbook,* 14th Edition; Brady, George S., Clauser, Henry R. and Vaccari, John A.; McGraw-Hill: New York, 1997

MII   *Morton International, Inc.,* Cepa, Ronald, personal communication; November 18, 1997

MS1   *Metal Statistics 1996,* 88th Edition; American Metal Market: New York, 1996

MSA   *Marine Survival and Rescue Systems;* House, D.J.; Cornell Maritime Press: Centreville, 1988

MST   *Matchmaking: Science, Technology and Manufacture;* Finch, C.A.; John Wiley & Sons: New York, 1983

MWD   *Medical Waste Disposal;* Brunner, Calvin R.; Incinerator Consultants: Reston, 1996

MWM   *Medical Waste Management and Disposal;* U.S Environmental Protection Agency; Noyes Data Corporation: Park Ridge, 1991

P   *Poisoning,* Arena, Jay M.; Charles C. Thomas; Springfield, 1979

PBP   *Portable Battery Powered Products: U.S. Markets and Opportunities,* GB-184; Saxman, D.; Business Communications Company: Norwalk

PF   *Plant Fibers;* Maiti, R.K.; Bishen Singh Mahendra Pal Singh: Dehra Dun, 1980

PFT   *Protocol for the Prohibition of the Use in War of Asphyxiating, Poisonous or Other Gases and of Bacteriological Methods of Warfare;* Geneva, 1925

PII   *Pyrotechnics in Industry;* Barbour, Richard T.; McGraw-Hill: New York, 1981

PME   *Penguin Medical Encyclopaedia;* Wingate, Peter, 2nd Edition; Penguin Books: Middlesex, 1978

PND   *Primadet® Nonelectric Delay Detonators,* product literature; The Ensign Bickford Company; 5/94 Rev., Stock #. 26519

POG   *Principles of Guided Missile Design;* Merrill, Grayson, ed.; Van Nostrand: New York, 1959

PSF   *Procurement Standards for Gas Turbines,* B133.1M - 83; American National Standards Committee, 1983

References

RBA   *Rock Blasting and Explosives Engineering;* Persson, Per-Anders, Holmberg, Roger, and Lee, Jaimin; CRC Press: Boca Raton, 1993

RDA   *RDX, HMX and PMX Oil Well Cartridge,* product literature; The Ensign Bickford Company; 5/94 Rev., Stock #. 26514

REI   *Rocket Encyclopedia Illustrated;* Herrick & Burgess: 1959

RFT   *Regulations for the Safe Transport of Radioactive Material,* 1996 Edition; IAEA Safety Standards Series; International Atomic Energy Agency (IAEA): Vienna, 1996

ROT   *Recommendations on the Transport of Dangerous Goods,* 10th Revised Edition; United Nations: Geneva, 1997

ROTS   *Recommendations on the Safe Use of Pesticides in Ships,* Supplement to the International Maritime Dangerous Goods Code, International Maritime Organization: London 1997

RW   *Recycler's World,* Internet site; RecycleNet Corporation: Guelph, 1999

SAS   *Survival at Sea, The Lifeboat and Liferaft;* Wright, C.H.; The James Laver Printing: 1973

SBV1   *Small Batteries, Volume 1, Secondary Cells;* Crompton, T.R.; John Wiley & Sons: New York, 1982

SBV2   *Small Batteries, Volume 2, Primary Cells;* Crompton, T.R.; John Wiley & Sons: New York, 1982

SDO   *Standard Definitions of Terms Relating to Pesticides,* E 609-81; American Society for Testing and Materials, 1991

SEO   *Safety Evaluation of Distress Flares and Smoke;* McHale, Edward T.; U.S Coast Guard, Office of Research and Development

SG   *Shippers Guide,* personal communication; Judge, John

SSD   *Stewart's Scientific Dictionary,* 4th Edition; Stewart, Jeffery R.; Stewart Research Laboratory: Franconia, 1953

SSF   *Sword® Safety Fuse,* product literature; Cia. Mexicana de Mecha para Minas, S.A. de C.V.; Stock # 26510

STRA   *Standard Terminology Relating to Asbestos,* D 2946-91; American Society for Testing and Materials, 1991

STRAC   *Standard Terminology Relating to Activated Carbon,* D 2652-94; American Society for Testing and Materials, 1995

STRCB   *Standard Terminology Relating to Carbon Black,* D 3053-96b; American Society for Testing and Materials, 1996

STRCC   *Standard Terminology Relating to Coal and Coke,* D121-95; American Society for Testing and Materials, 1995

STRF   *Standard Terminology Relating to Fibers,* D 123-96; American Society for Materials and Testing

STRP   *Standard Terminology Relating to Plastics,* D 883-96; American Society for Testing and Materials, 1997

STRPP   *Standard Terminology Relating to Petroleum, Petroleum Products, and Lubricants,* D 4175 96; American Society for Testing and Materials, 1996

STRPV   *Standard Terminology Relating to Paint, Varnish, Lacquer, and Related Products,* D 16-96a; American Society for Testing and Materials, 1996

STRR   *Standard Terminology Relating to Rubber,* D 1566-ENGL; American Society for Testing and Materials, 1997

SVO   *SAE Vehicle Occupant Restraint Systems and Components Standards Manual,* SAE HS-13, 1996 Edition; Society of Automotive Engineers: Warrendale

TAOE   *The Analysis of Explosives;* Yiron, Jehuda; Pergamon Press: 1981

TAOP   *The Analysis of Pigments, Paints and Varnishes;* Fox, J.J. and Bowles, T.H.; Ernest Benn: London, 1927

TAS   *The ACS Style* Guide, A Manual for Authors and Editors; Dodd, Janet S., ed.; American Chemical Society: Washington, 1986

TBO   *The Book of Rifles,* 2nd Edition; Smith, H.B.; The National Rifle Association; The Stockpole Co: Harrisburg, 1960

TCAOC   *The Chemistry and Technology of Coal;* Speight, J.G.; Marcel Dekker: New York, 1983

TCAOP   *The Chemistry and Technology of Petroleum,* 2nd Edition; Speight, James G.; Marcel Dekker: New York, 1991

TDO   *The Dictionary of Paper;* American Paper & Pulp Association: New York, 1940

TEA   *The Encyclopedia Americana, International Edition;* Grolier: Danbury, 1996

TFD   *The Food Drug and Cosmetics (FD&C) Colorants Market,* #A736; Frost & Sullivan: January, 1980

TGM   *Tear Gas Munitions;* Swearengen, Thomas F., Thomas, Charles C.; Springfield

THE   *The New Encyclopaedia Britannica,* 15th Edition; Encyclopaedia Britannica, Inc.

THO   *The Handbook of Solvents;* Scheflan & Jacobs, M.B.; Van Nostrand: New York 1953

THS   *The Handy Science Answer Book;* Science and Technology Department; The Carnegie Library of Pittsburgh; Visible Ink Press: Detroit, 1994

TI   *Technical Instructions for the Safe Transport of Dangerous Goods by Air,* 1999-2000; International Civil Aviation Organization (ICAO): Montreal, 1998

TIE   *The Illustrated Encyclopaedia of Ammunition;* Hogg, Ian V.; Apple Press: London, 1985

TIP   *The Illustrated Petroleum Reference Dictionary, 4th Edition;* Langenkamp, Robert D.; PennWell Books: Tulsa, 1980

TIR   *Tar in Road Construction;* International Road Tar Conference: Tokyo, 1967

TNP   *Trade Name Products,* Vol. 1.; Handbook of Paint and Coating Raw Materials; Ash, Michael and Irene; Gower Publishing: Aldershot, 1996

TOE   *The Oxford English Dictionary,* 2nd Edition; Oxford University Press: New York, 1989

TPD   *The Penguin Dictionary of Chemistry;* Sharp, D.W.A.; Penguin Books: Middlesex, 1983

TSM   *The Ship's Manual of the Inflatable Lifecraft;* Cory Brothers & Co: Cardiff, 1958

TWE   *The Water Encyclopedia,* 2nd Edition; van der Leeden, Fritz, Troise, Fred L., Todd, David, K.; Lewis Publishers: Chelsea, 1990

VAA   *Valve and Actuator Technology;* Ulanski, Wayne; McGraw-Hill: New York, 1991

VNS   *Van Nostrand's Scientific Encyclopedia,* 5th Edition; Considine, Douglas M., ed.; Van Nostrand Reinhold: New York, 1976

VOA   *Vegetable Oils and Agrichemicals;* Benedict, John H., Treacy, Michael F. and Kinard, David H.; The Cotton Foundation: Memphis, 1994

271

References

VSW   *Visual Sweep Width Determination for Three Visual Distress Signalling Devices*, Report No. CG-D-30-86, U.S. Department of Transportation; U.S. Coast Guard; Office of Research and Development: September, 1986

WAT   *Words and Terms, U.S. Army, Navy, Air Force Dictionary*, 1986-88; DCP: Washington

WDA   *Wood Deterioration and Its Prevention by Preservative Treatments, Volume II Preservatives and Preservative Systems;* Nicholas, Darrel D., ed.; Syracuse University Press: Syracuse, 1973

WDO   *Whittington's Dictionary of Plastics;* Carley, James F., ed.; Technomic Publishing: Lancaster, 1968

WNN   *Webster's Ninth New Collegiate Dictionary;* Merriam-Webster: Springfield, 1987

Index of UN Numbers

UN Number	Page	UN Number	Page
0000		0092	194
		0093	194
0005	8	0094	194
0006	8	0099	69
0007	8	0101	119
0009	8	0102	119
0010	8	0103	119
0012	8	0104	119
0014	8	0105	119
0015	194	0106	119
0016	194	0107	119
0018	19	0110	8
0019	19	0121	119
0020	19	0124	69
0021	19	0131	119
0027	74	0132	74
0028	74	0136	8
0029	119	0137	8
0030	119	0138	8
0033	8	0159	74
0034	8	0160	74
0035	8	0161	74
0037	194	0167	8
0038	194	0168	8
0039	194	0169	8
0042	119	0171	194
0043	8	0173	69
0044	119	0174	69
0048	69	0180	8
0049	194	0181	8
0050	194	0182	8
0054	194	0183	8
0055	69	0186	8
0056	8	0191	194
0059	69	0192	194
0060	119	0193	194
0065	119	0194	194
0066	119	0195	194
0070	69	0196	194
0073	119	0197	194
0081	74	0204	69
0082	74	0212	194
0083	74	0221	8
0084	74	0223	88

UN Number	Page	UN Number	Page
0225	119	0312	194
0237	69	0313	194
0238	8, 134	0314	119
0240	134	0315	119
0241	74	0316	119
0242	8	0317	119
0243	8	0318	8
0244	8	0319	119
0245	194	0320	119
0246	194	0321	8
0247	8	0322	8
0248	69	0323	8, 69
0249	69	0324	8
0250	8	0325	119
0254	194	0326	8
0255	119	0327	8
0257	119	0328	8
0267	119	0329	8
0268	119	0330	8
0271	8	0331	74
0272	8	0332	74
0275	69	0333	194
0276	8, 69	0334	194
0277	69	0335	194
0278	69	0336	194
0279	8	0337	194
0280	8	0338	8
0281	8	0339	8
0283	119	0344	8
0284	8	0345	8
0285	8	0346	8
0286	8	0347	8
0287	8	0348	8
0288	69	0349	8, 69
0289	119	0350	69
0290	119	0351	69
0291	8	0352	69
0292	8	0353	69
0293	8	0354	69
0294	8	0355	69
0295	8	0356	69
0296	69	0357	74
0297	194	0358	74
0299	194	0359	74
0300	8	0360	119
0301	19	0361	119
0303	194	0362	8
0305	194	0363	8
0306	194	0364	119

UN Number	Page	UN Number	Page
0365	119	0432	194
0366	119	0433	74
0367	119	0434	8
0368	119	0435	8
0369	8	0436	8
0370	8	0437	8
0371	8	0438	8
0372	8	0439	69
0373	194	0440	69
0374	69	0441	69
0375	69	0442	69
0376	119	0443	69
0377	119	0444	69
0378	119	0445	69
0379	69	0446	8
0380	226	0447	8
0381	69	0449	8
0382	119	0450	8
0383	119	0451	8
0384	119	0452	8
0395	8	0453	134
0396	8	0454	119
0397	8	0455	119
0398	8	0456	119
0399	8	0457	8
0400	8	0458	8
0403	194	0459	8
0404	194	0460	8
0405	194	0461	119
0408	119	0462	69
0409	119	0463	69
0410	119	0464	69
0412	8	0465	69
0413	8	0466	69
0414	8	0467	69
0415	8	0468	69
0417	8	0469	69
0418	194	0470	69
0419	194	0471	69
0420	194	0472	69
0421	194	0473	74
0424	8	0474	74
0425	8	0475	74
0426	8	0476	74
0427	8	0477	74
0428	194	0478	74
0429	194	0479	74
0430	194	0480	74
0431	194	0481	74

UN Number	Page	UN Number	Page
0482	74	1223	183
0485	74	1224	164
0486	69	1226	137
0487	194	1228	164
0488	8	1263	175
0491	8	1266	52
0492	194	1267	183
0493	194	1268	183
0494	69	1272	188
0495	8	1286	188
0497	8	1287	189
0498	8	1288	183
0499	8	1293	61
0500	119	1299	188
		1300	183
		1306	179
		1308	153

1000

UN Number	Page	UN Number	Page
1002	104	1309	153
1003	104	1323	137
1009	215	1324	161
1013	134	1325	61, 99
1018	215	1326	153
1020	215	1327	90
1021	215	1331	143
1022	215	1333	153
1023	44	1345	189
1028	215	1352	153
1029	215	1353	161
1030	215	1358	153
1043	88	1361	36
1044	44, 94	1362	36
1057	137	1363	90
1058	104	1364	90
1063	215	1365	90
1071	183	1373	90
1075	183	1374	226
1078	215	1376	44
1082	215	1378	39
1130	86	1379	90
1133	1	1383	151, 153, 226
1136	44	1386	90
1139	175	1389	151
1169	86	1390	156
1170	5	1391	153
1197	86	1392	151
1201	5	1393	151
1202	183	1396	153
1203	183	1409	156
1210	175	1418	151, 153

UN Number	Page
1420	151
1421	151
1422	151
1435	147
1436	147
1450	128
1461	128
1462	128
1477	128
1479	61, 170
1481	128
1482	128
1483	170
1544	7
1549	128
1556	128
1557	128
1562	147
1564	156
1566	156
1567	153
1583	164
1588	128
1598	128
1601	61
1602	63
1621	179
1649	23
1655	7
1693	19
1700	19
1707	156
1719	47
1740	128
1759	47, 61
1760	40, 42, 47, 61, 179
1774	94
1791	128
1796	47
1826	47
1845	104
1851	61
1854	151
1855	151
1856	90
1858	215
1863	183
1866	189
1869	151, 153

UN Number	Page
1903	61
1906	183
1908	128
1932	147
1935	128
1941	215
1944	143
1945	143
1950	3
1953	104
1954	104, 215, 218
1955	104
1956	104, 193
1958	215
1959	215
1964	104
1965	104
1967	179
1968	179
1971	183
1972	183
1973	215
1974	215
1976	215
1979	104
1980	104
1981	104
1982	215
1983	215
1984	215
1986	5
1987	5
1988	164
1989	164
1992	96
1993	42, 61, 96, 179, 183, 218
1999	32, 183

2000

2000	161
2002	161
2003	167
2006	161
2008	153
2009	153
2016	19
2017	19
2024	145

UN Number	Page	UN Number	Page
2025	145	2735	164
2026	145	2742	164
2028	194	2757	179
2035	215	2758	179
2037	104	2759	179
2067	88	2760	179
2068	88	2761	179
2069	88	2762	179
2070	88	2763	179
2071	88	2764	179
2072	88	2765	179
2076	44	2766	179
2193	215	2767	179
2206	164	2768	179
2211	189	2769	179
2212	25	2770	179
2216	226	2771	179
2217	90	2772	179
2254	143	2773	179
2291	156	2774	179
2315	111	2775	179
2319	86	2776	179
2422	215	2777	179
2424	215	2778	179
2430	164	2779	179
2445	167	2780	179
2453	215	2781	179
2454	215	2782	179
2478	164	2783	179
2517	215	2784	179
2545	153	2786	179
2546	153	2787	179
2570	156	2788	167
2571	47	2793	147
2583	47	2794	27
2584	47	2795	27
2585	47	2796	27
2586	47	2797	27
2588	179	2800	27
2590	25	2801	63
2599	215	2809	145
2600	104	2810	179, 255
2602	215	2811	255
2623	99	2813	58
2627	128	2814	115
2630	128	2837	128
2693	128	2845	226
2733	164	2846	226
2734	164	2856	128

UN Number	Page
2857	218
2858	153
2871	153
2878	153
2881	39
2900	115
2902	179
2903	179
2908	200
2909	200
2910	200
2911	200
2912	200
2913	200
2915	200
2916	200
2917	200
2918	200
2919	200
2920	47
2921	47
2922	47
2923	47
2924	96
2925	99
2926	99
2927	255
2928	255
2929	255
2930	255
2950	153
2956	99
2969	38
2974	200
2977	200
2978	200
2982	200
2985	128
2986	128
2987	128
2988	128
2990	134
2991	179
2992	179
2994	179
2995	179
2996	179
2997	179
2998	179

UN Number	Page
2999	179

3000

UN Number	Page
3000	179
3001	179
3002	179
3003	179
3004	179
3005	179
3006	179
3007	179
3008	179
3009	179
3010	179
3011	179
3012	179
3013	179
3014	179
3015	179
3016	179
3017	179
3018	179
3019	179
3020	179
3021	179
3024	179
3025	179
3026	179
3027	179
3028	27
3048	179
3049	167
3050	167
3051	167
3052	167
3053	167
3065	5
3066	175
3071	164
3072	134
3076	167
3077	67, 113, 158
3080	164
3082	67, 113, 158
3084	47
3085	170
3086	255
3087	170

UN Number	Page	UN Number	Page
3088	226	3140	7
3089	153	3141	128
3090	27	3142	61
3091	27	3143	63
3093	47	3144	7
3094	47	3145	164
3095	47	3146	167
3096	47	3147	63
3097	99	3148	58
3098	170	3150	104
3099	170	3151	111
3100	170	3152	111
3101	170	3156	104
3102	170	3157	104
3103	170	3158	104
3104	170	3159	215
3105	170	3160	104
3106	170	3161	104
3107	170	3162	104
3108	170	3163	104
3109	170	3164	193
3110	170	3165	27
3111	170	3166	219
3112	170	3167	40
3113	170	3168	40
3114	170	3169	40
3115	170	3170	147
3116	170	3171	27
3117	170	3172	115
3118	170	3175	99
3119	170	3176	99
3120	170	3178	74, 99
3121	170	3179	99
3122	255	3180	99
3123	255	3181	167
3124	255	3182	167
3125	255	3184	226
3126	226	3185	226
3127	226	3186	226
3128	226	3187	226
3129	58	3188	226
3130	58	3189	153
3131	58	3190	226
3132	58	3191	226
3133	58	3192	226
3134	58	3194	226
3135	58	3200	226
3137	170	3203	167
3139	170	3205	167

UN Number	Page	UN Number	Page
3206	167	3265	47
3207	167	3266	47
3208	156	3267	47
3209	156	3268	134
3210	128	3269	189
3211	128	3270	161
3212	128	3271	164
3213	128	3272	164
3214	128	3273	164
3215	128	3275	164
3216	128	3276	164
3218	128	3277	164
3219	128	3278	167
3220	215	3279	167
3221	99	3280	167
3222	99	3281	156
3223	99	3282	167
3224	99	3283	128
3225	99	3284	128
3226	99	3285	156
3227	99	3286	96
3228	99	3287	255
3229	99	3288	255
3230	99	3289	255
3231	99	3290	255
3232	99	3291	115
3233	99	3292	27
3234	99	3295	96
3235	99	3296	215
3236	99	3301	47
3237	99	3303	104
3238	99	3304	104
3239	99	3305	104
3240	99	3306	104
3243	255	3307	104
3244	47	3308	104
3245	115	3309	104
3248	61	3310	104
3249	61	3311	104
3252	215	3312	104
3256	65	3313	63
3257	65	3314	189
3258	65	3315	40
3259	164	3316	40
3260	47	3319	74
3261	47	3321	200
3262	47	3322	200
3263	47	3323	200
3264	47	3324	200

UN Number	Page	UN Number	Page
3325	200	3349	179
3326	200	3350	179
3327	200	3351	179
3328	200	3352	179
3329	200	3353	134
3330	200	3354	179
3331	200	3355	179
3332	200	3356	173
3333	200		
3334	19, 158		
3335	158		

8000

UN Number	Page
3337	215	8000	158
3338	215	8001	158
3339	215	8013	134
3340	215	8023	218
3343	74	8027	158
3344	74	8038	27
3345	179		
3346	179		
3347	179		
3348	179		

9000

UN Number	Page
9035	104

Index of Entries

The Index of Entries includes each of the entries extracted from the various lists of regulated dangerous goods and hazardous materials and covered in the *Glossary for the Worldwide Transportation of Dangerous Goods and Hazardous Materials*. These are listed in **bold**. The page numbers associated with these bolded regulatory entries indicate the title page of the chapter in which it is principally described.

The items listed here in *italicized* text include occurrences of the key words and phrases used to describe the regulatory entry.

Page

' (the prime mark), structural notation 248

A

a-, structural notation 248
α-, structural notation 248, 249, 250
Accumulators 106
Accumulators, electric 27
Accumulators, pressurized, hydraulic, 2.2 193
Accumulators, pressurized, pneumatic, 2.2 193
Acid 7, 11, 28, 29, 38, 42, 46, 47, 52, 53, 59, 62, 64, 76, 77, 86, 89, 92, 94, 99, 100
........129, 130, 132, 143, 148, 156, 161, 165, 166, 167, 168, 170, 177, 180, 181, 186
........188, 190, 191, 195, 205, 222, 224, 231, 234, 238, 239, 240, 242, 244, 247, 251
Acidic 47
Acid liquid, n.o.s. 47
Acid mixture, nitrating acid 47
Acid mixture, spent, nitrating acid 47
Acid Sludge 183
Acid sludges 186
Actinolite 25
Actinolite 25
Activated 231
Activated carbon 36
Activated carbon 36
Activated charcoal 36
Activated charcoal 36
Actuating cartridge, explosive 69
Actuating cartridges 69, 94
Adhesive 1, 40, 129, 162, 176, 189, 191, 224
Adhesives, containing a flammable liquid, 3, 3.1, 3.2, 3.3 1
Adhesives containing flammable liquid, 3 1
Aerial flares 196

Index of Entries

Page

Aeroplane flares...194
Aeroplane flares 196
Aerosols...3, 20, 53, 62, 74, 106, 224
Aerosols, 2...3
Aerosols, flammable, 2.1, 2.3, 6.1, 8...3
Aerosols, non-flammable, 2.2, 2.3, 6.1, 8..3
Agents, blasting type, B..74
Agents, blasting type, E..74
Air..105
Air bag inflator..135
Air bag inflators, 9..134
Air bag inflators, compressed gas, 2.2 134
Air bag inflators, pyrotechnic, 9..134
Air bag module 135, 219
Air bag modules, 9..134
Air bag modules, compressed gas, 2.2..134
Air bag modules, pyrotechnic, 9..134
Air bags 69, 219
Air, compressed, 2.2 104
Aircraft evacuation slides 134
Aircraft evacuation slides...134
Aircraft hydraulic power unit fuel tank 27
Aircraft hydraulics 29
Aircraft survival kits 134
Aircraft survival kits 134
Air, refrigerated liquid, 2.2...104
Air, refrigerated liquid, (cryogenic liquid), 2.2.............................104
Air, refrigerated liquid, (cryogenic liquid), non-pressurized, 2.2.......................104
Air, refrigerated liquid, low pressure or pressurised, 2.2, 5.1 104
Air, refrigerated liquid, non-pressurised, 2.2, 5.1..........................104
Alcohol, 3.2..5
Alcohol 5, 7, 40, 52, 53, 59, 61, 100, 132, 158, 167
....168, 169, 188, 190, 224, 241, 248, 251, 253, 256
Alcoholates 168
Alcoholates solution, n.o.s. in alcohol, 3, 3.1, 3.2, 8 167
Alcohol, denatured 5
Alcohol, denatured, 3.2 5
Alcohol, denatured solutions, 3.3 5
Alcoholic beverages, 3, 3.2, 3.3 5
Alcoholic beverages..5, 243
Alcohol, industrial...5
Alcohol, industrial, 3.2 5
Alcohol, industrial, solutions, 3.3 5
Alcohols, n.o.s. 3, 3.1, 3.2, 3.3, 6.1 5
Alcohol solutions........................1, 7, 22, 53, 58, 61, 86, 130, 162, 168, 176, 189
Alcohol solutions, 3.3..5
Aldehyde 164, 191, 248

bold: regulatory entries

Aldehydes, n.o.s., 3, 3.1, 3.2, 3.3, 6.1 164
Alkali 28, 42, 47, 52, 79, 99, 156, 161, 177, 195
Alkali metal alcoholates, n.o.s., 4.2, 8 167
Alkali metal alloy, liquid, n.o.s., 4.3 151
Alkali metal alloys 151
Alkali metal amalgam, liquid, 4.3 151
Alkali metal amalgams, 4.3 151
Alkali metal amalgam, solid, 4.3 151
Alkali metal amides, 4.3 156
Alkali metal amides 156
Alkali metal dispersion, 4.3 153
Alkali metals 242
Alkali metals 29, 59, 151, 153, 156
Alkaline 48
Alkaline caustic liquid, n.o.s. 47
Alkaline corrosive battery fluid 27
Alkaline corrosive battery fluid 28
Alkaline corrosive liquid, n.o.s. 47
Alkaline corrosive solid, n.o.s. 47
Alkaline earth metal 59, 151
Alkaline earth metal alcoholates, n.o.s., 4.2 167
Alkaline earth metal alloy, n.o.s., 4.3 151
Alkaline earth metal alloys 151
Alkaline earth metal amalgam, 4.3 151
Alkaline earth metal dispersion, 4.3 153
Alkaline earth metals 242
Alkaline metals 153
Alkaloid 7, 38, 61
Alkaloid pesticides 180
Alkaloid salts 7
Alkaloid salts, liquid, n.o.s., 6.1 7
Alkaloid salts, solid, n.o.s., 6.1 7
Alkaloids and alkaloid salts (pesticides) 179
Alkaloids, liquid, n.o.s., 6.1 7
Alkaloids, solid, n.o.s., 6.1 7
Alkyl aluminium halides 167
Alkylphenol 164, 239, 240
Alkylphenols, liquid, n.o.s. (including C_2-C_{12} homologues), 8 164
Alkylphenols, solid, n.o.s. (including C_2-C_{12} homologues), 8 164
Alkylsulphonic acids 49
Alkylsulphonic acids, liquid, 8 47
Alkylsulphonic acids, solid, 8 47
Alkylsulphuric acids, 8 47
Alkylsulphuric acids 49
Alloy 137, 139, 149, 151, 153, 205, 222, 242, 245
alpha-, structural notation 248
Aluminium alkyl 167

italics: key words and phrases

Page

Aluminium alkyl halides, 4.2, 4.3 167
Aluminium alkyl hydrides, 4.2, 4.3 167
Aluminium alkyls, 4.2, 4.3 167
Aluminium dross 147
Aluminium dross, hot 147
Aluminium dross, wet 147
Aluminium dross, wet or hot 147
Aluminium liquid 175
Aluminium liquid 175
Aluminium metal hydride 169
Aluminium paint..175
Aluminium paint 175
Aluminium phosphide pesticide, 6.1 179
Aluminium phosphide pesticides 180
Aluminium, powder..153
Aluminium powder 77, 78, 122
Aluminium powder, coated, 4.1 153
Aluminium powder, pyrophoric...153
Aluminium powder, uncoated, 4.3 153
Aluminium powder, uncoated, non-pyrophoric, 4.3....................................153
Aluminium remelting by-products, 4.3...147
Aluminium remelting by-products 148
Aluminium residue..148
Aluminium residues...147
Aluminium skimmings 147
Aluminium smelting by-products...148
Aluminium smelting by-products, 4.3..147
Amalgams 146, 152, 158
Amatol 77
Amatols...74
Amine...................11, 20, 48, 64, 75, 77, 164, 167, 180, 240, 249, 250, 251
Amines, liquid, n.o.s., 3, 8 164
Amines, n.o.s., 3, 8, 3.1, 3.2, 3.3 164
Amines, solid, n.o.s., 8 164
Ammonium nitrate explosive 11, 76, 89
Ammonium nitrate explosives 74
Ammonium nitrate fertilizer, 5.1, 1.1D, 9..88
Ammonium nitrate fertilizer, n.o.s., 5.1...88
Ammonium nitrate fertilizers 88
Ammonium nitrate mixed fertilizers, 5.1..88
Ammunition 8, 20, 70, 71, 75, 77, 121
Ammunition, blank 8
Ammunition, fixed, semi-fixed or separate loading..................................8
Ammunition, illuminating with or without burster, expelling
   charge or propelling charge, 1.2G, 1.3G, 1.4G.............................194
Ammunition, incendiary liquid or gel, with burster, expelling
   charge or propelling charge, 1.3J..8

bold: regulatory entries

Ammunition, incendiary (water-activated contrivances)......................................8
Ammunition, incendiary (water-activated contrivances) with
  burster, expelling charge or propelling charge......................................8
Ammunition, incendiary, white phosphorus with burster,
  expelling charge or propelling charge, 1.2H, 1.3H 8
Ammunition, incendiary with or without burster, expelling charge
  or propelling charge, 1.2G, 1.3G, 1.4G..8
Ammunition, industrial..69
Ammunition, lachrymatory 19
Ammunition, practice, 1.3G, 1.4G..8
Ammunition, proof, 1.4G..8
Ammunition, rocket 8
Ammunition, SA (small arms) 8
Ammunition, smoke (water-activated contrivances) 194
Ammunition, smoke (water-activated contrivances), white
  phosphorus, with burster, expelling charge or propelling charge.........194
Ammunition, smoke (water-activated contrivances), with or
  without white phosphorus or phosphides, with burster,
  expelling charge or propelling charge..194
Ammunition, smoke, white phosphorus, with burster, expelling
  charge or propelling charge, 1.2H, 1.3H..194
Ammunition, smoke, with or without burster, expelling charge or
  propelling charge, 1.2G, 1.3G, 1.4G, 8..194
Ammunition, sporting 8
Ammunition, tear-producing, non-explosive with neither burster
  nor expelling charge, 6.1, 8 19
Ammunition, tear-producing, non-explosive with neither burster
  nor expelling charge, non-fuzed, 6.1 19
Ammunition, tear-producing, non-explosive, without burster,
  expelling charge or propelling charge, non-fuzed, 6.1, 8......................19
Ammunition, tear-producing, with burster, expelling charge or
  propelling charge, 1.2G, 1.3G, 1.4G, 6.1, 8..19
Ammunition, toxic, non-explosive, without burster or expelling
  charge, non-fuzed, 6.1 19
Ammunition, toxic (water-activated contrivances), with burster,
  expelling charge or propelling charge..19
Ammunition, toxic with burster, expelling charge or propelling
  charge, 1.2K, 1.3K, 6.1 19
Amorce 197
Amorces..194
Amorces (caps, toy) 194
Amorphous..231
Amosite..25
Amosite 25
Anaesthetic ether 61
Anaesthetic ether 62
Anhydrous..231

italics: key words and phrases

Page

Animal fabrics, oily 90

Animal fibres 90

Animal fibres, burnt, wet or damp...90

Animal fibres, oily 90

Anthophyllite 25

Anthophyllite 25

Antifreeze..22, 94

Antifreeze, liquid 22

Antiknock compound, mixture..23

Antiknock compounds...23, 151

Antimony compound...128, 232, 234

Antimony compound, inorganic, liquid, n.o.s., 6.1 128

Antimony compound, inorganic, solid, n.o.s., 6.1 128

Antimony powder, 6.1 153

Apparatus 158

Aqueous 231

Aromatic extracts...86

Aromatic liquids 86

Aromatic liquids 86

Aromatic nitro-derivatives..76

Arsenates 128

Arsenates, liquid, n.o.s., inorganic 128

Arsenates, n.o.s. 128

Arsenates, solids, n.o.s., inorganic...128

Arsenical dust, 6.1 147

Arsenical flue dust 147

Arsenical pesticide, liquid, 3, 3.2, 6.1 179

Arsenical pesticides 180

Arsenical pesticide, solid, 6.1 179

Arsenic compound..128, 149, 180, 181, 197, 245

Arsenic compound, liquid, n.o.s. 128

Arsenic compound, liquid, n.o.s., inorganic, 6.1 128

Arsenic compound, solid, n.o.s., inorganic, 6.1 128

Arsenic compounds (pesticides)..179

Arsenic sulphides..129

Arsenic sulphides, liquid, n.o.s., inorganic 128

Arsenic sulphides, n.o.s. 128

Arsenic sulphides, solid, n.o.s., inorganic 128

Arsenites 129

Arsenites, liquid, n.o.s., inorganic 128

Arsenites, n.o.s. 128

Arsenites, solid, n.o.s., inorganic 128

Articles, EEI, 1.6N...69

Articles, explosive, extremely insensitive, 1.6N.....................................69

Articles, explosive, n.o.s., 1.1C, 1.1D, 1.1E, 1.1F, 1.1L, 1.2C, 1.2D,
1.2E, 1.2F, 1.2L, 1.3C, 1.3L, 1.4B, 1.4C, 1.4D, 1.4E, 1.4F, 1.4G,
1.4S...69

bold: regulatory entries

Articles, pressurized, hydraulic, 2.2...193
Articles, pressurized, pneumatic, 2.2 193
Articles, pyrophoric, 1.2L...226
Articles, pyrotechnic for technical purposes, 1.1G, 1.2G, 1.3G,
   1.4G, 1.4S...194
Asbestos 25
Asbestos 25, 90, 241, 255, 256
Asbestos, blue, 9 25
Asbestos, brown 25
Asbestos, white, 9 25
Asphalt...183
Asphalt...................................11, 12, 25, 32, 44, 65, 185, 240, 244
Asphalt, at or above its flashpoint, 3...183
Asphalt, cut back 32
Atomic number 231
Atomic weight 231
Automobile...219
Automobile 28, 139, 184, 219, 254
Automobile, motorcycle, tractor, or other self-propelled vehicle,
   engine, or other mechanical apparatus......................................219
Aviation gasoline 183
Aviation regulated liquid, n.o.s., 9 158
Aviation regulated materials 159
Aviation regulated solid, n.o.s., 9 158
Aviation turbine engine fuels..185
Azeotropic 232

B

b-, structural notation 249
Bag charges 8
Bag charges..10, 123, 162
Ballistite..74
Ballistite 77
Bangalore torpedoes..8
Bangalore torpedoes 12
Barium alloys..151
Barium alloys, non-pyrophoric 151
Barium alloys, pyrophoric, 4.2 151
Barium amalgams...151
Barium compound, n.o.s., 6.1 156
Barium compounds..129, 141, 156, 195
Barium dispersions...153
Barium, powder 153
Barium powder, pyrophoric, 4.2 153
Basic..48
Batteries 27, 71, 134, 158, 219

italics: key words and phrases

Batteries, containing lithium 27
Batteries, containing sodium, 4.3 27
Batteries, dry......................27
Batteries, dry, containing potassium hydroxide, solid, electric
   storage, 8 27
Batteries, electric storage 27
Batteries, lithium type 27
Batteries, wet, filled with acid, electric storage, 8 27
Batteries, wet, filled with alkali, electric storage, 8......................27
Batteries, wet, non-spillable, electric storage, 8 27
Batteries, wet, without electrolyte and fully discharged......................27
Battery acid 27
Battery acid 28
Battery fluid......................27
Battery fluid, acid, 8 27
Battery fluid, alkali, 8......................27
Battery-operated......................29
Battery-powered equipment......................29, 137
Battery-powered equipment, 9 27
Battery-powered vehicle, 9......................27
Battery-powered vehicles 29
Battery, wet, filled with acid or alkali with automobile (or named
   self-propelled vehicle or mechanical equipment containing
   internal combustion engine) 27
Battery, wet, with wheelchair 27
Benzoic derivative pesticide, liquid, 3, 6.1 179
Benzoic derivatives pesticides 180
Benzoic derivative pesticide, solid, 6.1 179
Beryllium compound, n.o.s., 6.1 156
Beryllium compounds......................156
Beryllium powder, 4.1, 6.1 153
beta-, structural notation 249
Beverage extract (concentrate)......................86
Beverage extracts 86
Bhusa, 4.1 90
Bhusa......................91
bi-, structural notation 249
Bifluorides 130
Bifluorides, n.o.s. 128
Bifluorides, solid, n.o.s. 128
Bifluorides, solution, n.o.s.......................128
Biological products 115
Biological products 115
Biomedical waste......................115
Biomedical waste, n.o.s., 6.2......................115
(Bio)medical waste, n.o.s., 6.2 115
Bipyridilium pesticides 180

bold: regulatory entries

Bipyridilium pesticide, liquid, 3, 3.2, 6.1 179
Bipyridilium pesticide, solid, 6.1 179
bis-, structural notation..249
Bisulphates 130, 170
Bisulphates, aqueous solution, 8...128
Bisulphites 34, 130
Bisulphites, aqueous solution, n.o.s., 8..128
Bitumen 32
Bitumen..1, 32
Black powder...11, 76, 121, 123
Black powder, compressed, 1.1D..74
Black powder for small arms, 4.1..74
Black powder granular or as a meal, 1.1D..74
Black powder in pellets, 1.1D 74
Blank ammunition 13
Blasting agent...70, 78, 120, 124
Blasting agent, n.o.s. 74
Blasting cap assemblies 119
Blasting cap assembly 122
Blasting caps 120
Blasting caps, electric 119
Blasting caps, non-electric 119
Blasting gelatine...79
Blau gas 44
Blau gas 45
Bleach 34
Bleach..34, 42, 52, 130, 132, 159, 170
Bleaching powder 34
Bleaching powders 34
Bleach liquor 34
Bleach liquor 34
Bleach solution...34
Bleach solutions 34
Block matches...143
Blue asbestos 25
Blue asbestos 25
Blue asbestos (crocidolite), 9..25
Boiling Point...232
Bombs..12, 20, 123, 195, 201, 202
Bombs, illuminating 194
Bombs, photo-flash, 1.1D, 1.1F, 1.2G, 1.3G.....................................194
Bombs, smoke, non-explosive, with corrosive liquid, without
    initiating device, 8..194
Bombs, target identification...194
Bombs with bursting charge, 1.1D, 1.1F, 1.2D, 1.2F 8
Bombs with flammable liquid with bursting charge, 1.1J, 1.2J 8
Boosters...11, 124

italics: key words and phrases

Page

Boosters with detonator, 1.1B, 1.2B 119
Boosters without detonator, 1.1D, 1.2D 119
Borings 149
Box toe gum 161
Box toe gum 162
Briquettes 36, 44
Bromates 129, 170
Bromates, inorganic, aqueous solution, n.o.s., 5.1 128
Bromates, inorganic, n.o.s., 5.1 128
Brown asbestos 25
Brown asbestos (amosite, mysorite), 9 25
Burster 9, 20, 194, 196
Bursters, explosive, 1.1D 8
Bursting charge 9, 197
Butter 232
Butylphenols 164
Butylphenols, liquid, n.o.s., 8 164
Butylphenols, solid, n.o.s., 8 164
By Mass 232
By Volume 233
By Weight 233

C

Cable cutter 69
Cable cutters, explosive 69
Cadmium compound, 6.1 156
Cadmium compounds 156
Caesium alloy (liquid) 151
Caesium amalgams 151
Caesium dispersions 153
Caesium, powder 153
Caesium powder, pyrophoric 153
Calcined 233
Calcined 147, 233
Calcined pyrites 222
Calcined pyrites (pyritic ash, fly ash) 221
Calcium alloy, non-pyrophoric 151
Calcium alloys 151
Calcium alloys, pyrophoric, 4.2 151
Calcium amalgams 151
Calcium dispersions 153
Calor gas 104
Calor gas 105
Camphor oil 86
Camphor oil, 3.3 86
Camping gas 104

bold: regulatory entries

Camping gas..105
Candles, gas...137
Cannon primers...119
Caps, blasting..119
Caps, primer..119
Caps, toy..194
Carbamate pesticide, liquid, 3, 3.2, 6.1 179
Carbamate pesticides 180
Carbamate pesticide, solid, 6.1 179
Carbon................36, 44, 46, 76, 92, 148, 151, 200, 215, 222, 231, 233, 240, 244, 247
Carbon black...36, 92, 176, 190, 231
Carbon black (animal or vegetable origin) 36
Carbon paper..90
Carbon paper 36, 92
Carbon, activated, 4.2 36
Carbon, animal or vegetable origin, 4.2 36
Carbon, non-activated, mineral origin 36
Cargo transport units under fumigation..181
Cargo transport unit under fumigation, 9...179
Cartridge case 9, 69
Cartridge cases..69
Cartridge cases, empty, primed...69
**Cartridges, actuating, for aircraft ejector seat catapult, fire
   extinguisher, canopy removal or apparatus**..69
**Cartridges, actuating, for fire extinguisher or apparatus or
   apparatus valve**...69
Cartridges for weapons 122
Cartridges for weapons, blank, 1.1C, 1.2C, 1.3C, 1.4C, 1.4S.........................8
Cartridges for weapons, inert projectile, 1.2C, 1.3C, 1.4C, 1.4S 8
**Cartridges for weapons, with bursting charge, 1.1E, 1.1F, 1.2E,
   1.2F, 1.4E, 1.4F** 8
Cartridges, explosive 69
Cartridges, flash, 1.1G, 1.3G 194
Cartridges, illuminating..194
Cartridges, oil well, 1.3C, 1.4C 69
Cartridges, power device, 1.2C, 1.3C, 1.4C, 1.4S...69
Cartridges, safety...8
Cartridges, safety, blank...8
Cartridges, signal, 1.3G, 1.4G, 1.4S 194
Cartridges, small arms, 1.2C, 1.3C, 1.4C, 1.4S 8
Cartridges, small arms, blank, 1.3C, 1.4C, 1.4S 8
Cartridges, sporting 8
Cartridges, starter, jet engine..69
Case oil 183
Case oil...184
Cases, cartridge, empty, with primer, 1.4C, 1.4S..69
Cases, combustible, empty, without primer, 1.3C, 1.4C................................69

italics: key words and phrases

Page

Casinghead gasoline 183
Casinghead gasoline 183
Castor 100, 175, 244
Castor bean 38, 159, 221
Castor beans, 9..38
Castor flake 38
Castor flake, 9 38
Castor meal 38
Castor meal, 9 38
Castor pomace..38
Castor pomace, 9 38
Catalysts 1, 11, 36, 39, 59, 93, 121, 145, 152, 153, 161
.................................165, 167, 168, 169, 170, 185, 188, 226, 241, 245
Caustic..48, 156, 222
Caustic alkali liquid, n.o.s., 8..47
Cells..27, 219
Cells containing sodium, 4.3 27
Celluloid..161
Celluloid, in blocks, rods, rolls, sheets, tubes, etc. (except scrap),
   4.1..161
Celluloid scrap, 4.2 161
Cement..1, 25, 189
Cement, flammable..1
Cement, liquid..1
Cerium powder, pyrophoric 153
Cerium slabs, ingots, or rods, 4.1 153
Cerium, turnings or gritty powder, 4.3....................................147
Cer Mischmetall..137
Cer mischmetall..137
Charcoal 36, 221
Charcoal..36, 76, 78, 221
Charcoal, activated..36
Charcoal briquettes, shell, screenings, wood, etc , 4.2 36
Charcoal, non-activated 36
Charcoal screenings, wet..36
Charcoal, wet 36
Charges, bursting, plastics bonded, 1.1D, 1.2D, 1.4D, 1.4S..........8
Charges, demolition, 1.1D..69
Charges, depth, 1.1D..8
Charges, expelling, explosives, for fire extinguishers 69
Charges, explosives, commercial, without detonator, 1.1D, 1.2D,
   1.4D, 1.4S..69
Charges, propelling, 1.1C, 1.2C, 1.3C, 1.4C..............................8
Charges, propelling, for cannon, 1.1C, 1.2C, 1.3C 8
Charges, shaped, flexible, linear, 1.1D, 1.4D............................69
Charges, shaped, without detonator, 1.1D, 1.2D, 1.4D, 1.4S........69
Charges, supplementary, explosive, 1.1D 119

bold: regulatory entries

Chemical kit, 8, 9 40
Chemical kits 40
Chemical sample, liquid, 6.1 40
Chemical samples 40
Chemical sample, solid, 6.1 40
Chemical sample, toxic 40
Chemical sample, toxic, liquid or solid, 6.1 40
Chlorates 76, 129, 143, 170, 173, 196, 197
Chlorates, inorganic, aqueous solution, n.o.s., 5.1 128
Chlorates, inorganic, n.o.s., 5.1 128
Chlorinated paraffins 165
Chlorinated paraffins (C_{10} - C_{17}) 164
Chlorites 129
Chlorites, inorganic, n.o.s., 5.1 128
Chlorocarbonate 165
Chlorocarbonates, n.o.s., 3, 6.1, 8 164
Chloroformate 165
Chloroformates, n.o.s., 3, 6.1, 8 164
Chloropicrin 165
Chloropicrin mixture, n.o.s., 6.1 164
Chlorosilanes 59, 129
Chlorosilanes, n.o.s., 3, 3.2, 4.3, 8 128
Chrysotile 25
Chrysotile 25
Cigar and cigarette lighter fluid 137
Cigar and cigarette lighters, charged with fuel 137
Cinnabar 145
Cinnabar 145
cis-, structural notation 249
Cleaning compounds 42
Cleaning fluid or liquid 42
Cleaning fluids 42
Cleaning liquids 42
Clinical waste 115
Clinical waste, unspecified, n.o.s., 6.2 115
Coal 44, 221
Coal 36, 39, 44, 63, 75, 78, 79, 222, 238, 240, 244
Coal briquettes 44
Coal briquettes, hot 44
Coal gas 44
Coal gas, 2.3 44
Coal gas, compressed, 2.1, 2.3 44
Coal tar 32, 36, 45, 63, 181
Coal tar, crude and solvent 44
Coal tar distillates 45
Coal tar distillates, flammable, 3, 3.2, 3.3 44
Coal tar dye, corrosive, liquid, n.o.s. 63

italics: key words and phrases

Page

Coal tar dyes 63
Coal tar naphtha..44
Coal tar naphtha..45
Coal tar oil..44
Coal tar oil..45
Coal tar solvents...45
Coatings 3, 26, 34, 129, 161, 162, 184, 188, 189, 191, 224
Coating solution, 3, 3.1, 3.2, 3.3...175
Coating solutions..175
Cobalt catalyst 39
Cobalt catalysts 39
Cocculus 115
Cocculus 115
Coir 90
Coir 91
Coke...36, 44
Coke, hot 44
Collodion 161
Collodion..162
Collodion cotton...162
Collodion cottons 161
Cologne spirits 52
Cologne spirits 53
Combination fuzes 123
Combustible cases 10, 71
Combustible liquid 86
Combustible liquid, n.o.s....96
Combustible liquids...96, 188
Combustion 233
Combustion..........................10, 11, 13, 23, 36, 39, 76, 77, 96, 99, 105, 129, 150, 162
.................................164, 168, 170, 184, 185, 194, 195, 237, 238, 245, 246
Commercial charges..70
Complex..234
Complex...63, 168, 180, 242
Components, explosive train, n.o.s., 1.1B, 1.2B, 1.4B, 1.4S...............119
Compound..234
Compound, anti-freeze liquid...22
Compound, cleaning liquid, 3, 8...42
Compound, enamel..175
Compounds, tree killing, liquid, 3, 6.1, 8..................................179
Compounds, weed killing, liquid, 3, 6.1, 8................................179
Compressed air..3, 11, 94, 105, 254
Compressed gas...................3, 12, 41, 94, 104, 134, 135, 173, 183, 193, 219, 254
Compressed gas, n.o.s., 2.1, 2.2, 2.3, 5.1, 8...............................104
Concentrates 234
Concentrates...147, 156, 221, 222
Consumer commodities 158

bold: regulatory entries

Consumer commodity, ORM-D, 9...158
Container under fumigation...179
Containers under fumigation...181
Contrivances, water-activated with burster, expelling charge or
    propelling charge, 1.2L, 1.3L..69
Copper based pesticide, liquid, 3, 3.2, 6.1..179
Copper-based pesticides..180
Copper based pesticide, solid, 6.1...179
Copper compounds...156
Copper compounds..156, 180, 181, 197
Copper metal powder...153
Copper metal powder...141
Copra...91, 221
Copra, 4.2..90
Cordeau detonant cord..121
Cordeau detonant fuse..119
Cord, detonating, flexible, 1.1D, 1.4D...119
Cord, detonating, metal clad, 1.1D, 1.2D...119
Cord, detonating, mild effect, metal clad, 1.4D;..............................119
Cord, igniter, 1.4G...119
Cordite...74
Cordite..77
Corrosive battery fluid..27
Corrosive battery fluid..28
Corrosive liquid.......................................22, 30, 42, 45, 48, 130, 146, 195
Corrosive liquid, acidic, inorganic, n.o.s., 8....................................47
Corrosive liquid, acidic, organic, n.o.s., 8.......................................47
Corrosive liquid, basic, inorganic, n.o.s., 8.....................................47
Corrosive liquid, basic, organic, n.o.s., 8...47
Corrosive liquid, n.o.s., 3, 4.2, 4.3, 5.1, 6.1, 8................................47
Corrosive solid, acidic, inorganic, n.o.s., 8......................................47
Corrosive solid, acidic, organic, n.o.s., 8...47
Corrosive solid, basic, inorganic, n.o.s., 8.......................................47
Corrosive solid, basic, organic, n.o.s., 8...47
Corrosive solid, n.o.s., 4.1, 4.2, 4.3, 5.1, 6.1, 8..............................47
Corrosive solids...48
Cosmetics...52
Cosmetics..3, 26, 38, 52, 86, 145, 158
Cosmetics, corrosive, liquid, n.o.s. 52
Cosmetics, corrosive, solid, n.o.s. 52
Cosmetics, flammable, liquid, n.o.s. 52
Cosmetics, flammable, solid, n.o.s. 52
Cosmetics, n.o.s. 52
Cosmetics, oxidizing material, liquid, n.o.s. 52
Cosmetics, oxidizing material, solid, n.o.s. 52
Cotton..40, 91, 121, 143, 161, 226
Cotton, 9 90

italics: key words and phrases

Page

Cotton, dry 90
Cottonseed..91, 132
Cotton seed, cut linters, hull fibres, pulp, waste, and shavings,
   with animal or vegetable oil..90
Cotton waste 91
Cotton waste, oily, 4.2 90
Cotton, wet, 4.2 90
Coumarin derivative pesticide, liquid, 3, 3.2, 6.1................................179
Coumarin derivative pesticides 180
Coumarin derivative pesticide, solid, 6.1 179
Creosote...44
Creosote 45, 181
Creosote (coal tar or wood tar) 44
Creosote salts 44
Creosote salts 45
Cresol 45, 61, 164
Cresols (o-, m-, p-), 6.1, 8 44
Cresols (ortho-; meta-) 44
Crocidolite..25
Crocidolite..25
Crude coal tar 44
Crude naphtha...183
Crude naphtha...184
Crude oil...70, 75, 79, 183
Cryogenic liquid 104
Cryogenic liquids 105
Cutback asphalt..32
Cut-backs 32
Cut linters 91
Cutters, cable, explosive, 1.4S...69
Cuttings 149
Cyanide mixture, inorganic, solid, n.o.s. 128
Cyanide or cyanide mixture, dry...128
Cyanides............................19, 46, 67, 128, 129, 130, 166, 181, 245, 256
Cyanides, inorganic, solid, n.o.s., 6.1 128
Cyanide solution, n.o.s., 6.1 128
Cyanides, organic, flammable, toxic, n.o.s. 164
Cyanides, organic, toxic, flammable, n.o.s. 164
Cyanides, organic, toxic, n.o.s. 164
cyclo-, structural notation 249

D

D-, structural notation 249
Dangerous goods...54, 67, 158, 221, 248
Dangerous goods in apparatus 158
Dangerous goods in machinery 158

bold: regulatory entries

Dead oil...44

Dead oil...45

Deflagrating metal salts of aromatic nitro-derivatives, n.o.s., 1.3C...................74

De-icing fluids..22

Delay electric igniter...119

Delay electric igniter..120

Demolition charges..70, 75, 78, 120

Denatured alcohol..6

Denatured alcohol, 3..5

Depleted uranium..12, 202

Depth charges..12

Depth charges...8

Derivative..234

Derivatives..1, 63, 75, 115, 128, 164, 176, 181, 244, 246

Desensitized...234

Desensitized..100

Detonating cord..121

Detonating fuse...121

Detonating fuze...123

Detonating relays...119

Detonating relays..122

Detonator...75, 78, 79, 120, 197

Detonator assemblies, non-electric, for blasting, 1.1B, 1.4B, 1.4S....................119

Detonator assembly...122

Detonators for ammunition, 1.1B, 1.2B, 1.4B, 1.4S...119

Detonators, electric for blasting, 1.1B, 1.4B, 1.4S..119

Detonators, non-electric for blasting, 1.1B, 1.4B, 1.4S....................................119

Devices, small, hydrocarbon gas powered, 2.1..104

Devices, small, hydrocarbon gas powered with release device, 2.1..................104

di-, structural notation...249

Diagnostic specimens...115

Diagnostic specimens..115

Diesel fuel...77, 79, 185

Diesel fuel, 3, 3.3..183

Dimer..235

Dimers..189

Direct reduced iron..221

Direct reduced iron...222

Disinfectants...3, 34, 45, 61, 145, 180, 188

Disinfectants, liquid, n.o.s., 6.1, 8..61

Disinfectants, solid, n.o.s., 6.1..61

Dispersant gas...104

Dispersant gases...106

Dispersant gases, n.o.s....104

Dispersants...64, 175

Dispersion...235

Dispersions..106, 153, 181, 224, 256

italics: key words and phrases

Page

Dithiocarbamate pesticide 179
Dithiocarbamate pesticide, liquid, 3, 6.1..179
Dithiocarbamate pesticides 180
Dithiocarbamate pesticide, solid, 6.1..179
Double-base propellant 9
Double-base propellants 77
Dressing, leather 132
Driers, paint or varnish, liquid, n.o.s...175
Driers, paint or varnish, solid, n.o.s....175
Dross 148, 149, 150
Drugs..3, 7, 40, 61, 63, 86, 113
Drugs, corrosive, liquid, n.o.s..61
Drugs, corrosive, solid, n.o.s...61
Drugs, flammable, liquid, n.o.s....61
Drugs, flammable, solid, n.o.s...61
Drugs, n.o.s....61
Drugs, oxidizing, liquid, n.o.s...61
Drugs, oxidizing, solid, n.o.s...61
Drugs, toxic, liquid, n.o.s....61
Drugs, toxic, solid, n.o.s...61
Dry 235
Dry..91, 94, 162, 246
Dry batteries..28
Dusts...153
Dye and dye intermediate, n.o.s....63
Dye intermediate, liquid, n.o.s., 6.1, 8 63
Dye intermediates 63
Dye intermediate, solid, n.o.s., 6.1, 8 63
Dye, liquid, n.o.s., 6.1, 8 63
Dyes..22, 34, 50, 52, 63, 111, 132, 158, 175, 176, 196
Dye, solid, n.o.s., 6.1, 8 63
Dynamite 74
Dynamite 78

E

Electric accumulators...28
Electric squibs..119
Electric squibs 121
Electric storage batteries...27
Electric storage batteries...28, 219
Electric wheelchairs 29
Electrolyte 27, 71
Electrolyte (acid) for batteries..27
Electrolyte (acid or alkali) for batteries..27
Electrolyte (alkali) for batteries 27
Electron tubes containing mercury..145

bold: regulatory entries

Elements 235
Elevated temperature liquid, n.o.s., 3, 3.3, 9 65
Elevated temperatures 32, 65, 96
Elevated temperature solid, n.o.s., 9 65
Enamel...175
Enamels 176
Engines...11, 22, 23, 29, 39, 69, 70, 106, 184, 185, 219
Engines, internal combustion (flammable gas powered) including
    where fitted in machinery or vehicles, 9 219
Engines, internal combustion (flammable liquid powered)
    including where fitted in machinery or vehicles, 9 219
Engines, internal combustion, including when fitted in machinery
    or vehicles, 9...219
Engines, rocket...8
Engine starting fluid..219
Engine starting fluids 106, 219
Engine starting fluid, with flammable gas, 2.1...219
Environmentally hazardous substance 67, 113, 141
Environmentally hazardous substance, liquid, n.o.s., 9 67
Environmentally hazardous substance, solid, n.o.s., 9 67
Esters..11, 42, 59, 75, 77, 92, 132, 165, 168, 180
Esters, n.o.s., 3, 3.2, 3.3..164
Etching acid...49
Etching acid, liquid, n.o.s..47
Ethers 1, 7, 23, 42, 62, 86, 113, 132, 162, 165, 167, 215, 219
Ethers, n.o.s., 3, 3.1, 3.2, 3.3...164
Etiologic agent 115
Etiologic agents...115
Excepted packages...205
Expelling charge...9, 20, 122, 196
Explosive article 69, 75, 197
Explosive articles 69
Explosive, blasting, type A, 1.1D 74
Explosive, blasting, type B, 1.1D, 1.5D 74
Explosive, blasting, type C, 1.1D 74
Explosive, blasting, type D, 1.1D 74
Explosive, blasting, type E, 1.1D, 1.5D 74
Explosive cartridge..69, 134, 195, 196
Explosive emulsions...79
Explosive, emulsion 74
Explosive fracturing devices...70
Explosive rivets...70
Explosive samples...40
Explosive, seismic 74
Explosive, slurry 74
Explosive substances 74
Explosive substances 74

italics: key words and phrases

Page

Explosive train..119
Explosive, water gel...74
Extract, aromatic or flavouring 86
Extracts...7, 20, 61, 90, 188, 224
Extracts, aromatic, liquid, 3, 3.2, 3.3 86
Extracts, flavouring, liquid, 3, 3.2, 3.3....................86

F

Fabrics 25, 42, 90, 121, 162, 181
Fabrics, animal, n.o.s. with oil, 4.2..........................90
Fabrics impregnated with weakly nitrated nitrocellulose, n.o.s.,
   4.1..161
Fabrics impregnated with weakly nitrated nitrocellulose, n.o.s.
   (including toe puffs, nitrocellulose base), 4.1............161
Fabrics, synthetic, n.o.s. with oil, 4.2 90
Fabrics, vegetable, n.o.s. with oil, 4.2.....................90
Fermentation amyl alcohol 5
Fermentation amyl alcohol...6
Ferrocerium..137
Ferrocerium, 4.1 137
Ferrophosphorus 222
Ferrophosphorus (including briquettes).....................221
Ferrosilicon..221, 222
Ferrosilicon, with 25% to 30% silicon, or 90% or more silicon
   (including briquettes) 221
Ferrous metal borings in a form liable to self-heating, 4.2 147
Ferrous metal cuttings in a form liable to self-heating, 4.2....147
Ferrous metals 242
Ferrous metal shavings in a form liable to self-heating, 4.2....147
Ferrous metal turnings in a form liable to self-heating, 4.2....147
Fertilizer ammoniating solution with free ammonia, 2.2 88
Fertilizers 38, 50, 88, 186, 226
Fertilizers containing ammonium nitrate, n.o.s......................88
Fertilizer with ammonium nitrate, n.o.s......................88
Fibre........................25, 26, 34, 38, 69, 90, 121, 161, 162, 235, 239, 255
Fibreglass repair kit 189
Fibreglass repair kits 191
Fibres, animal, burnt, wet, or damp, 4.2 90
Fibres, animal, n.o.s. with oil, 4.2.............................90
Fibres impregnated with weakly nitrated nitrocellulose, n.o.s., 4.1 161
Fibres impregnated with weakly nitrated nitrocellulose, n.o.s.
   (including toe puffs, nitrocellulose base), 4.1............161
Fibres, synthetic, n.o.s. with oil, 4.2 90
Fibres, vegetable, burnt, wet or damp, 4.2 90
Fibres, vegetable, dry, 4.1 90
Fibres, vegetable, n.o.s. with oil, 4.2........................90

bold: regulatory entries

Filler, liquid...175
Film scrap...161
Film scrap 162
Films, nitrocellulose base from which gelatin has been removed......................161
Films, nitrocellulose base, gelatin coated, except scrap, 4.1.............................161
Fire extinguisher 69, 94, 106, 219
Fire extinguisher charge 94
Fire extinguisher charges, corrosive liquid, 8..94
Fire extinguisher charges, expelling, explosive 69
Fire extinguishers, containing compressed or liquefied gas.................................94
Fire extinguishers with compressed or liquefied gas, 2.2 94
Firelighters...99
Firelighters, solid with flammable liquid, 4.1...99
Fire Point...238
Fireworks..196
Fireworks, 1.1G, 1.2G, 1.3G, 1.4G, 1.4S 194
First aid kit..40, 134
First aid kit, 9...40
Fischer-Tropsch gas 44
Fischer-Tropsch gas 45, 244
Fischer-Tropsch gas compressed, 2.2..44
Fish meal 226
Fish meal (fish scrap), unstabilized, 4.2...226
Fish meal, stabilized, 9 226
Fish meal, unstabilized, 4.2 226
Fish scrap 226
Fish scrap, stabilized, 9 226
Fish scrap, unstabilized, 4.2...226
Fissile excepted 205
Fissile material..202
Fixed ammunition..10
Flammable gas 104
Flammable gas 30, 47, 49, 59, 74, 137, 150, 167, 219, 221
Flammable gas in lighters 137
Flammable gas (small receptacles not fitted with a dispersion
   device, not refillable) 104
Flammable liquid, n.o.s., 3, 3.1, 3.2, 3.3, 6.1, 8...96
Flammable liquid preparation 96
Flammable liquid preparations, n.o.s. 96
Flammable liquids...53, 65, 74, 96, 99, 165, 184, 190, 237
Flammable range..237
Flammable solid, inorganic, n.o.s., 4.1, 6.1, 8..99
Flammable solid, n.o.s., 4.1, 5.1..99
Flammable solid, organic molten, n.o.s., 4.1..99
Flammable solid, organic, n.o.s., 4.1, 6.1, 8...99
Flammable solids 53, 74, 86, 99
Flares 71, 134, 196

italics: key words and phrases

Page

Flares, aerial, 1.1G, 1.2G, 1.3G, 1.4G, 1.4S 194
Flares, aeroplane 194
Flares, distress, small...194
Flares, highway or railway 194
Flares, surface, 1.1G, 1.2G, 1.3G................................194
Flares, water-activated..194
Flash cartridges 195
Flash point...237
Flash powder..197
Flash powder, 1.1G, 1.3G...194
Flavouring extracts 7, 86
Flavouring liquids 86
Flavouring liquids 86
Flax 91
Flax, dry 90
Flowers 238
Flue dust..149
Flue dusts, arsenical 147
Flue dusts, poisonous...147
Flue dusts, toxic 147
Fluid..238
Fluorine compounds (pesticides) 179
Fluorine pesticides 180
Fluorosilicates...130, 156
Fluorosilicates, n.o.s., 6.1 128
Fluorspar..88, 222
Fluorspar (calcium fluoride)......................................221
Forbidden dangerous goods..56
Fracturing devices, explosive, for oil wells, without detonators,
　1.1D..69
Freon...215
Freon...215
Fuel 10, 238
*Fuel 6, 10, 11, 12, 21, 23, 29, 30, 44, 70, 76, 78, 79, 106, 135, 137, 153, 170
　　　.........173, 183, 184, 195, 196, 202, 203, 219, 222, 233, 235, 237, 240, 244*
Fuel, aviation, turbine engine, 3, 3.1, 3.2, 3.3 183
Fuel oil 183
Fuel oil 77, 89, 96, 185
Fuel oil No. 1 183
Fuel oil (No. 1, 2, 4, 5, or 6), 3.................................183
Fulminating...238
Fumigant 179
Fumigant 7, 181
Fuming 195
Fuming..239
Fungicide..179
Fungicides 145, 179

bold: regulatory entries

Fuse, detonating, metal clad, 1.1D, 1.2D...119
Fuse, detonating, mild effect, metal clad, 1.4D...119
Fusee, matches...143
Fusee matches..143
Fusees (railway or highway)..194
Fusees, railway or highway, explosive..194
Fuse, igniter, tubular, metal clad, 1.4G..119
Fuse lighters..121
Fusel oil..6
Fusel oil, 3, 3.2, 3.3...5
Fuse, non-detonating, 1.3G..119
Fuses...21, 77, 121, 197
Fuse, safety, 1.4S...119
Fuses, tracer..194
Fuzes...10, 11, 12, 20, 123
Fuzes, combination, percussion or time..119
Fuzes, detonating, 1.1B, 1.2B, 1.4B, 1.4S..119
Fuzes, detonating, with protective features, 1.1D, 1.2D, 1.4D.....................119
Fuzes, igniting, 1.3G, 1.4G, 1.4S..119

G

gamma-, structural notation..249
Gas..239
Gas 3, 7, 40, 41, 44, 45, 46, 59, 65, 70, 71, 74, 76, 88, 94, 104, 128, 129, 135
............154, 156, 159, 164, 165, 166, 169, 170, 181, 184, 185, 190, 193, 194, 197
......215, 218, 222, 223, 224, 226, 232, 233, 238, 240, 241, 243, 248, 252, 253, 255
Gas candle..106, 137, 173
Gas candles, charged with flammable gas..104
Gas cartridge..106
Gas cartridges, 2.3, 5.1..104
**Gas cartridges, without a release device, non-refillable, 2.1, 2.2,
   2.3, 5.1, 8**...104
Gas, compressed...104
Gas drips..184
Gas drips, hydrocarbon..183
Gas generator assemblies...134
**Gas generator assemblies for aircraft escape slides containing a
   non-flammable, non-toxic gas and a propellant cartridge, 2.2**.....................134
Gas identification set, 2.3...40
Gas liquefied..104
Gasohol...184
Gasohol, 3...183
Gas oil..185
Gas oil, 3, 3.3...183
Gasoline 12, 39, 184, 219
Gasoline, 3, 3.1...183

italics: key words and phrases

Page

Gasoline, casinghead 183
Gas-powered devices 106
Gas, refrigerated liquid, n.o.s., 2.1, 2.2, 5.1 104
Gas sample, non-pressurised, n.o.s., not refrigerated liquid, 2.1,
   2.3 40
Gas samples 40, 106
Gel 239
Gelatin, blasting 74
Gelatin dynamite 74
Gelatine, blasting 74
Gelatine dynamite 74
Gelatine dynamite 79
Gels 12, 28, 40, 53, 77, 79, 92, 162
Genetically modified microorganisms 115
Genetically modified micro-organisms, 9 115
Glacial 239
Granules 32, 39, 45, 76, 88, 153, 189
Grenades 12, 20, 21, 123, 195
Grenades, empty primed, 1.4S 8
Grenades, hand or rifle, with bursting charge, 1.1D, 1.1F, 1.2D,
   1.2F 8
Grenades, illuminating 194
Grenades, practice, hand or rifle, 1.2G, 1.3G, 1.4G, 1.4S 8
Grenades, smoke 194
Grignard 167
Grignard solution 167
Guided missiles 11
Gunpowder 9, 76, 100, 246
Gunpowder, compressed, 1.1D 74
Gunpowder granular or as a meal, 1.1D 74
Gunpowder in pellets, 1.1D 74
Gutta percha 190
Gutta percha solution 189

H

Hafnium powder, dry, 4.2 153
Hafnium powder, wetted, 4.1 153
Hair 52, 90, 106, 132
Hair, wet 90
Hay 91
Hay, 4.1 90
Hazardous materials 54, 67
Hazardous substances 67
Hazardous substances, liquid or solid, n.o.s. 67
Hazardous waste, liquid, n.o.s., 9 113
Hazardous wastes 67, 113

bold: regulatory entries

Hazardous waste, solid, n.o.s., 9 113
Heating oil 185
Heating oil, light, 3, 3.3 183
Heat producing article, battery operated equipment, 9 27
Heat producing articles 29
Heavy 239
Heavy metal 242
Hemp 91
Hemp, dry 90
High explosives 74
High explosives 9, 10, 12, 70, 75, 119, 121, 122, 123
Highway flare 196
Homologues 23, 165
Homologues 239
Hull fibre 91
Hydrides, metal, water-reactive, n.o.s., 4.3 156
Hydrocarbon gas 59, 105
Hydrocarbon gases, compressed, n.o.s., 2.1 104
Hydrocarbon gas, liquefied, n.o.s., 2.1 104
Hydrocarbon gas mixture, compressed, n.o.s., 2.1 104
Hydrocarbon gas mixture, liquefied, n.o.s., 2.1 104
Hydrocarbon gas-powered small devices 104
Hydrocarbon gas refills for small devices with release device, 2.1 104
Hydrocarbons 240
Hydrocarbons 1, 29, 32, 36, 39, 42, 44, 45, 63, 86, 105, 106, 128, 164
....183, 184, 185, 189, 190, 218, 219, 222, 224, 225, 244
Hydrocarbons, liquid, n.o.s., 3.1, 3.2, 3.3 96
Hydrogendifluorides 130
Hydrogendifluorides, n.o.s., 8 128
Hydrogendifluorides, solid, n.o.s., 8 128
Hydrogendifluorides, solution, n.o.s., 8 128
Hypergolic 240
Hypochlorites 34, 62, 129
Hypochlorites, inorganic, n.o.s., 5.1 128

I

Igniter cord 121
Igniter fuse, metal clad 119
Igniter fuses 121
Igniters 121
Igniters, 1.1G, 1.2G, 1.3G, 1.4G, 1.4S 119
Igniting fuze 123
Ignition element for lighter, containing pyrophoric liquid 137
Illuminating ammunition 194
Illuminating bombs 194
Illuminating cartridges 194

italics: key words and phrases

Page

Illuminating grenades..194
Illuminating projectiles 194
Immiscible 240
Incendiary ammunition..12
Indiarubber...189
Indiarubber..189
Industrial alcohol 6
Industrial ammunition 69
Inert 240
Inert 3, 9, 11, 29, 78, 106, 134, 143, 148, 154, 170, 176, 223, 224, 226
Infectious substance, affecting animals only, 6.2 115
Infectious substance, affecting humans, 6.2 115
Infectious substances 115, 223
Ingots..154
Inhibited 241
Ink, printer's, flammable..175
Inorganic 241
Inorganic...1, 39, 42, 44, 47, 63, 79, 128, 129, 148, 151
....156, 164, 165, 166, 167, 183, 218, 224, 234, 244
Inorganic acids..47
Inorganic base...48
Insecticide...179
Insecticide gas, n.o.s., 2.1, 2.2, 2.3..179
Insecticides 7, 130, 179, 188
Insecticide solid or liquid 179
Iron oxide, spent (obtained from coal gas purification), 4.2 44
Iron powder..153
Iron powder 173
Iron powder, pyrophoric..153
Iron sponge..46
Iron sponge, spent, 4.2..44
Iron sponge, spent (obtained from coal gas purification), 4.2 44
Iron swarf..147
Irritating agents 19
Irritating agents..20
Irritating material 19
Irritating material...20, 86, 130, 159, 165, 221
iso-, structural notation...249
Isocyanates 165
Isocyanates, n.o.s., 3, 3.1, 3.2, 3.3, 6.1 164
Isocyanate solution, n.o.s., 3, 3.1, .3.2, 3.3, 6.1 164
Isomers 241
Isomers 6, 45, 105, 164, 190, 240, 249, 250, 251, 252

bold: regulatory entries

J

JATO..11
Jet engine starter cartridges..70
Jet fuel 183
Jet fuel 185
Jet perforating guns, charged, oil well, with detonator, 1.1D, 1.4D 69
Jet perforating guns, charged, oil well, without detonator, 1.1D, 1.4D 69
Jet perforator guns 70
Jet perforators 69
Jet tappers 70
Jet tappers, without detonator 69
Jet thrust igniters...121
Jet thrust igniters, for rocket motors or Jato 119
Jet thrust unit (Jato)..8
Jet thrust units 11
Jute 90
Jute 91
Jute, dry..90

K

Kapok 90
Kapok 91
Kapok, dry 90
Kerosene..99, 154, 185, 235
Kerosene, 3, 3.3 183
Ketones 165
Ketones, liquid, n.o.s., 3, 3.2, 3.3 164

L

L-, **structural notation** 249
Lachrymatory ammunition 19
Lacquer 175
Lacquer base, liquid 175
Lacquer base or lacquer chips, nitrocellulose, dry 161
Lacquer base or lacquer chips, plastic, wet with alcohol or solvent..........161, 175
Lacquer base, solution, 3.1, 3.2, 3.3..........................175
Lacquer, liquid..161, 175
Lacquers...162, 176
Lamp black 36
Lamp black...36, 176
Lead compound 23, 75, 76, 113, 119, 120, 148, 150, 156, 175
Lead compound, soluble, n.o.s., 6.1..........................156
Lead dross 147

italics: key words and phrases

Page

Lead dross 148
Leather bleaches..132
Leather bleach or dressing..132
Leather dressing..132
Life rafts 134
Life rafts 71, 134
Life saving appliance, not self-inflating, 9..134
Lifesaving appliances 71, 134
Life-saving appliances, not self-inflating containing dangerous
    goods as equipment, 9..134
Life-saving appliances, self-inflating, 9...134
Lighter flint..137
Lighter flints 137
Lighter fluid 137
Lighter fluid..106, 137
Lighter refills (cigarettes) containing flammable gas, 2.1 137
Lighter refills containing flammable gas, 2.1 137
Lighter refills, 2.1 137
Lighter replacement cartridges containing liquefied petroleum
    gases..137
Lighters...106, 137
Lighters, 2.1 137
Lighters (cigarettes), containing flammable gas, 2.1 137
Lighters (cigarettes), containing pyrophoric liquid...................................137
Lighters containing flammable gas, 2.1 137
Lighters for cigars, cigarettes, etc., with lighter fluids, 3 137
Lighters, fuse, 1.4S..119
Lighters, with lighter fluids (cigarettes) 137
Ligroin..183
Ligroin..184
Lime...1, 88, 180, 222, 247
Lime (unslaked) (calcium oxide, quicklime, dolomitic quicklime)....................221
Line-throwing rockets..134
Liquefied gas 3, 11, 41, 104, 137, 173, 238, 240
Liquefied gases, non-flammable charged with nitrogen, carbon
    dioxide 104
Liquefied gas, n.o.s., 2.1, 2.2, 2.3, 5.1, 8 104
Liquefied hydrocarbon gas...104
Liquefied natural gas 183
Liquefied petroleum gas....183
Liquefied petroleum gases...184
Liquid 241
Liquid 3, 7, 11, 20, 27, 28, 32, 36, 44, 45, 49, 53, 65, 74, 76, 92, 94, 96, 99
    ....105, 106, 107, 114, 121, 130, 137, 145, 151, 152, 153, 159, 162, 164
    ....165, 167, 175, 181, 183, 184, 185, 193, 204, 215, 218, 219, 224, 231
    ....232, 233, 234, 235, 237, 238, 239, 240, 243, 246, 247, 252, 253, 255, 256
Liquid fillers 176

bold: regulatory entries

Liquid lacquer base 176
Liquids, other than those classified as flammable, corrosive, or
  toxic, charged with nitrogen, carbon dioxide, or air 104
Liquor 5
Liquor 5
Lithium alkyl 167
Lithium alkyls, 4.2, 4.3 167
Lithium alloy (liquid) 151
Lithium amalgams 151
Lithium batteries 28
Lithium batteries, 9 27
Lithium batteries contained in equipment, 9 27
Lithium batteries packed with equipment, 9 27
Lithium cartouches 154
Lithium cartridges 154
Lithium dispersions 153
Lithium in cartouches 153
Lithium in cartridges 153
LNG 183
LNG 183
London purple 180
London Purple, 6.1 179
Low-specific activity (LSA) 204
LPG 183
LPG 105, 106, 184
LSA-I 204
LSA-II 204
LSA-III 204
Lythene 183

M

m-, structural notation 249
M86 fuel 29
M86 fuel, 3.2 27
Machinery 158
Magnesia 222
Magnesia (unslaked) (Lightburned magnesia, calcined magnesite,
  caustic calcined magnesite) 221
Magnesium alkyl 12, 167
Magnesium alkyls, 4.2, 4.3 167
Magnesium alloys, 4.1 151
Magnesium alloys, powder, 4.2, 4.3 151
Magnesium amalgams 151
Magnesium dispersions 153
Magnesium dross 148
Magnesium dross, hot 147

italics: key words and phrases

Page

Magnesium dross, wet 147
Magnesium granules, coated, 4.3 153
Magnesium in pellets, turnings, or ribbons, 4.1 153
Magnesium powder 195, 197
Magnesium powder, 4.2, 4.3 153
Magnesium scrap 147
Magnetized material 56, 139, 219
Magnetized material, 9 139
Marine pollutants 141
Marine pollutants, liquid or solid, n.o.s. 141
Match 99, 121, 134, 143
Matches, block 143
Matches, fusee, 4.1 143
Matches, safety (book, card or strike on box), 4.1 143
Matches, 'strike anywhere,' 4.1 143
Matches, trick 194
Matches, wax 'vesta,' 4.1 143
Matting acid 47
Matting acid 49
Meal 90
Meal, oily, 4.2 90
Medical waste 115
Medical waste, n.o.s., 6.2 115
Medicinal preparation 61
Medicinal preparation 7, 96
Medicinal tinctures 61
Medicine 3, 7, 61, 203
Medicine, liquid, n.o.s., 3, 3.2, 3.3, 6.1, 8 61
Medicine, n.o.s. 61
Medicine, solid, n.o.s., 4.1, 5.1, 6.1, 8 61
Mercaptan mixture, liquid, n.o.s., 3, 3.1, 3.2, 3.3, 6.1 164
Mercaptans 165
Mercaptans, liquid, n.o.s., 3, 3.1, 3.2, 3.3, 6.1 164
Mercuric salt 145
Mercuric salt 145
Mercurous compound 145
Mercurous compounds 145
Mercury based pesticide, liquid, 3, 3.2, 6.1 179
Mercury-based pesticides 180
Mercury based pesticide, solid, 6.1 179
Mercury compound, liquid, n.o.s, 6.1 145
Mercury compounds 52, 61, 75, 120, 141, 145, 180, 181, 238
Mercury compound, solid, n.o.s., 6.1 145
Mercury contained in manufactured articles 40, 158
Mercury contained in manufactured articles, 8 145
Mercury electron tubes 145
Mercury vapour tubes 145

bold: regulatory entries

Mercury vapour tubes..145

meta-, structural notation 249

Metal.. ....1, 9, 10, 12, 13, 30, 39, 42, 44, 47, 48, 49, 58, 59, 62, 65
..................................69, 70, 75, 94, 111, 121, 122, 128, 129, 130, 137, 139, 145
..........................146, 147, 148, 149, 150, 151, 153, 154, 155, 156, 164, 167, 168
.....................173, 175, 176, 183, 188, 197, 201, 222, 224, 234, 247, 248, 252, 254

Metal alkyl halide 167

Metal alkyl halides, n.o.s., 4.2, 4.3..167

Metal alkyl hydrides 168

Metal alkyl hydrides, n.o.s., 4.2, 4.3...167

Metal alkyls 12, 146, 167

Metal alkyls, n.o.s., 4.2, 4.3 167

Metal alkyl, solution, n.o.s., 3 167

Metal aryl halide 167

Metal aryl halides, n.o.s., 4.2, 4.3 167

Metal aryl hydrides..168

Metal aryl hydrides, n.o.s., 4.2, 4.3...167

Metal aryls...167

Metal aryls, n.o.s., 4.2, 4.3...167

Metal carbonyl 23, 113, 156

Metal carbonyls, n.o.s., 6.1..156

Metal catalyst..39

Metal catalyst, dry, 4.2 39

Metal catalysts...39

Metal catalyst, wetted with a visible excess of liquid, 4.2 39

Metal catalyst, wetted without a visible excess of liquid 39

Metal clad detonating cord 122

Metal clad detonating fuse 122

Metal hydrides 59, 156

Metal hydrides, n.o.s., 4.1, 4.3 156

Metallic substance, n.o.s., 4.2, 4.3..156

Metallic substances 156

Metalloids..242

Metalloids..128, 167

Metal powder, n.o.s., 4.1, 4.2 153

Metal powders 129, 206

Metals 242

Metals 39

Metal salts of organic compounds...42, 59, 76, 168, 247

Metal salts of organic compounds, n.o.s., 4.1 167

Metal sulphide concentrates 221

Metal sulphides...156, 222

Metal sulphides concentrates...156

Metal turnings 149

Methylated spirit 5

Methylated spirits 6

Mine rescue equipment...134

italics: key words and phrases

Page

Mine rescue equipment containing carbon dioxide 134
Mines 12
Mines with bursting charge, 1.1D, 1.1F, 1.2D, 1.2F 8
Mischmetall 137
Mischmetall 137
Miscible 243
Miscible 5, 96, 151, 224, 248
Missiles, guided 8
Mixed acid 47
Mixed acid 50, 161
Mixed acid, spent 47
Mixture 243
Mobility aids 27
Mobility aids 29
Model rocket motor, 1.4C, 1.4S 8
Model rocket motors 11
Molecular weight 243
Molecular weight 32, 165, 189, 239, 246
mono-, structural notation 249
Motorcycle 219
Motorcycles 219
Motor fuel anti-knock compounds 23
Motor fuel anti-knock mixture, 3, 6.1 23
Motor spirit 184
Motor spirit, 3, 3.1 183
Musk xylene 100
Musk xylene, 4.1 99
Mysorite 25
Mysorite 25

N

n-, structural notation 250
Naphtha 183
Naphtha 45, 114, 154, 184, 235
Naphtha, petroleum 183
Naphtha, solvent 183
Naphtha solvent 1, 176, 184, 188
Natural 243
Natural gas 183, 184, 240
Natural gas, compressed with high methane content, 2.1 183
Natural gases 183
Natural gasoline 183
Natural gasoline 183, 244
Natural gas, refrigerated liquid with high methane content, 2.1 183
Natural uranium 202
New explosive 74

bold: regulatory entries

New explosive device 69
Nickel catalyst..39
Nickel catalysts...39
Nicotine 7, 180
Nicotine compound, liquid, n.o.s., 6.1 7
Nicotine compounds 7
Nicotine compound, solid, n.o.s., 6.1 7
Nicotine preparation, liquid, n.o.s., 6.1...7
Nicotine preparations...7
Nicotine preparation, solid, n.o.s., 6.1 7
Nitrated papers..162
Nitrated paper (unstable)..161
Nitrates ...7, 11, 13, 75, 76, 77, 78, 79, 88, 89, 130, 143, 161, 162, 170, 195, 197, 221
Nitrates of diazonium compounds..63
Nitrates of diazonium compounds 64
Nitrates, inorganic, aqueous solution, n.o.s., 5.1 128
Nitrates, inorganic, n.o.s., 5.1 128
Nitrating acid mixture 50
Nitrating acid mixture, 5.1, 8..47
Nitrating acid mixture, spent, 5.1, 8...47
Nitrating acid, mixture, spent, all concentrations, unstable 47
Nitriles...20, 129, 165
Nitriles, liquid, n.o.s., 6.1...164
Nitriles, n.o.s., 3, 3.1, 3.2, 6.1..164
Nitriles, solid, n.o.s., 6.1..164
Nitrites 130
Nitrites, inorganic, aqueous solution, n.o.s., 5.1 128
Nitrites, inorganic, n.o.s., 5.1 128
Nitrocellulose films..162
Nitrocellulose membrane filters 162
Nitrocellulose membrane filters, 4.1 161
Nitroglycerine..75
Nitroglycerin mixture, desensitized, liquid, n.o.s., 3.....................................74
Nitroglycerin mixture, desensitized, solid, n.o.s., 4.1 74
Non-activated 244
Non-activated carbon 36
Non-activated charcoal 36
Nonflammable gas 134
Non-flammable gas, n.o.s. 104
Non-liquefied gas 104
Nonliquefied gas..104, 193
Non-liquefied hydrocarbon gas 104
Non-volatile..244
Nonvolatile 253
Numerals, structural notation 250

italics: key words and phrases

O

o-, structural notation 250
Oil 244
Oil cake...90
Oil cake 90
Oil gas 184
Oil gas, compressed, 2.3 183
Oils 1, 7, 11, 32, 34, 36, 38, 42, 44, 45, 70, 75, 79, 86, 90,
....91, 92, 99, 100, 132, 141, 150, 168, 175, 176, 181, 183
................................184. 185, 186, 188, 190, 191, 193, 224, 226, 235, 243, 253
Oil well cartridges...70
Oil well sampling device, charged 183
Oily rags...92
omega-, structural notation 250
Organic...244
Organic..........5, 7, 12, 32, 34, 36, 39, 42, 44, 45, 47, 48, 49, 50, 63, 75, 111, 128, 129
..............151, 161, 164, 165, 167, 168, 170, 173, 175, 176, 177, 180, 186, 189, 190
......191, 205, 215, 224, 225, 233, 234, 238, 239, 240, 241, 248, 249, 250, 251, 253
Organic acid...7, 47, 168. 188. 246. 247
Organic bases..48
Organic compounds of arsenic...168
Organic compounds of arsenic, n.o.s., 6.1 167
Organic cyanides..165
Organic peroxides 76, 170, 234
Organic peroxide types B, C, D, E, F, liquid, 5.2 170
Organic peroxide types B, C, D, E, F, liquid, temperature
   controlled, 5.2..170
Organic peroxide types B, C, D, E, F, solid, 5.2 170
Organic peroxide types B, C, D, E, F, solid, temperature
   controlled, 5.2..170
Organic pigments, self-heating, 4.2...63
Organoarsenic compound, n.o.s., 6.1 167
Organoarsenic compounds...168
Organochlorine pesticide, liquid, 3, 3.2, 6.1 179
Organochlorine pesticides..180
Organochlorine pesticide, solid, 6.1 179
Organometallic compound dispersion, n.o.s., 3, 4.3 167
Organometallic compound, n.o.s., 3, 4.3, 6.1......................................167
Organometallic compounds..12, 23, 39, 145, 156, 167
Organometallic compound solution, n.o.s., 3, 4.3 167
Organophosphorus compound, n.o.s., 3, 6.1 167
Organophosphorus compounds...168
Organophosphorus pesticide, liquid, 3, 3.2, 6.1 179
Organophosphorus pesticides 180
Organophosphorus pesticide, solid, 6.1 179

bold: regulatory entries

Organotin compound, liquid, n.o.s., 6.1...167
Organotin compounds 168, 180
Organotin compound, solid, n.o.s., 6.1..167
Organotin pesticide, liquid, 3, 3.2, 6.1 179
Organotin pesticides...180
Organotin pesticide, solid, 6.1 179
ORM-D 158
ortho-, structural notation 250
Other regulated substance, aromatic extract or flavouring.................................86
Other regulated substance, liquid, n.o.s., 9 158
Other regulated substances 87, 159
Other regulated substance, solid, n.o.s., 9 158
Oxidation..245
Oxidation....................................29, 34, 36, 39, 47, 49, 59, 65, 92, 93, 129, 130, 148
....170, 173, 175, 221, 222, 226, 233, 235, 241, 245, 251
Oxidation Number..245
Oxidizer 10, 11, 47, 62, 76, 78, 113, 129, 130, 132, 135, 143, 155, 197, 233, 234, 240
Oxidizing liquid..34, 52, 170
Oxidizing liquid, n.o.s., 5.1, 6.1, 8..170
Oxidizing solid...13, 170
Oxidizing solid, n.o.s., 4.1, 4.2, 4.3, 5.1, 6.1, 8......................................170
Oxygen generator, chemical, 5.1 173
Oxygen generators...173

P

p-, structural notation 251
Paint driers..175
Paint driers..175, 188
Paint (including paint, lacquer, enamel, stain, shellac, varnish,
   polish, liquid filler and liquid lacquer base), 3, 3.1, 3.2, 3.3, 8...................175
Paint related material (including paint thinning or reducing
   compounds), 3, 3.1, 3.2, 3.3, 8..175
Paints...3, 38, 63, 113, 145, 158, 175, 184, 188, 224
Paint thinners 175
Paper stock...91, 92
Paper stock, wet..90
Paper waste 92
Paper waste, wet..90
Paper, unsaturated oil treated, incompletely dried, 4.2 90
Paper, unsaturated oil treated, incompletely dried (includes
   carbon paper), 4.2...90
para-, structural notation 251
Paraffin..183
Paraffin 132, 143, 164, 184, 240
Passenger restraint systems...135
P.c.b.s..111

italics: key words and phrases

Page

P.c.b.s 111
PCBs 111
PCB's 111
PCB's 111
PCBs...111, 141, 256
Peat 32, 36, 44, 99, 222
Peat moss 221
Pellets 32, 76, 78, 124, 135, 153, 189
Pencil pitch..32
Pencil pitch..32, 222
penta-, **structural notation** 251
Pentaerythrite tetranitrate 75
Pentaerythrite tetranitrate mixture, desensitized, solid, n.o.s., 4.1 74
Pepper spray 19
Pepper spray...20
per, structural notation 251
Perchlorates 11, 79, 129, 173
Perchlorates, inorganic, aqueous solution, n.o.s., 5.1 128
Perchlorates, inorganic, n.o.s., 5.1..128
Percussion cap 9, 122
Percussion caps 119
Percussion fuze..123
Perfumery products 3, 52, 159, 252
Perfumery products in small inner packagings 52
Perfumery products, with flammable liquid, 3.2, 3.3 52
Perfumery products with flammable solvents, 3......................................52
Permanganates...130, 170
Permanganates, inorganic, aqueous solution, n.o.s., 5.1 128
Permanganates, inorganic, n.o.s., 5.1..128
Permeation devices...41
Permeation devices, containing dangerous goods, for calibrating
    air quality monitoring..40
Peroxide, organic...170
Peroxides..11, 34, 52, 62, 132, 170, 173
Peroxides, inorganic, n.o.s., 5.1 170
Persulphates 130, 170
Persulphates, inorganic, aqueous solution, n.o.s., 5.1...........................128
Persulphates, inorganic, n.o.s., 5.1 128
Pesticide, liquid, n.o.s., 3, 3.2, 6.1 179
Pesticide, solid, n.o.s., 6.1 179
Pesticide, toxic, under compressed gas, n.o.s........................................179
Pesticides...3, 7, 141, 156, 165, 179
Petrol...184
Petrol, 3, 3.1 183
Petroleum 32, 39, 238, 240, 244
Petroleum coke 44, 186, 222, 233
Petroleum coke, calcined or uncalcined..183, 221

bold: regulatory entries

Petroleum crude oil, 3, 3.1, 3.2, 3.3 183
Petroleum distillates 32, 181, 184, 224
Petroleum distillates, n.o.s., 3, 3.1, 3.2, 3.3 183
Petroleum ether 183
Petroleum ether 184
Petroleum gases, liquefied, 2.1 183
Petroleum naphtha 183
Petroleum naphtha 184
Petroleum oil 183
Petroleum oil, 3 183
Petroleum products 183
Petroleum products 6, 12, 32, 45, 63, 183
Petroleum products, n.o.s., 3, 3.1, 3.2, 3.3 183
Petroleum raffinate 183
Petroleum raffinate 185
Petroleum spirit 183
Petroleum spirits 184
Phenoxyacetic acid pesticide, liquid, 3, 3.2, 6.1 179
Phenoxyacetic acid pesticides 180
Phenoxyacetic acid pesticide, solid, 6.1 179
Phenoxy pesticides, liquid, 3, 6.1 179
Phenoxy pesticides, solid, 6.1 179
Phenylmercuric compound, n.o.s., 6.1 145
Phenylmercuric compounds 145
Phenyl urea pesticide, liquid, 3, 6.1 179
Phenyl urea pesticides 180
Phenyl urea pesticide, solid, 6.1 179
Phlegmatized 246
Photoflash bombs 195
Phthalimide derivative pesticide, liquid, 3, 6.1 179
Phthalimide derivative pesticides 181
Phthalimide derivative pesticide, solid, 6.1 179
Picrotoxin 115
Picrotoxin 115
Pigments 36, 52, 63, 86, 92, 145, 156, 161, 175, 176, 224
Pine oil 188
Pine oil, 3, 3.3 188
Pitch 32, 45
Pitch prill 32
Pitch prill 32, 222
Pitch prill, prilled coal tar, pencil pitch 221
Plastic explosives 74
Plastic explosives 78
Plasticized 246
Plasticizers 1, 11, 79, 161, 165, 168, 188, 190
Plastic moulding compounds 190

italics: key words and phrases

Page

Plastics moulding compound in dough, sheet or extruded rope
　form evolving flammable vapour, 9 189
Plastics, nitrocellulose-based, self-heating, n.o.s., 4.2 161
Plastic solvent, n.o.s. 224
Plastic solvents 224
Poisonous 255
Poisonous 255
Poisonous by inhalation 107, 256
Poisonous gas 107
Poisonous gases, n.o.s. 104
Polish 175
Polishes 26, 38, 52, 132, 176
Polyamines 164
Polyamines, n.o.s., 3.1, 3.2, 3.3, 8 164
Polychlorinated biphenyls 111
Polychlorinated biphenyls, 9 111
Polychlorinated biphenyls, liquid, 9 111
Polychlorinated biphenyls, solid, 9 111
Polyester resin kit, 3, 3.2, 3.3 189
Polyester resin kits 191
Polyhalogenated biphenyls 111
Polyhalogenated biphenyls, liquid, 9 111
Polyhalogenated biphenyls, solid, 9 111
Polyhalogenated terphenyls 111
Polyhalogenated terphenyls, liquid, 9 111
Polyhalogenated terphenyls, solid, 9 111
Polymeric bead 190
Polymeric beads, expandable evolving flammable vapour, 9 189
Polystyrene beads 190
Polystyrene beads, expandable 189
Polystyrene beads, expandable, evolving flammable vapour 189
Potassium alloys, metal 151
Potassium amalgams 151
Potassium dispersions 153
Potassium metal alloys, 4.3 151
Potassium metal, liquid alloy 151
Potassium sodium alloys, 4.3 151
Powder 246
Powder 34, 36, 39, 52, 76, 99, 135, 149, 151
　..................................153, 154, 155, 173, 189, 197, 221, 238, 247
Powder cakes 76
Powder cake, wetted, 1.1C, 1.3C 74
Powder paste 76
Powder paste, wetted, 1.1C, 1.3C 74
Powder, smokeless, 1.1C, 1.3C 74
Power device cartridges 69
Power device, explosive 69

bold: regulatory entries

Power devices 69
Practice ammunition 13
Pressurized accumulators 193
Pressurized articles 106, 193
Pressurized products 3
Pressurized products 3
Prilled coal tar 44
Prilled coal tar 45, 222
Primer caps 122
Primers 9, 10, 77, 122, 196, 197
Primers, cap type, 1.1B, 1.4B, 1.4S 119
Primers, small arms 119
Primers, tubular, 1.3G, 1.4G, 1.4S 119
Printing ink, 3, 3.1, 3.2, 3.3 175
Printing inks 176
Projectile ammunition 9, 20, 123, 195
Projectiles, illuminating 194
Projectiles, inert with tracer, 1.3G, 1.4G, 1.4S 8
Projectiles with burster or expelling charge, 1.2D, 1.2F, 1.2G,
   1.4D, 1.4F, 1.4G 8
Projectiles with bursting charge, 1.1D, 1.1F, 1.2D, 1.2F, 1.4D 8
Proof ammunition 13
Propellant 3, 9, 10, 74, 195, 196, 215
Propellant, liquid, 1.1C, 1.3C 8
Propellant, solid, 1.1C, 1.3C 8
Propelling charge 11
Propelling charge 9, 20, 122, 194, 196
Pyrethroid pesticide, liquid, 3, 3.2, 6.1 179
Pyrethroid pesticides 181
Pyrethroid pesticide, solid, 6.1 179
Pyrophoric 12, 137, 167, 226, 233
Pyrophoric alloy 137
Pyrophoric alloy, n.o.s., 4.2 151
Pyrophoric articles 226
Pyrophoric articles 226
Pyrophoric liquid, inorganic, n.o.s., 4.2 226
Pyrophoric liquid, organic, n.o.s., 4.2 226
Pyrophoric metal, n.o.s., 4.2 226
Pyrophoric organometallic compound (liquid), n.o.s., 4.2, 4.3 167
Pyrophoric organometallic compound, n.o.s., 4.2, 4.3 167
Pyrophoric organometallic compound (solid), n.o.s., 4.2, 4.3 167
Pyrophoric solid, inorganic, n.o.s., 4.2 226
Pyrophoric solid, organic, n.o.s., 4.2 226
Pyrotechnic articles 71, 134, 158, 197
Pyrotechnics 9
Pyrotechnics 75, 77, 121, 123, 135, 153, 162, 226
Pyroxylin 52, 161

italics: key words and phrases

Page

Pyroxylin cement 1
Pyroxylin cements 1
Pyroxylin plastic 161
Pyroxylin solution 161
Pyroxylin solution or solvent 161
Pyroxylin solvent, n.o.s. 161

Q

Quickmatch 119
Quickmatches 121
Quicksilver 145
Quicksilver 145

R

R11 215
R12 215
R13 215
R14 215
R12B1 215
R12B2 215
R13B1 215
R21 215
R22 215
R23 215
R32 215
R40 215
R41 215
R114 215
R115 215
R116 215
R124 215
R125 215
R133a 215
R134a 215
R142b 215
R143 215
R143a 215
R152a 215
R161 215
R218 215
R227 215
R404A 215
R407A 215
R407B 215
R407C 215

bold: regulatory entries

R500 215
R502 215
R503 215
R1113 215
R1132a 215
R1216 215
R1318 215
Radioactive material, excepted package, articles, 7 200
Radioactive material, excepted package, articles manufactured
    from depleted uranium, 7 200
Radioactive material, excepted package, articles manufactured
    from natural or depleted uranium or natural thorium, 7 200
Radioactive material, excepted package, articles manufactured
    from natural thorium, 7 200
Radioactive material, excepted package, articles manufactured
    from natural uranium, 7 200
Radioactive material, excepted package - articles manufactured
    from natural uranium or depleted uranium or natural thorium 200
Radioactive material, excepted package - empty packaging, 7 200
Radioactive material, excepted package, instruments, 7 200
Radioactive material, excepted package - instruments or articles, 7 200
Radioactive material, excepted package - limited quantity of
    material, 7 200
Radioactive material, fissile, n.o.s., 7 200
Radioactive material, low specific activity (LSA), n.o.s., 7 200
Radioactive material, low specific activity (LSA-I) non fissile or
    fissile excepted, 7 200
Radioactive material, low specific activity (LSA-II) fissile, 7 200
Radioactive material, low specific activity (LSA-II) non fissile or
    fissile excepted, 7 200
Radioactive material, low specific activity (LSA-III) fissile, 7 200
Radioactive material, low specific activity (LSA-III) non fissile or
    fissile excepted, 7 200
Radioactive material, low specific activity, n.o.s., 7 200
Radioactive material, LSA, n.o.s., 7 200
Radioactive material, n.o.s., 7 200
Radioactive materials 55, 201, 223
Radioactive material, SCO, 7 200
Radioactive material, special form, n.o.s., 7 200
Radioactive material, special form, Type A package, non fissile or
    fissile excepted 200
Radioactive material, surface contaminated object, 7 200
Radioactive material, surface contaminated objects (SCO), 7 200
Radioactive material, surface contaminated objects (SCO-I or
    SCO-II) fissile, 7 200
Radioactive material, surface contaminated objects (SCO-I or
    SCO-II) non fissile or fissile excepted, 7 200

italics: key words and phrases

Page

Radioactive material, transported under special arrangement,
  fissile, 7 200
Radioactive material, transported under special arrangement, non
  fissile or fissile excepted, 7...200
Radioactive material, Type A package, fissile non-special form, 7 200
Radioactive material, Type A package, non-special form, non
  fissile or fissile excepted, 7...200
Radioactive material, Type A package, special form, fissile, 7.........................200
Radioactive material, Type B(M) package fissile, 7...200
Radioactive material, Type B(M) package non fissile or fissile
  excepted, 7 200
Radioactive material, Type B(U) package, fissile, 7 200
Radioactive material, Type B(U) package, non fissile or fissile
  excepted, 7 200
Radioactive material, Type C package, fissile, 7...200
Radioactive material, Type C package, non fissile or fissile
  excepted, 7 200
Radioactive material, uranium hexafluoride, fissile, 7....................................200
Radioactive material, uranium hexafluoride non fissile or fissile
  excepted, 7 200
Rags, oily, 4.2 90
Rags, wet 90
Railway fusees..194
Railway fusees 196
Railway torpedo..194
Railway torpedo 196
Railway track signal..196
Rare-earth metals 243
Rare gases...104
Rare gases 106, 240
Rare gases, mixture, compressed, 2.2..104
RC318 215
Reaction..246
Reaction...1, 23, 27, 28, 29, 30, 39, 40, 47, 49, 50, 55, 58, 59, 64
  ....65, 71, 74, 76, 78, 92, 94, 99, 106, 122, 135, 150, 153, 154
  ....155, 159, 162, 165, 169, 170, 173, 183, 195, 200, 201, 221, 224
  ....226, 233, 234, 235, 238, 240, 241, 242, 243, 245, 247, 251, 255
Receptacles, small, containing gas without a release device, non-
  refillable, 2.1, 2.2, 2.3, 5.1, 8...104
Reducing compounds...175
Reduction 247
Reduction.............................27, 29, 34, 44, 48, 49, 147, 148, 154, 222, 245, 248
Refrigerant gases..215, 218
Refrigerant gas, n.o.s., 2.1, 2.2..215
Refrigerant gas, n.o.s., (e.g., non-flammable halocarbons), 2.2.....................215
Refrigerated air 105
Refrigerated liquids 104

bold: regulatory entries

Refrigerating machines 218
Refrigerating machines, 2.1, 2.2, 3 218
Regulated medical waste, n.o.s., 6.2 115
Release device 3, 69, 94, 106
Release devices, explosive, 1.4S 69
Resinates..188
Resinates, liquid 188
Resinates, solid 188
Resin solution, flammable, 3, 3.1, 3.2, 3.3............................189
Resin solutions..191
Ribbons..154
Rifle grenade 8
Rifle powder...74
Rifle powder 77
Rivets, explosive, 1.4S...69
Road asphalt 32
Road asphalt..32, 185
Road asphalt or tar liquids 32
Road oil 32, 65
Rocket engine 10
Rocket motor..10
Rocket motor..121
Rocket motors, 1.1C, 1.2C, 1.3C...8
Rocket motors, liquid fuelled, 1.2J, 1.3J................................8
**Rocket motors with hypergolic liquids with or without expelling
    charge, 1.2L, 1.3L**...8
Rockets, line-throwing, 1.2G, 1.3G, 1.4G 134
Rockets, liquid fuelled with bursting charge, 1.1J, 1.2J.......8
Rockets with bursting charge, 1.1E, 1.1F, 1.2E, 1.2F 8
Rockets with expelling charge, 1.2C, 1.3C, 1.4C 8
Rockets with inert head, 1.3C...8
Rods..70, 154, 161, 203
Rosin oil 188, 191
Rosin oil, 3, 3.2, 3.3..188
Rubber scrap 189
Rubber scrap, powdered or granulated, 4.1.........................189
Rubber shoddy..189
Rubber shoddy, powdered or granulated, 4.1.......................189
Rubber solution 189
Rubber solution, 3, 3.2, 3.3 189
Rubidium alloy (liquid)...151
Rubidium amalgam 151
Rubidium dispersion 153

italics: key words and phrases

S

s-, structural notation 251
SA ammunition 9
Safety cartridges 13
Safety fuse 119
Safety fuse 121
Safety matches 143
Safety squibs 119
Safety squibs 121
Salts 247
*Salts *7, 28, 42, 45, 47, 48, 52, 58, 59, 62, 64, 65, 76, 77, 88, 94
..................................100, 128, 129, 130, 145, 147, 148, 180, 181, 188, 195, 196, 224
Salt slags 148
Samples, explosive other than initiating explosives, 1 40
Sawdust 90, 221
Sawdust 78, 91, 99, 223
SCO-I 205
SCO-II 205
Scrap 149
Seatbelt modules 135
Seat-belt modules, 9 134
Seatbelt pretensioner 135
Seat-belt pretensioners, 9 134
Seat-belt pretensioners, compressed gas, 2.2 134
Seat-belt pretensioners, pyrotechnic, 9 134
sec-, structural notation 251
secondary-, structural notation 251
Security-type attaché cases 158
Security type attaché cases incorporating dangerous goods, for
    example lithium batteries or pyrotechnic material.................158
Seed cake 38, 90, 235
Seed cake, 4.2 90
Seed cake, containing vegetable oil mechanically expelled seeds,
    4.2...90
Seed cake containing vegetable oil, solvent extractions, 4.2.........90
Seed cake, containing vegetable oil, solvent extractions and
    expelled seeds, 4.2 90
Seed expeller 90
Seed expellers 90
Seed expellers, oily 90
Seismic explosives 70, 79
Selenates 130
Selenates, 6.1 128
Selenites, 6.1 128
Selenium compounds 130

bold: regulatory entries

Selenium compounds, n.o.s., 6.1 128
Self-defense spray 20
Self-defense spray, aerosol 19
Self-defense spray, non-pressurized, 9 19
Self-heating 36, 65, 221, 222, 223, 226
Self-heating liquid, inorganic, n.o.s., 4.2, 6.1, 8 226
Self-heating liquid, organic, n.o.s., 4.2, 6.1, 8 226
Self-heating solid, inorganic, n.o.s., 4.2, 6.1, 8 226
Self-heating solid, n.o.s., 4.2, 5.1 226
Self-heating solid, organic, n.o.s., 4.2, 6.1, 8 226
Self-inflating passenger restraint systems (air bags) for motor
   vehicles 134
Self-propelled vehicle 27, 219
Self-propelled vehicles 29, 219
Self-reactive 99, 107
Self-reactive liquid type B, C, D, E, F, 4.1 99
Self-reactive liquid type B, C, D, E, F, temperature controlled, 4.1 ...99
Self-reactive solid type B, C, D, E, F, 4.1 99
Self-reactive solid type B, C, D, E, F, temperature controlled, 4.1 ...99
Semifixed ammunition 10
Sensitized 247
Sensitizer 76, 79, 119
Separate loading ammunition 10
Shale oil 186, 240
Shale oil, 3, 3.2, 3.3 183
Shaped charges 69
Shaped charges 70
Shaped charges, commercial 69
Shavings 149
Sheets 154
Shellac 175
Shellac 1, 176, 191, 197
Shellac solution 175
Signal cartridges 196
Signal devices, hand, 1.4G, 1.4S 194
Signal, highway 194
Signals 75, 134, 196
Signals, distress, ship, 1.1G, 1.3G 194
Signals, distress, ship, water-activated 194
Signals, railway track, explosive, 1.1G, 1.3G, 1.4G, 1.4S 194
Signals, smoke, 1.1G, 1.2G, 1.3G, 1.4G 194
Silicofluorides 130
Silicofluorides, n.o.s., 6.1 128
Silicomanganese 222
Silicomanganese with a silicon content of 25% or more 221
Single-base propellant 9, 77
Sisal 90

italics: key words and phrases

Page

Sisal 91

Sisal, dry 90

Skimmings 148

Slabs 154

Slaked 247

Slaked 222

Sludge acid, 8 183

Slurry explosives 74

Slurry explosives 79

Small arms ammunition 9, 20, 195

Small arms cartridges 9, 122

Smoke ammunition 195

Smoke bombs 195

Smoke grenades 195

Smokeless powder 74

Smokeless powder 9, 77

Smokeless powder for small arms, 4.1 74

Sodium alloys (liquid) 151

Sodium amalgam 151

Sodium batteries 28

Sodium dispersion 153

Sodium metal, liquid alloy 151

Solid 247

Solids. 1, 3, 7, 11, 13, 20, 28, 29, 34, 36, 39, 40, 42, 44, 48, 53, 55

........................59, 65, 74, 76, 77, 89, 94, 99, 106, 114, 130, 151, 152, 153, 164

........................173, 175, 176, 181, 183, 190, 204, 221, 222, 224, 231, 233, 234

........................235, 238, 239, 240, 241, 242, 246, 247, 252, 253, 254, 255, 256

Solids containing corrosive liquids 49

Solids containing corrosive liquids, n.o.s., 8 47

Solids containing flammable liquid, n.o.s., 4.1 99

Solids containing toxic liquid, n.o.s., 6.1 255

Solutions 247

Solutions 1, 3, 5, 7, 22, 28, 34, 42, 47, 48, 49, 58, 61, 62, 63, 79, 88, 94, 105

........................130, 141, 147, 156, 162, 176, 181, 222, 224, 231, 232, 242, 247

Solvent 1, 2, 6, 22, 32, 38, 42, 47, 49, 52, 53, 58, 62, 86, 90, 105, 111, 132, 161

......162, 167, 168, 175, 176, 177, 181, 184, 185, 188, 189, 190, 191, 224, 231, 247

Solvents, flammable, n.o.s. 224

Solvents, flammable, toxic, n.o.s. 224

Sounding devices 71

Sounding devices, explosive, 1.1D, 1.1F, 1.2D, 1.2F 69

Special arrangement 205

Special form radioactive materials 204

Spent iron oxide 46

Spent iron sponge 46

Sponge 248

Sponge 154

bold: regulatory entries

Spontaneous combustion..12, 36, 46, 89, 92, 93, 99, 162
....170, 195, 221, 222, 223, 226, 233, 240
Sporting ammunition...9
Squibs..119
Squibs..121
Stabilized..248
Stabilizers...1, 11, 53, 57, 64, 77, 132, 168, 226, 227
Stain..175
Stains...132, 176
Steel swarf..147
Storage batteries, wet...27
Storage battery..28
Straw...91, 99
Straw, 4.1..90
Strike-anywhere matches...134, 143
Strips...154
Strontium alloy, non-pyrophoric..151
Strontium alloy, pyrophoric...151
Strontium amalgam..151
Strontium dispersion..153
Strontium, powder...153
Structural notation..248
Substances, EVI, n.o.s., 1.5D..74
Substances, explosive, n.o.s., 1.1A, 1.1C, 1.1D, 1.1G, 1.1L, 1.2L,
1.3C, 1.3G, 1.3L, 1.4C, 1.4D, 1.4G, 1.4S...74
Substances liable to spontaneous combustion...226
Substances liable to spontaneous combustion, n.o.s....226
Substances which in contact with water emit flammable gases..........................58
Substances, explosive, very insensitive, n.o.s., 1.5D...74
Substituted...251
Substituted..164, 251
Substituted nitrophenol pesticide, liquid, 3, 3.2, 6.1.......................................179
Substituted nitrophenol pesticides..181
Substituted nitrophenol pesticide, solid, 6.1..179
Supplementary explosive charges..123, 124
Surface-contaminated objects (SCO)...204
Swarf...149, 150
sym-, structural notation..251
Symmetrical..252
Symmetrical...241
Synthesis gas..44
Synthesis gas...45
Synthesis gas, compressed...44
Synthetic..1, 2, 32, 45, 61, 63, 86, 94, 128, 175
....176, 188, 189, 190, 191, 201, 243, 244
Synthetic..252
Synthetic fabrics, oily..90

italics: key words and phrases

	Page
Synthetic fibres	38, 90
Synthetic fibres, oily	90

T

t-, structural notation	251
Talcum	26, 241
Talcum with tremolite and/or actinolite	25
Tankage	88
Tankage	88, 89, 223
Tankage (garbage tankage (containing 8% or more moisture)), (rough ammonia tankage (containing 7% or more moisture)), (tankage fertilizer (containing 8% or more moisture))	221
Tar	32, 45
Target-identification bombs	195
Tars, liquid, 3.2, 3.3	32
Tars, liquid, including road asphalt and oils, bitumens and cut backs, 3	32
Tear gas	9, 12, 19, 158, 165, 168
Tear gas candles	21
Tear gas candles, 4.1, 6.1	19
Tear gas candles, non-explosive, 4.1, 6.1	19
Tear gas cartridges	19
Tear gas devices	19
Tear gas devices containing tear gas substances	19
Tear gas grenades	19
Tear gas grenades, non-explosive, 4.1, 6.1	19
Tear gas substance, liquid, n.o.s., 6.1	19
Tear gas substance, solid, n.o.s., 6.1	19
Tear-producing ammunition	19
Tellurium compounds	129
Tellurium compounds, n.o.s., 6.1	128
Terpene hydrocarbons, n.o.s., 3.3	86
Terpenes	86, 188, 190, 244
Terpenes, n.o.s.	86
tert-, structural notation	251
Tertiary alcohol	5
Tertiary alcohols	5, 251
tertiary-, structural notation	251
tetra-, structural notation	251
Textile waste	91, 111
Textile waste, wet, 4.2	90
Thallium compound, n.o.s., 6.1	156
Thallium compounds	156, 158, 181
Thiocarbamate pesticide, liquid, 3, 3.2, 6.1	179
Thiocarbamate pesticides	180
Thiocarbamate pesticide, solid, 6.1	179

bold: regulatory entries

Time fuzes 12, 123, 196
Tinctures, medicinal, 3, 3.2, 3.3 61
Tinning Flux 252
Tire assemblies inflated, above maximum rated pressure, 2.2 254
Tire assemblies inflated, unserviceable, damaged, 2.2 254
**Tire assemblies inflated, unserviceable, damaged or above
   maximum rated pressure, 9** 254
**Tire assemblies serviceable, inflated to pressure not greater than
   their rated inflation pressure** 254
Tires 254
Titanium powder, dry, 4.2 153
Titanium powder, wetted, 4.1 153
Titanium sponge 154, 248
Titanium sponge granules, 4.1 153
Titanium sponge powders, 4.1 153
Toe puffs 162
Toe puffs, nitrocellulose base 161
Torpedoes 11
Torpedoes, liquid fuelled with inert head, 1.3J; 8
Torpedoes, liquid fuelled with or without bursting charge, 1.1J 8
Torpedoes with bursting charge, 1.1D, 1.1E, 1.1F 8
Toxic 3, 6, 7, 19, 20, 21, 22, 23, 30, 32, 38, 42, 44, 45, 46, 52, 59, 61, 62, 78
   ....86, 99, 106, 107, 111, 113, 114, 115, 128, 129, 130, 141, 145, 146
   ....150, 156, 159, 164, 165, 166, 168, 180, 181, 221, 222, 223, 225, 255
Toxic ammunition 19
Toxic liquid, inorganic, n.o.s., 6.1, 8 255
Toxic liquid, n.o.s., 4.3, 5.1, 6.1 255
Toxic liquid, organic, n.o.s., 3, 6.1, 8 255
Toxic solid, inorganic, n.o.s., 6.1, 8 255
Toxic solid, n.o.s., 4.2, 4.3, 5.1, 6.1 255
Toxic solid, organic, n.o.s., 4.1, 6.1, 8 255
Toxins, liquid, extracted from living sources, n.o.s., 6.1 115
Toxins, solid, extracted from living sources, n.o.s., 6.1 115
Toy caps 197
Toy caps, 1.4S 194
Tracer fuses 195
Tracers 9, 195
Tracers for ammunition, 1.3G, 1.4G 194
Tractors 219
Tractors 219
trans-, structural notation 251
Tree killers 179
Tremolite 25
Tremolite 25
tri-, structural notation 251
Triaryl phosphates 168
Triaryl phosphates, n.o.s. 167

italics: key words and phrases

Page

Triazine pesticide, liquid, 3, 3.2, 6.1 179
Triazine pesticides 181
Triazine pesticide, solid, 6.1 179
Trick matches 197
Triple-base propellant 9, 77
tris-, structural notation 251
Tubular primers 123
Turpentine 176, 184, 188
Turpentine, 3, 3.3 188
Turpentine substitute, 3, 3.2, 3.3 183
Turpentine substitutes 184
Type A packages 204
Type B(M) packages 204
Type B(U) packages 204
Type C packages 204
Tyre assemblies inflated, above maximum rated pressure, 2.2 254
Tyre assemblies inflated, unserviceable, damaged, 2.2 254
Tyre assemblies inflated, unserviceable, damaged or above maximum rated pressure, 9 254
Tyre assemblies serviceable, inflated to pressure not greater than their rated inflation pressure 254
Tyres 254

U

Uncalcined 252
Uninhibited 252
Uninhibited 241
Unphlegmatized 252
uns-, structural notation 251
Unstabilized 252
Unstabilized 248
Unsymmetrical 252

V

Vanadium compound, n.o.s., 6.1 156
Vanadium compounds 156
Vanadium ore 221
Vanadium ore 223
Varnish 175
Varnish 38, 176
Varnish drier, liquid 175
Varnish driers 175
Varnish drier, solid 175
Vegetable fabrics, oily 90
Vegetable fibre 90

bold: regulatory entries

Vegetable fibres, burnt, wet or damp 90
Vegetable fibres, dry 90
Vegetable fibres, oily 90
Vehicle (flammable gas powered) including where containing an
    internal combustion engine, 9 219
Vehicle (flammable liquid powered) including where containing
    an internal combustion engine, 9 219
Vehicles 29, 106, 218, 254
Vehicles, self-propelled 27, 219
Vehicles, self-propelled including internal combustion engines or
    other apparatus containing an internal combustion engine or
    electric storage battery 219
Very signal cartridge 194
Very signal cartridge 196
Volatile 252
Volatile 20, 32, 159, 181, 184, 185, 190, 219, 233

W

Warheads 11
Warheads for guided missiles 8
Warheads, rocket with burster or expelling charge, 1.4D, 1.4F 8
Warheads, rocket with bursting charge, 1.1D, 1.2D, 1.1F 8
Warheads, torpedo, with bursting charge, 1.1D 8
Wastes 40, 44, 67, 91, 93, 111, 113, 115, 116, 145, 149, 162, 203, 205
Water-activated contrivances 69
Water-activated contrivances 71
Water gas 44
Water gas 45
Water gas, compressed 44
Water gels 74
Water gels 79
Water-reactive ..28, 59, 71, 99, 129, 150, 151, 154, 156, 165, 167, 169, 221, 242, 247
Water-reactive liquid, n.o.s., 4.3, 6.1, 8 58
Water-reactive solid, n.o.s., 4.1, 4.2, 4.3, 5.1, 6.1, 8 58
Wax 'vesta' matches 143
Weed killers 179
Wet batteries 28
Wet rags 93
Wetted 253
Wetted 39, 100, 162
Wheelchair, electric (spillable or non-spillable type batteries), 9 27
Wheelchair, electric with batteries 27
White asbestos 25
White asbestos 25
White asbestos (chrysotile, actinolite, anthophyllite, tremolite), 9 25
White spirit 183

italics: key words and phrases

Page

White spirit 99, 184
Wires 154
Woodchips 221
Woodchips 91, 223
Wood chips 90
Wood preservative 45, 113, 145, 181
Wood preservatives, liquid, 3, 3.2, 3.3 179
Wood pulp 78, 92, 162
Wood pulp pellets 221
Wood pulp pellets 223
Wood pulp, pellets 90
Wool waste, wet, 4.2 90

Z

Zinc ash 149
Zinc ashes, 4.3 147
Zinc dross 148, 149
Zinc dross, 4.3 147
Zinc dust 148
Zinc dust, 4.2, 4.3 147
Zinc dust, pyrophoric 147
Zinc powder, 4.2, 4.3 153
Zinc powder, non-pyrophoric, 4.3 153
Zinc powder, pyrophoric 153
Zinc residue 148, 149
Zinc residue, 4.3 147
Zinc skimmings 149
Zinc skimmings, 4.3 147
Zirconium, dry, coiled wire, finished metal sheets, strip, 4.1 153
Zirconium, dry, finished sheets, strip or coiled wire, 4.2 153
Zirconium powder 12
Zirconium powder 153
Zirconium powder, dry, 4.2 153
Zirconium powder, wetted, 4.1 153
Zirconium scrap, 4.2 147
Zirconium suspended in a flammable liquid, 3, 3.1, 3.2, 3.3 153

bold: regulatory entries